CIRCUITS AND SOFTWARE
FOR ELECTRONICS ENGINEERS

Edited by Howard Bierman,
Managing Editor, Technical,
Electronics

CIRCUITS & SOFTWARE
FOR ELECTRONICS ENGINEERS

Library of Congress Cataloging in Publication Data

Main entry under title:

Circuits and software for electronics engineers.

1. Electronics circuits. 2. Electronic circuit design
--Computer programs. I. Bierman, Howard, 1925-
TK7867.C498 1983   621.3815'3   83-9806
ISBN 0-07-005243-3 (McGraw-Hill Book Co.)
ISBN 0-07-606870-6 (McGraw-Hill Publications Co. : pbk.)

A McGraw-Hill Publication

CONTENTS

I. AMPLIFIERS

1. Pair of pnp/npn transistors form high-voltage amplifier
2. Dual-function amp chip simplifies many circuits
5. Feedback reduces offset in wideband video amplifiers
5. Power-sharing bridge circuit improves amplifier efficency
7. Knowing gate-charge factor eases power MOS FET design
8. Dynamic depletion circuits upgrade MOS performance
9. Exploiting the full potential of an rf power transistor
11. Fm receiver mixes high gain with low power
12. Dual-function amplifier eases circuit design
15. Antilog amplifiers improve biomedical signal's S/N ratio
16. Differential amp cancels integrator's crosstalk

II. AUDIO CIRCUITS

17. OTAs and op amp form voltage-controlled equalizer
18. Digital processor improves speech-signal quality
20. Electronic-music generator retriggers waveforms
21. Agc prevents noise build-up in voice-operated mike
22. Switching circuit fulfills three functions
23. Parametric equalizer improves Baxandall tone control
25. Audio-visual controller synchronizes museum display
26. Two-chip generator shapes synthesizer's sounds
27. On-chip transistors extend audio amp's design flexibility
29. Audio-range mixer calibrates intermod-distortion analyzers
30. Angle-modulated signals suffer less a-m distortion

III. CLOCK CIRCUITS

32. Making a clock chip keep better time
34. Recovering the clock pulse from NRZ-inverted data
35. Sync clock, counter improve programmable-width generator
35. Counter-time resolves l us over extremely wide range

IV. COMPARATORS

37. Frequency comparator uses synchronous detection
38. Comparator, one-shot produce pulse-width inversion for servos
39. Digital comparator minimizes serial decoding circuitry

V. CONTROL CIRCUITS

40. One-chip tachometer simplifies motor controller
41. Thermocouple simplifies temperature controller
42. Stepper resolves motor's angular position to 0.1
43. Current-biased transducer linearizes its response
44. Pulse generator has independent phase control
44. C-MOS circuit controls stepper motor
46. Low-cost controller stabilizes heater-type cryostats
47. Resistor-based servo replaces mechanical governors in motors
48. Interface lets microprocessor control stepper motor
50. Sequencer plus PROM step three motors independently
52. Thermistor probe helps regulate liquid levels
53. Octal register links dissimilar a-d converters
54. Conductive foam forms reliable pressure sensor
55. Positive pulse triggers 555 integrated-circuit timer
56. Speeding floppy data transfer under program control
58. Variable duty-cycle circuit controls valve mixer
59. Appliance-controller interface aids the paralyzed
60. PROM forms flexible stepper-motor controller

VI. CONVERTERS

63. D-a converter also performs a-d translation
64. 5-V converter powers EE-PROMs, RS-232-C drivers
66. Interfacing 10-bit a-d converter with a 16-bit microprocessor
68. Access control logic improves serial memory systems
69. Voltage-controlled resistance switches over preset limits
70. Current-biased transducer linearizes its response
71. Low-power f-V converter turns portable tachometer
72. Frequency comparator uses synchronous detection
73. Voltage-detector chip simplifies V-f converter
74. Photocurrent-to-frequency converter notes light levels
75. Delay circuit produces precision time intervals
76. OTA multiplier converts to two-quadrant divider
77. Current mirror linearizes remote-controlled timer
78. Analog multiplexer and op amp unite for precise d-a converter
79. A-d converter linearizes 100-ohm temperature detector
80. Low-cost coordinate converter rotates vectors easily

VII. ENCODERS/DECODERS

82. Two-chip VCO linearly controls ramp's amplitude and frequency
82. Interleaving decoder simplifies serial-to-parallel conversion
84. Decoder reworks binary input into hexadecimal or octal form
86. Simplified multiplier improves standard shaft encoder
87. Decoder logs signals' order of arrival
88. RC oscillator linearizes thermistor output
89. Multiplier increases resolution of standard shaft encoders
90. Interfacing digital circuitry with plastic rotary encoders

VIII. FILTERS

92. Delay line eases comb-filter design
93. Sample-and-hold devices ease tracking-filter design
94. High-pass Chebyshev filters use standard-value capacitors
96. State-variable filter has high Q and gain
97. State-variable filter trims predecessor's component count
98. Selecting the right filters for digital phase detectors
99. Logic-gate filter handles digital signals
101. Standard C, L-input filters stabilize hf transistor amplifiers

IX. FUNCTION GENERATORS

103. Canceling cusps on the peaks of shaped sine waves
104. TTL can generate composite video signals
105. Generating triangular waves in accurate, adjustable shapes
107. Doubling the resolution of a digital waveform synthesizer

X. INSTRUMENT CIRCUITS

108. Measuring irregular waveforms by detecting amplitude changes
109. Four chips generate pseudorandom test data
110. Wide-range capacitance meter employs universal counter
111. Building a low-cost opitical time-domain reflectometer
112. Low-cost thermometer has high resolution
113. Adapter equips HP analyzer for general ROM tracing
114. Instrumentation-meter driver operates on 2-V supply
115. Ac voltmeter measures FET dynamic transconductance
116. One-chip DVM displays two-input logarithmic ratio
117. Meter measures processor's dynamic utilization capacity
118. Pulse-width meter displays values digitally

119. Reading a dual-trace scope with a +/-0.025% accuracy
120. Tester checks functionality of static RAMs
123. Continuity tester buzzes out breadboards's flaws
124. Hall-effect tachometer senses speed, direction of rotation
125. Probing chip nodes with minimum disturbance
125. Meter measures coatings magnetically
128. Charge integrator measures low-level currents
130. Thermistor gives thermocouple cold-junction compensation
132. Five-chip meter measures impedances ratiometrically
134. Hall compass points digitally to headings
136. Power-up relays prevent meter from pinning
137. 8048 system generates alphanimerics for CRT display
139. Processed seismometer signals yield ground acceleraion
140. Latch grabs glitches for waveform recorder
142. Transistor probe simplifies solid-state gaussmeter
143. Digital phase meter updates measurement each cycle
144. Phase-locked loops replaced precision component bridge
145. Four-chip meter measures capacitance to within 1%
146. Chip changes the colors of light-emitting diodes
148. 'Dithering' display expands bar graph's resolution
150. Light pen generates plotter signals
151. Versatile circuit measures pulse width accurately
152. Digital weighing scale resolves quarter counts

XI. LOGIC CIRCUITS

155. Treating the three-state bus as a transmission line
156. LED indicates timing error in emitter-coupled-logic one-shot
157. Address checker troubleshoots memory drive, logic circuitry
158. Protecting TTL gates from electrostatic-discharge pulses
158. Interfacing TTL with fast bipolar drivers
160. Signal chip solves MC6809 timing problems

XII. MEMORY CIRCUITS

162. Chip computer gains I/O lines when adding memory
163. Stepper checks state of E-PROM's memory
164. Transferring data reliably between core and CPU RAM
166. RAM makes programmable digital delay circuit
167. Programmable source sets voltage of E-PROMs
168. Bipolar PROMs make versatile Camac instruction decoders
169. Access control logic improves serial memory systems
169. Deselected RAM refuses all data during power failure
170. Subroutine tests RAM nondestructively
171. Random-access memories form E-PROM emulator
173. Nonvolatile RAM provides on-board storage for computer
176. Simple interface links RAM with multiplexed processors

XIII. MICROPROCESSORS

177. Redundancy increases microprocessor reliability
178. Macroinstruction for Z80 guarantees valid output data
179. Interfacing C-MOS directly with 6800 and 6500 buses
180. Static RAM relocates addresses in PROM
181. 8X300 microcontroller performs fast 8-by-8-bit multiplication
182. Octal latches extend bus hold time
183. Simple patch reconciles parity flags in Z80, 8080
184. Mapping an alterable reset vector for the MC68000
185. Reset circuit reacts to fast line hits
186. Z80B controller waits for slower memory
187. Enabling a processor to interact with peripherals using DMA
189. Power-fail detector uses chip's standby mode

XIV. MODULATORS/DEMODULATORS

191. Switching regulator performs multiple analog division
192. Tracking filters demodulte two audio-band fm signals
193. Integrator improves 555 pulse-width modulator
194. Linear one-chip modulator eases TV circuit design
196. Achieving linear control of a 555 timer's frequency
197. Ultrafast hybrid counter converts BCD into binary
198. Pulse modulator provides switched-mode amplification

XV. MULTIPLEXERS

199. HP64000 emulates MC6801/6803 using bidirectional multiplexer
200. Multiplexers compress data for logarithmic conversion
202. Multiplexer does double duty as dual-edge shift register
203. Time-slot assigner chip cuts multiplexer parts count

XVI. MULTIPLIERS

204. Frequency multiplier uses digital technique
205. Joining a PLL and VCO forms fractional frequency multiplier
206. Simplified multiplier improves standard shaft encoder
207. Enhanced multiplier cuts parts count

XVII. OPERATIONAL AMPLIFIERS

209. Current booster drives low-impedance load
210. Diode plus op amp provide double-threshold function
211. Op-amp summer forms simple high-speed phase generator
212. Absolute-value amplifier uses just one op amp
213. Bi-FET op amps invade 741's general-purpose domain
215. Extending the range of a low-cost op amp
216. Bi-FET op amps simplify AGC threshold design
217. Bi-FETs expand applications for general-purpose op amps

XVIII. OPTOELECTRONIC CIRCUITS

219. Manchester decoder optimizes fiber-optic receiver
220. Fiber-optic link taps time-division multiplexing
221. Optical coupler isolates comparator inputs
222. Improving the LM395 for low-level switching
223. Twin optocouplers raise serial transmission speed
223. Programmed comparator finds loss in optical fiber
225. Optocouplers clamp spikes fast over wide range
225. Swapable fiber-optic parts ease isolation problems
227. Opto-isolated line monitor provides fall-safe control
229. Opto-isolated RS-232 interface achieves high data rate
230. Optical agc minimizes video measurement errors
231. Fiber-optic transmitter measures high voltage safely

XIX. OSCILLATORS

233. Divider sets tuning limits of C-MOS oscillator
234. C-MOS IC achieves triggered phase-locked oscillations
235. Stable sinusoidal oscillator has multiple phased outputs
236. Stable and fast PLL switches loop bandwidths
237. Phase shifters simplify frequency-multiplier design
239. V-MOS oscillator ups converter's switching frequency
240. Programmable sine generator is linearly controlled

XX. PERIPHERAL INTERFACING

242. Low-cost interface unites RAM with multiplexed processors
243. Enabling a processor to interact with peripherals using DMA
245. Interface unites Z8000 with other families of peripheral devices
247. Adapting a home computer for data acquisition
248. Serial-communication link controls remote displays
250. Transferring data reliably between core and CPU RAM
252. Interfacing C-MOS directly with 6800 and 6500 buses
253. Simple patch reconciles parity flags in Z80, 8080
253. Current loop supports remote distributed processing
255. Serial-data interface eases remote use of terminal keyboard
256. 8-bit DMA controller handles 16-bit data transfers
257. Driver simplifies design of display interface

XXI. POWER SUPPLIES

259. Stabilizer boosts current of +/-dc-dc converter
260. Milliampere current source is voltage-controlled
261. Low-cost timers govern switched-mode regulator
262. Ni-Cd battery charger has wide range of features
263. External pass FET boosts regulated output voltage
264. Comparator circuit regulates battery's charging current
265. Current mirror stabilizes zener-diode voltage
266. Reed-coil relay is behind flexible fault detection
267. Voltage translator switches auxiliary voltages when needed
267. Three-level inverter conserves battery power
269. Capacitive voltage doubler forms +/-12 to +/-15-V converter
270. External transistor boosts load current of voltage regulator
270. Linear controller attenuates switching-supply ripple
271. Stacked voltage references improve supply's regulation
273. High frequencies, winding setup improve voltage conversion

XXII. PROTECTION CIRCUITS

274. One-chip alarm scares auto thieves
275. Noise-immuinized annunciator sounds change-of-state alarm
277. Phase-reversal protector trips main contactor

XXIII. PULSE GENERATORS

278. Deglitcher delay circuit serves also as pulse generator
279. Two-chip pulse generator operates at 75 MHz
280. Three-chip circuit produces longer one-shot delays
281. Synchronous one-shot has integral pulse width
281. Synchronous pulse catcher snares narrow glitches
283. Dual one-shot keeps firmware on track
284. Parallel power MOS FETs increase circuit current capacity
286. 'Demultiplexing' pulses of different widths
287. Pulse-width discriminator eliminates unwanted pulses
289. Low-cost generator delivers all standard bit rates
290. Generator has independent duty cycle and frequency
291. Thumbwheel switch programs retriggerable one-shot

XXIV. SOFTWARE/CALCULATORS

294. TI-59 program tracks satellites in elliptical orbits
298. Calculator plots time response of inverse Laplace transform
299. HP-41C generates a pseudorandom sequence
300. Macroinstruction for Z80 guarantees valid output data
301. HP-67 calculates maximum nonlinearity error
302. Codec program compands samples for u-law simulation
304. HP-41C calculator analyses resistive attenuators
305. TI-59's reverse-Polish routine simplifies complex arithmetic

XXV. SOFTWARE/COMPUTERS

307. Interface program links a-d chip with microprocessor
308. Push-pop program aids 6800's register swapping
310. Nystrom integration gives dynamic system's response
312. Pocket computer solves for LC resonance using Basic
314. TRS-80 program helps to load cassette data twice as fast
316. TRS-80 program simplifies design of PROM decoders
318. 8048 program transmits messages in ASCII
320. Pocket computer tackles classical queuing problems
322. Interface, software form smart stepper controller
326. GPIB software helps to provide automatic test switching
328. Interactive software controls data-acquisition process
331. Tracing out program bugs for Z80A processor
332. Writing relocatable code for 8-bit microprocessors
334. Universal E-PROM controller eases computer linkup
336. Home computer displays inverse Laplace transforms
338. 'Surgical' program speeds 6909 debugging process
340. Very efficient 8080 program multiplies and divides
342. Hardware-software integration eases E-PROM programming

PREFACE

☐ Growth and change are two characteristics long associated with the electronics industry. As applications of ever more sophisticated integrated circuits and other solid-state devices extend farther into such markets as appliances, photography, automobiles, and home entertainment, the electronics engineer is faced with a growing challenge to innovate and develop more advanced products with higher performance and quality.

The task is awesome, but this volume of novel circuits and software should help. It is a collection of the creative solutions developed by readers of *Electronics* for specific design problems and offered for public consumption in Designers' Casebooks and Software Notebooks from mid-1980 through the end of 1982. Over 300 pages of thought-provoking ideas contain speedy answers for short-term projects and stimulating new wrinkles for long-term programs. ☐

1. AMPLIFIERS

Pair of pnp/npn transistors form high-voltage amplifier

by H. F. Nissink
Electrical Engineering Department, University of Adelaide, Australia

This simple high-voltage amplifier circuit provides a large output voltage swing with low-current consumption and uses only a few components. Its 280-volt regulated supply produces an unclipped output of up to 260 v peak to peak. In addition, rise and fall times of the output for a square-wave input are 150 nanoseconds, and the no-load supply current is only 4 milliamperes.

The principle behind this circuit is just a simple transistor amplifier (a) employing collector feedback through resistor R_F. The dc output is approximately $V_{be} \times R_F/R_1$. The circuit has an active pull-down action, with pull-up through R_2. However, if R_2 is replaced with a pnp transistor in a similar circuit, the pull-up and pull-down are through the transistor.

This substitution is the basis for the circuit in (b). Its output-voltage level is theoretically determined by the 300-kilohm and 1.3-kΩ resistors and thus the ac circuit gain is approximately 300 kΩ/10 kΩ = 30. The power supply (±10 v dc) for the current amplifier A_1 driving 2N5416 is isolated from the supply for A_2, which is driving the 2N3439.

The circuit has an input impedance of 5 kΩ and an output impedance of 2.4 kΩ. For the component values shown, the actual gain measures about 27, and the output over the frequency range of 1 kilohertz to 300 kHz is 260 v peak to peak (without clipping) and 100 v peak to peak at 1 megahertz. Because the amplifier is not short-circuit protected at the output, the regulated power supply is limited by the current. This high-voltage amplifier may drive capacitive-type transducers and be used for several other applications. □

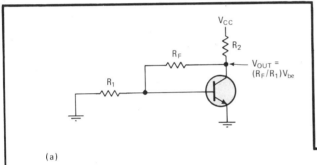

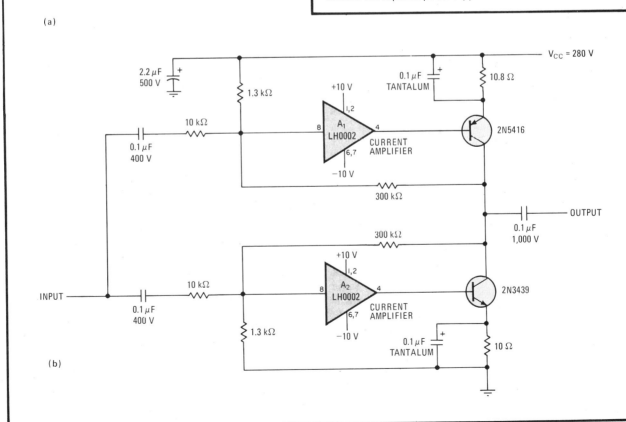

High-voltage amplifier. A simple transistor amplifier (a) employs collector feedback with R_F. Resistor R_2 is replaced by a circuit similar to (a) but with a pnp transistor to form the basis for a high-voltage amplifier (b). The current amplifiers A_1 and A_2 driving 2N5416 and 2N3439 use separate power supplies of ± 10 V dc.

Dual-function amp chip simplifies many circuits

by Jim Williams
National Semiconductor Corp., Santa Clara, Calif.

Various circuits that combine low cost, single- or dual-supply operation, and ease of use can easily be built with comparators and operational amplifiers like National Semiconductor's LM339 and LM324 because of their general applicability to a wide range of design problems. Now circuit complexity can be reduced even further with up-and-coming dual-function devices like the LM392, which put both a comparator and an op amp on one chip. Besides allowing a degree of flexibility in circuit function not readily implemented with separate chips, this device retains simplicity at low cost. The building of such circuits as a sample-and-hold circuit, a feed-forward low-pass filter, and a linearized platinum thermometer is

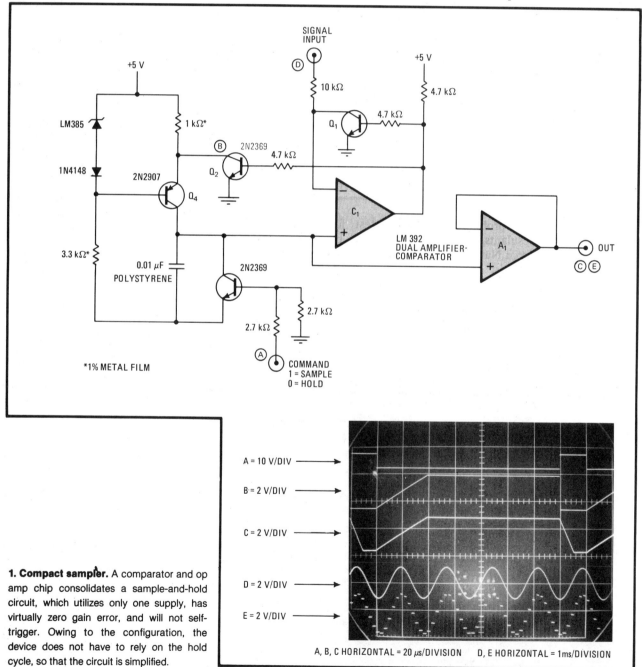

1. Compact sampler. A comparator and op amp chip consolidates a sample-and-hold circuit, which utilizes only one supply, has virtually zero gain error, and will not self-trigger. Owing to the configuration, the device does not have to rely on the hold cycle, so that the circuit is simplified.

discussed here in the first of two articles.

The circuit in Fig. 1 is an unusual implementation of the sample-and-hold function. Although its input-to-output relationship is similar to standard configurations, its operating principle is different. Key advantages include no hold-step glitch, essentially zero gain error and operation from a single 5-volt supply.

When the sample-and-hold command pulse (trace A) is applied to transistor Q_3, it turns on, causing Q_4's collector to go to ground. Thus the output sits at ground. When the command pulse drops to logic 0, however, Q_4 drives a constant current into the 0.1-microfarad capacitor (trace B). At the instant the capacitor ramping voltage equals the signal input voltage, comparator C_1 switches, thereby causing transistor Q_2 to turn off the current source. Thus the voltage at Q_4's collector and A_1's output (trace C) will equal the input.

Q_1 ensures that the comparator will not self-trigger if the input voltage increases during a hold interval. If a dc-biased sine wave should be applied to the circuit (trace D), a sampled version of its contents will appear at the output (trace E). Note that the ramping action of the current source, Q_4, will just be visible at the output during sample states.

In Fig. 2, the LM392 solves a problem common to filters used in multiplexed data-acquisition systems, that of acquiring a signal rapidly but providing a long filtering time constant. This characteristic is desirable in electronic scales where a stable reading of, for example, an infant's weight is desired despite the child's motion on the scale's platform.

When an input step (trace A) is applied, C_1's negative input will immediately rise to a voltage determined by the setting of the 1-kilohm potentiometer. C_1's positive input, meanwhile, is biased through the 100 K $-$ 0.01 F time constant, and phase lags the input. Under these conditions, C_1's output will go low, turning on Q_1.

This action causes the capacitor (trace B) to charge rapidly up to the input value. When the voltage across the capacitor equals the voltage at C_1's positive input, C_1's output will go high, turning off Q_1. Now, the capacitor can only charge through the 100-kΩ resistor and the time constant must therefore be long.

The point at which the filter switches from the short to the long time constant is adjustable with the potentiometer. Normally, this pot will be set so that switching occurs at 90% to 98% of the final value (note that the trip point is taken at about the 70% point in the photo so that circuit operation may be easily seen). A_1 provides a buffered output. When the input returns to zero, the

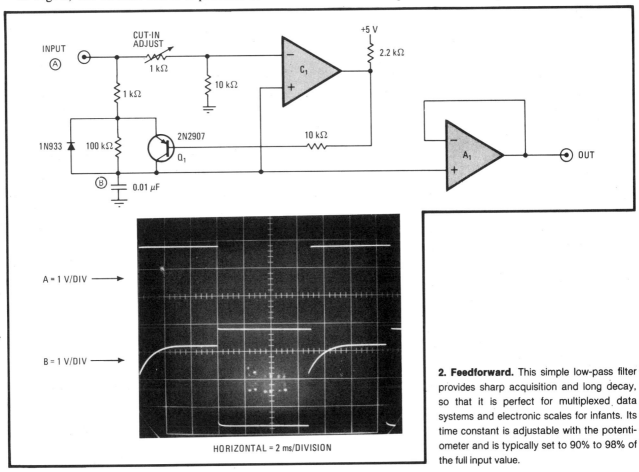

2. Feedforward. This simple low-pass filter provides sharp acquisition and long decay, so that it is perfect for multiplexed data systems and electronic scales for infants. Its time constant is adjustable with the potentiometer and is typically set to 90% to 98% of the full input value.

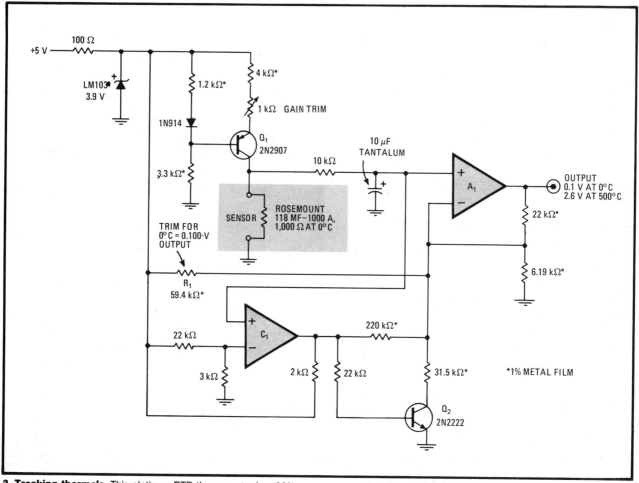

3. Tracking thermals. This platinum RTD thermometer has 99% accuracy over the 0°-to-500°C range. C_1 derives the breakpoint change in A_1's gain for sensor outputs exceeding 250°C, compensating for the sensor's nonlinearity. Current through the 220-kΩ resistor shifts A_1's offset voltage, in effect preventing glitches at the breakpoint. The instrument is calibrated only at two points with a decade resistor box.

1N933 diode (a low forward-drop type), provides rapid discharge for the capacitor.

In Fig. 3, the LM392 is used to provide gain and linearization for a platinum resistor-temperature device in a single-supply thermometer circuit. This one measures from 0°C to 500°C with ±1° accuracy.

Q_1 functions as a current source that is slaved to the 3.9-v reference. The constant-current–driven platinum sensor consequently yields a voltage drop that is proportional to its temperature. A_1 amplifies the signal and provides the circuit output.

Normally, the slightly nonlinear response of the sensor would limit the circuit accuracy to about ±3°C. C_1 compensates for this error by generating a breakpoint change in A_1's gain at sensor outputs corresponding to temperatures exceeding 250°C. Then, the potential at the comparator's positive output exceeds the potential at the negative input and C_1's output goes high. This turns on Q_2, which shunts A_1's 6.19-kΩ feedback resistor and causes a change in gain that compensates for the sensor's slight loss of gain from 250° to 500°C. Current through the 220-kΩ resistor shifts the offset voltage of A_1 so no discernible glitch will occur at the breakpoint.

A precision decade box should be used to calibrate this circuit. Once inserted in place of the sensor, it is adjusted for a value of 1,000 ohms and a 0.10-v output by means of resistor R_1. Next, its resistance is set to 2,846 Ω (500°C) and its gain trim control adjusted for an output of 2.6 V. These adjustments are repeated until the zero and full-scale readings remain fixed at these points. □

Feedback reduces offset in wideband video amplifiers

by Alan Cocconi
California Institute of Technology, Pasadena, Calif.

Wideband video amplifiers such as the LM733 generally have large input offset voltages that, when multiplied by their gain, can result in unacceptably high dc offset at the output. This undesirable effect can be reduced by feedback by means of a low–input-offset integrator.

As shown in (a), summing resistors R_1 and R_2 are selected so that the input to the integrator is proportional to the video amplifier's input offset voltage. The integral feedback drives the video amp's input offset to zero, leaving only the low offset of the integrator (which can be trimmed to zero) to appear at the amplifier output.

A practical implementation of the approach is given in (b). The integrating operational amplifier, a CA3140, was chosen for its low input offset voltage. Here, the 1N4371 zener diode and the 2N2222 transistor, in an emitter-follower configuration, are required to ensure that the output can go down to 0 volt, since the 733 video amp suffers from the restriction of a minimum positive output voltage. □

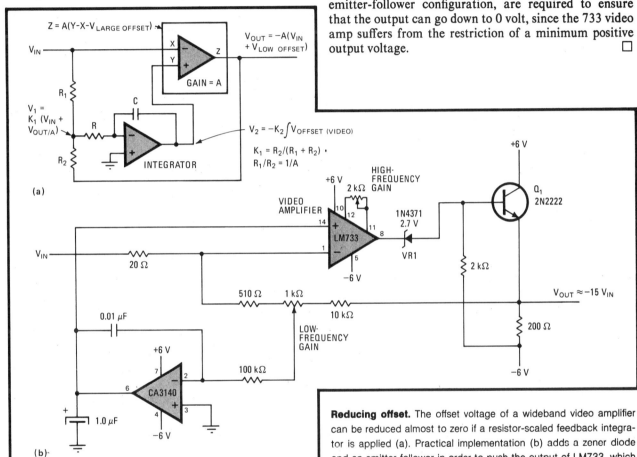

Reducing offset. The offset voltage of a wideband video amplifier can be reduced almost to zero if a resistor-scaled feedback integrator is applied (a). Practical implementation (b) adds a zener diode and an emitter follower in order to push the output of LM733, which has a minimum positive output voltage, down to zero.

Power-sharing bridge circuit improves amplifier efficiency

by Jim Edrington
Texas Instruments Inc., Austin, Texas

This linear bridge amplifier offers several advantages in driving motors and servo systems, including obtaining maximum efficiency with a single power supply and with dc coupling, which as a result reduces circuit complexity. Most notable, however, is that the four transistors in the amplifier will equally share load currents, as well as simplifying the drive requirements. These factors permit lower-cost transistors to be applied and allow their heat-sink requirements to be reduced.

Shown in (a) is one half of the bridge-type circuit, which illustrates the amplifier's operation. Positive input excursions from the driver turn on current sink Q_2, with a portion of Q_2's collector current passing through transistor Q_3. Q_3's current flow causes source transistor Q_1 to turn on.

Because the majority of the flow must pass through Q_1 and Q_2, the collector-to-emitter voltage of both transistors must be equal to ensure equal power dissipation. This voltage-matching requirement is achieved by clamping the gain of Q_1 to the voltage at the center of the load with a zener diode. Thus the virtual center of the load will be maintained at $V_{cc}/2$ and $V_{Q_1ce} = V_{Q_2ce}$, provided $R_1 = R_2$. The zener diode, D_1, must have a value of $V_z = (V_{cc}/2) - 1.4$ to meet the requirement for the reference potential.

Two of these circuits may be readily incorporated into a full-bridge arrangement, as shown in (b), that is suitable for driving electromechanical devices. Adding diodes D_2 through D_5 isolate one branch's functions from the other. With this configuration, each branch conducts for half of the input cycle thereby eliminating virtually all crossover difficulties.

The isolation diodes will alter the divider's center voltage by 0.7 volts, however, and so the value of the zener voltage must be slightly changed. In this case, it will be $V_z = (V_{cc}/2) - 1.4 + 0.7 = 11.3$ v. In most applications, selecting the nearest standard zener value will suffice. □

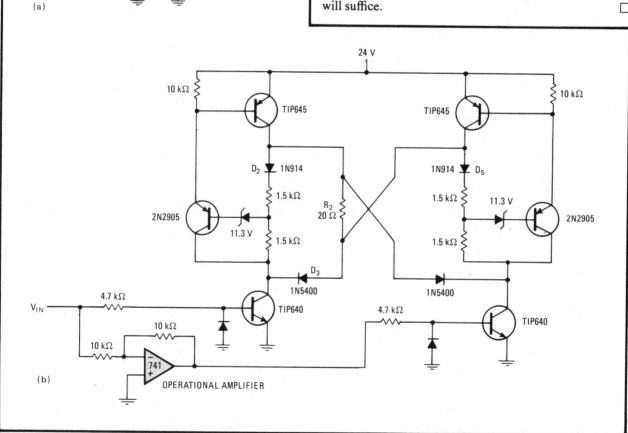

Divided driver. A rudimentary amplifier (a) may be designed so that Q_1 and Q_2 carry equal load on a positive excursion of an input signal, using a zener diode of suitable value for biasing a load center to cause $V_{Q_1ce} = V_{Q_2ce}$. Combining two such sections in a balanced bridge arrangement (b) builds a dc-coupled amplifier that is simple, can run from one supply, and can ensure that all amplifiers may handle a proportionate share of the power. This combination reduces electrical specifications of individual transistors, thereby reducing their cost.

Knowing gate-charge factor eases power MOS FET design

by Brian Pelly
International Rectifier Corp., El Segundo, Calif.

Unlike bipolar transistors, power MOS field-effect transistors are essentially voltage-controlled devices whose drive circuits are best designed around their gate-charge factor. Obtaining a measurement of this factor with this circuit (a) will ease switching-time calculations and, as a result, reduce drive-component selection to a series of simple Ohm's Law equations.

Gate charge comprises both gate-to-source and gate-to-drain (Miller) capacitances. To measure this charge, a constant current is supplied to the gate of the device under test from capacitor C_1 through regulator diode D_1. In addition, a constant current is established in the drain circuit by setting the voltage on the gate of power MOS FET Hexfet 1. The net charge consumed by the gate is related to the given current and voltage that is in the source-to-drain path.

The graph in (b) represents gate voltage versus gate charge in nanocoulombs. It shows exactly when the gate-to-source and gate-to-drain capacitances take on charge. The first voltage rise charges the gate-to-source capacitance and the flat portion charges the gate-to-drain capacitance. At the second voltage rise, both capacitances are charged to a level that can switch the given voltage and current.

Although the second voltage rise indicates the point at which the switching operation is completed, the design safety margin requires that the drive-voltage level applied to the gate be slightly higher than the voltage that is required to switch the given drain current and voltage. Since gate charge is the product of the gate input current and the switching time, a designer can quickly develop a drive circuit that is appropriate for the switching time required. □

Charge. The gate-charge factor, measured with test circuit (a), is the total charge that must be supplied to the gate to switch a given drain current and voltage. The gate voltage versus gate charge (b) for Hexfet IRF 131 shows that the total charge consumed by the gate must be higher than the minimum required to initiate switching.

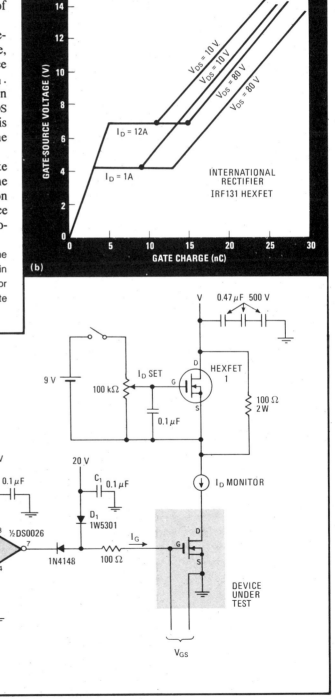

Dynamic depletion circuits upgrade MOS performance

by Clay Cranford, *IBM Corp.*,
System Communications Division, Research Triangle Park, N.C.

In the design of MOS integrated circuits, the need frequently arises for an efficient, low-power driver to charge and discharge high-capacitance loads, be they on chip or off. Standard driver circuits include either enhancement-depletion inverters or inverters with push-pull output stages. However, both suffer from high input capacitance, and with a push-pull driver the high-state output voltage is limited to a threshold-voltage drop below the power-supply potential. Clocked driver circuits cut power dissipation, but chip area must be provided for clock-signal generation or routing or both.

Two new circuit solutions include the dynamic depletion-mode driver (Fig. 1) and the active bootstrap driver (Fig. 2). The first takes advantage of the high conductance of a depletion-mode device under high gate bias. The output can be charged to the full power-supply voltage, V_{DD}, and dc power is reduced by limiting the low output-level current drain. The idea behind the approach is to charge a bootstrap capacitor, C_B, and then redistribute that capacitor's charge when the output is being driven to its high level.

In Fig. 1, transistor Q_6 serving as the bootstrap capacitor is charged to V_{DD} when the input is low. Q_4 is in a low-conductivity state and Q_3 and Q_5 are turned on, causing the gate of Q_7 to be held near ground. As the input rises, the charge on C_B is redistributed between C_B and the gate of Q_7 via Q_4. At this point, Q_3 and Q_5 turn off (Q_3 has functioned as the dynamic depletion-mode device, switching between conductive and nonconductive states). Device Q_7 is switched to its linear region and Q_8 has turned off, charging the output to V_{DD}.

In the active bootstrap technique (Fig. 2), a voltage-bootstrapping circuit and a power-down feature provide a large amount of overdrive and a reduced output-low power dissipation, respectively. The operation of this circuit also has several steps.

With the input low, node 1 is high. Q_6 is turned off and Q_7 turned on; consequently, node 3 is low and driver Q_3 shuts off. Since Q_8 can be made physically long, its current can be limited to a negligible amount. This accounts for the minimal output-low current.

When the input is raised, Q_6 turns on, and after one inverter delay, Q_7 turns off. The bootstrap capacitor—Q_5 in this circuit—is then charged to approximately a threshold voltage below the input, since node 2 is heavily loaded. Node 2 is held near ground by Q_4 during part of the time that Q_6 is turned on because of the inverter delay between the input and the gate of Q_4.

If node 2 begins to move upward during this precharge period because of different loading conditions or because Q_6 is given a smaller width-to-length ratio, Q_6 will dynamically precharge node 3, being bootstrapped by the rising voltage at node 2 and the bootstrap capacitor, and it will turn off when node 3 reaches a threshold voltage below the level of the input signal. Q_4 is not conducting while node 2 is being charged through Q_3.

Since the bootstrap capacitor is precharged, it will

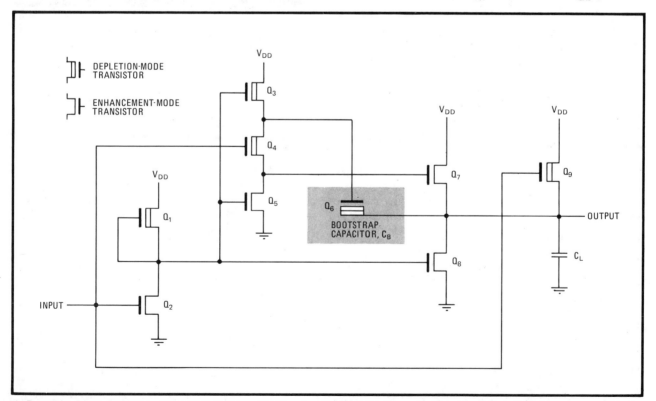

1. Dynamic driver. Bootstrap capacitor C_B is charged to V_{DD} when the input is low, causing the gate of Q_7 to be held near ground. As the input rises, the charge on C_B is redistributed between C_B and the gate of Q_7. The output is charged to V_{DD} as Q_7 is switched to its linear region.

boost node 3 to a voltage higher than a threshold drop below the input. This provides increased on-drive for Q_3 and, in turn, a faster rising output transition than might otherwise be possible. The actual voltage to which node 3 is bootstrapped is determined by the ratio of the bootstrap capacitance to that of Q_3 plus the contribution made by parasitic capacitances.

When the input falls, Q_6 turns off, Q_7 turns on, and node 3 is pulled near ground. Q_3 enters a nonconducting state, resulting in a rapidly falling response, since Q_4 need sink current only from the load capacitance. This action helps to reduce the down-level power consumption as well.

Unlike the dynamic depletion-mode driver, this configuration provides for dynamic precharging of the bootstrap capacitor directly from the power supply (through Q_6). A detailed analysis shows that to obtain a given amount of bootstrap voltage, a bootstrap capacitor less than half the size of that necessary for other configurations is required. For the typical layout, it will be considerably less than half.

The active bootstrap technique can be applied wherever high speed and low power are prime considerations — if the extra chip area required is acceptable. The circuit of Fig. 2 has been designed and tested using n-channel silicon-gate technology. □

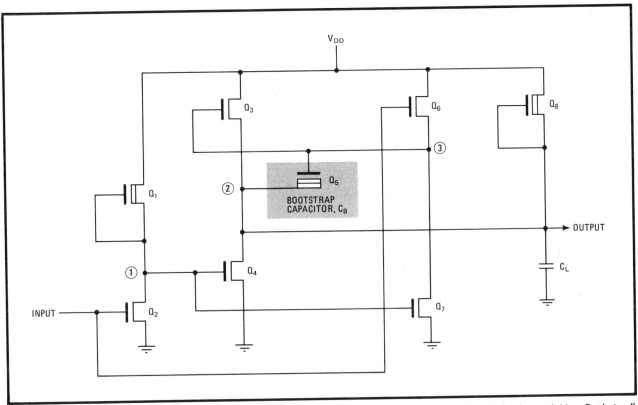

2. Better bootstrap. With the input low, node 1 is high, Q_6 is turned off, and Q_7 is turned on. As a result, node 3 is low and driver Q_3 shuts off. Q_8 can be made physically long, limiting its current and reducing overall power consumption.

Exploiting the full potential of an rf power transistor

by Dan Moline and Dan Bennett
Motorola Semiconductor Products Sector, Phoenix, Ariz.

With improved packaging and appropriate circuit design, the new MRF630 radio-frequency power transistor can be used out to its design limits — the generation of 3 watts with 9.5 decibels of gain at ultrahigh frequencies when assembled with an all-gold metal system.

Good heat sinking enables Motorola's low-cost grounded-emitter TO-39 package for rf transistors to perform like a stripline opposed-emitter type. In this package, the MF630, also from Motorola, shows impressive boardband response, excellent heat dissipation, and high reliability.

So that heat can flow directly away from the transistor die, a flange is soldered to the bottom of the TO-39 can and secured to a heat sink by one or two screws (Fig. 1a). This assembly method maximizes heat dissipation while minimizing space requirements. Also, electrical grounding is better as the package is now connected mechanically to the chassis ground.

The broadband uhf amplifier circuit in Fig. 1b uses a distributed-element design to optimize the gain and bandwidth of the MRF630. The transmission lines are simulated by epoxy fiberglass G-10 board, whose high dielectric constant and low cost keep the circuit small

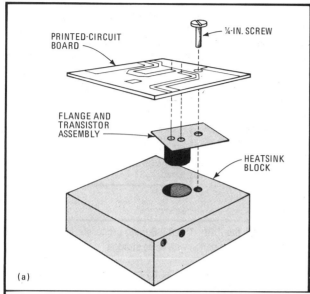

and inexpensive. (In contrast, the commonly used glass Teflon board offers a low dielectric constant at a relatively high price.) To further cut the cost and exploit readily available components, mica capacitors are chosen for the matching network.

Broadband circuit performance (Fig. 2a) shows that the amplifier can furnish more than 3.0 w at frequencies in the range of 450 to 512 megahertz. The high power output can be extended above 490 MHz by optimizing the input and output–impedance-matching networks. With the addition of the copper flange in the circuit assembly, the thermal resistance of the transistor can be expected to be only 12° to 13°C/W. The gain curve (Fig. 2b) demonstrates typical performance of the transistor at ultrahigh frequencies. □

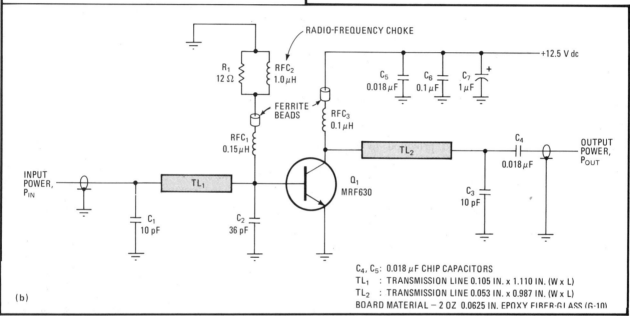

1. High power. Utilizing the construction technique outlined in (a), common emitter TO-39 package for Motorola's MRF630 provides excellent heat dissipation and reliability at high power levels. The amplifier (b) uses the distributed element design to obtain high power at uhf.

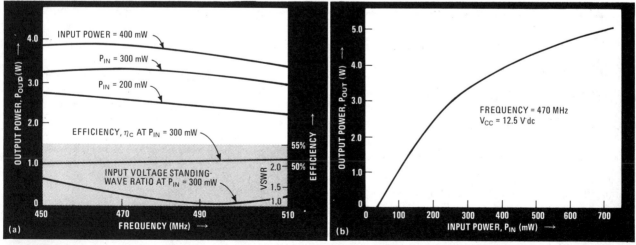

2. Broadband. Its high-frequency performance (a) shows that this amplifier can provide an output of more than 3.0 W above 490 MHz. The gain roll-off above 490 MHz is minimized by optimizing matching networks. The amplifier gain curve at 470 MHz is depicted in (b).

Fm receiver mixes high gain with low power

by Peter Whatley
Motorola Inc., Phoenix, Ariz.

Scanning receivers, ham transceivers, and other narrow-band frequency-modulated systems that receive voice or digital information need as much gain as they can get to pick up weak signals. Normally such gain is expensive. But the Motorola MC3359P chip, which has an oscillator-mixer, limiting amplifier, quadrature discriminator, operational amplifier, squelch, scan control, automatic frequency control, and mute switch neatly combines low cost, high gain, and as a bonus, low power consumption. It requires only a front-end tuner and a few other components to form a complete narrow-band scanning receiver (see figure).

A typical application of the MC3359P is as a narrow-band fm scanning receiver for voice communication. As shown, the input to pin 18 (typically 10.7 megahertz) is converted down by a mixer-oscillator combination to 455 kilohertz, and most of the amplification is done by the chip at this frequency. The mixer is doubly balanced to reduce the fm receiver's spurious responses. Its output at pin 3 has a 1.8-kilohm impedance to match an external ceramic filter. For its part, the oscillator is a Colpitts design that can readily be controlled by a crystal.

After limiting, the fm signal is demodulated using a quadrature detector. The recovered audio is filtered through R_1 and C_1 to remove noise and is then coupled via C_2 to a volume control. The recovered audio is 800 millivolts peak to peak at the junction of R_2 and C_2. The unfiltered recovered audio at pin 10 is fed through R_4 to an internal inverting operational amplifier that, with R_5, C_4, C_5, and R_6, forms an active bandpass filter in the 6-kHz range. Therefore any noise or tone frequency, which may be present above the normal audio range, can be selected, amplified, and then detected by the C_6 and D_1 combination. This detected signal is, in turn, sent to the squelch control at pin 14. Squelch sensitivity may be adjusted by R_9, which provides a bias to the squelch input.

If pin 14 is raised to 0.7 volt by the detected noise, tone, or dc bias, the squelch detector will be activated. This causes pin 15 to act as an open circuit and pin 16 to be shorted to ground via pin 17. Pin 16 is thus connected to the input of the audio amplifier and mutes the audio signal during squelching. Pin 15 can be used for scan control and may be connected to a frequency synthesizer in the receiver's front end. An afc connection is also available at pin 11. In this crystal-controlled application, an afc is not required, so that pin 11 can be grounded or tied to pin 9. With this last connection, the recovered audio is doubled in amplitude. □

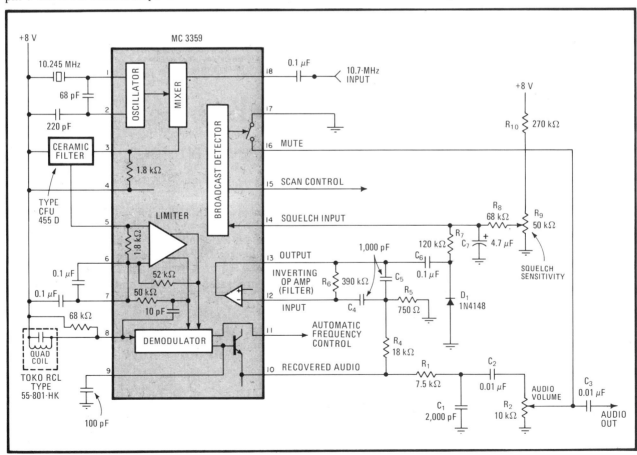

Narrow band. In the scanning-receiver circuit shown, the MC3359P provides an audio output voltage of 800 mV peak to peak. Current drain from a 6-V supply is 3 mA, and the sensitivity is 2 μV for −3 dB of input limiting. Only one crystal and some passive components are needed.

Dual-function amplifier eases circuit design

by Jim Williams
National Semiconductor Corp., Santa Clara, Calif.

To simplify and cut the cost of the myriad of general-purpose and specialized circuits, chips like National's LM392 combine both amplifier and comparator functions on a single substrate. As has already been noted [*Electronics*, May 5, p. 142], it can be used to build a sample-and-hold circuit, a feed-forward low-pass filter and a linearized–platinum-resistor thermometer. This article will present designs for its use in the construction of a variable-ratio digital divider, an exponential voltage-to-frequency converter for electronic music, and a temperature controller for quartz-crystal stabilization.

Figure 1 shows a divider whose digital-pulse input can be divided by any number from 1 to 100 by means of a single-knob control. This function is ideal for bench-type work where the ability to set the division ratio rapidly is advantageous.

With no input signal, transistors Q_1 and Q_3 are off and Q_2 is on. Thus, the 100-picofarad capacitor (C_1) at the junction of Q_2 and Q_3 accumulates a charge equal to $Q_{cap} = C_1 V_0$, where V_0 is the potential across the LM385 zener diode (1.2 volts), minus the saturated collector-to-emitter potential across Q_2.

When the input signal to the circuit goes high (see trace A, in the photograph), Q_2 goes off and Q_1 turns on Q_3. As a result, the charge across C_1 is displaced into A_1's summing junction. A_1 responds by jumping to the value required to maintain its summing junction at zero (trace B).

This sequence is repeated for every input pulse. During this time, A_1's output will generate the staircase waveshape shown as the 0.02-microfarad feedback capacitor (C_2) is pumped by the charge-dispensing action to the A_1 summing junction. When A_1's output is

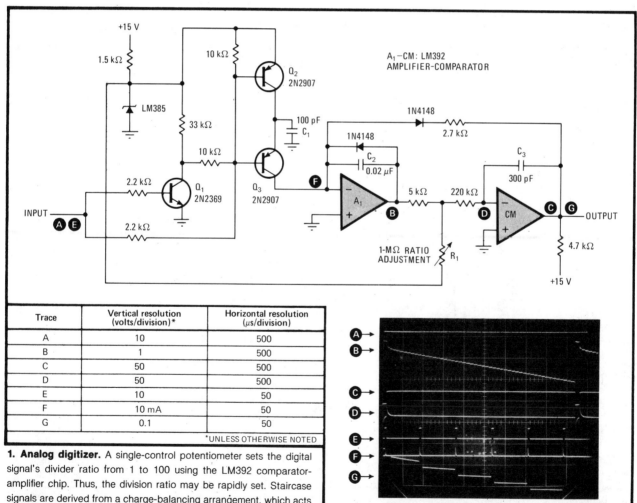

1. Analog digitizer. A single-control potentiometer sets the digital signal's divider ratio from 1 to 100 using the LM392 comparator-amplifier chip. Thus, the division ratio may be rapidly set. Staircase signals are derived from a charge-balancing arrangement, which acts to maintain A_1's summing junction at a voltage null.

Trace	Vertical resolution (volts/division)*	Horizontal resolution (µs/division)
A	10	500
B	1	500
C	50	500
D	50	500
E	10	50
F	10 mA	50
G	0.1	50

*UNLESS OTHERWISE NOTED

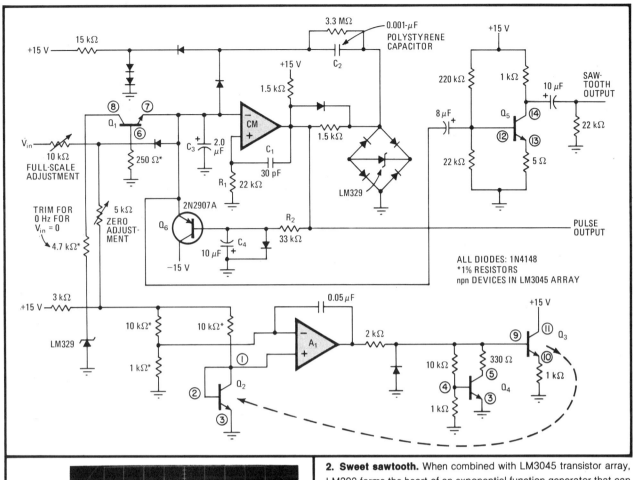

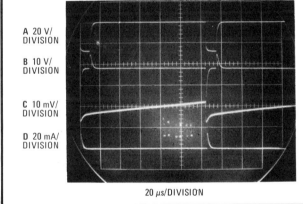

2. Sweet sawtooth. When combined with LM3045 transistor array, LM392 forms the heart of an exponential function generator that can easily be built. Waveform conformity to a pure exponential is excellent—±0.25% over the 20-Hz-to-15-kHz range. Thermal drift is minimized with a simple servo loop. Provision is made for eliminating servo lock-up under virtually all conditions.

age. The one shown in Fig. 2 provides conformity within 0.25% over the range from 20 hertz to 15 kHz using a single LM392 and an LM3045 transistor array. These specifications will be adequate for all but the most demanding of applications.

The exponential function is generated by Q_1, whose collector current varies exponentially with its base-emitter voltage in accordance with the well-known relationship between that voltage and current in a bipolar transistor. An elaborate and expensive compensation scheme is usually required because the transistor's operating point varies widely with temperature. Here, Q_2 and Q_3, located in the array, serve as a heater-sensor pair for A_1, which controls the temperature of Q_2 by means of a simple servo loop. As a consequence, the LM3045 array maintains its constant temperature, eliminating thermal-drift problems in the operation of Q_1. Q_4 is a clamp, preventing the servo from locking up during circuit start-up.

In operation, Q_1's current output is fed into the summing junction of a charge-dispensing current-to-frequency converter. The comparator's output state is used to switch the 0.001-μF capacitor between a reference voltage and the comparator's inverting input, the reference

just great enough to bias the noninverting input of the comparator (CM) below ground, the output (trace C) goes low and resets A_1 to zero. Positive feedback to the comparator (trace D) is applied through the 300-pF capacitor (C_3), ensuring adequate reset time for A_1.

Potentiometer R_1 sets the number of steps in the ramp required to trip the comparator. Thus the circuit's input-to-output division ratio may be conveniently set. Traces E through G expand the scope trace to show the dividing action in detail. When the input E goes high, charge is deposited into A_1's summing junction F, and the resultant waveform G takes a step.

Professional-grade electronic-music synthesizers require voltage-controlled frequency generators whose output frequency is exponentially related to the input volt-

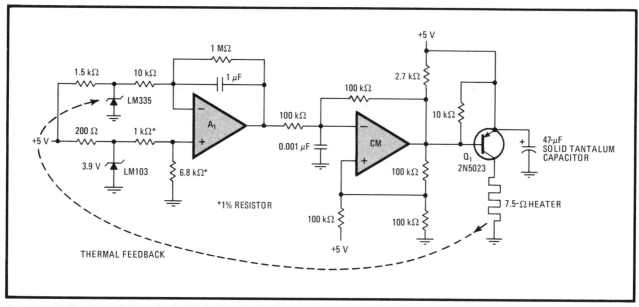

3. Oven cut. Quartz crystals are maintained at 75°C with this temperature controller, thus stabilizing output frequency of these sources. Switched-mode servo loop simplifies circuitry considerably. Long-term temperature accuracy is estimated at 10 parts per million.

being furnished by the LM329.

The comparator drives the capacitor C_1 and resistor R_1 combination, this network providing regenerative feedback to reinforce the direction of its output. Thus, positive feedback ceases when the voltage across the R_1C_1 combination decays, and any negative-going amplifier output will be followed by a single positive edge after the time constant R_1C_1 (see waveforms A and B in the photograph).

The integrating capacitor C_3 is never allowed to charge beyond 10 to 15 millivolts because it is constantly reset by charge dispensed from the switching of C_2 (trace C). If the amplifier's output goes negative, C_2 dumps a quantity of charge into C_3, forcing it to a lower potential (trace D). When a short pulse is transferred through to the comparator's noninverting input, C_2 is again able to charge and the cycle repeats. The rate at which this sequence occurs is directly related to the current into the comparator's summing junction from Q_1. Because this current is exponentially related to the circuit's input voltage, the overall current-to-frequency transfer function is exponentially related to the input voltage.

Any condition that allows C_3 to charge beyond 10 to 20 mV will cause circuit lock-up. Q_6 prevents this by pulling the inverting input of A_1 towards -15 v. The resistor and capacitor combination of R_2 and C_4 determines when the transistor comes on. When the circuit is running normally, Q_6 is biased off and is in effect out of the circuit.

The circuit is calibrated by simply grounding the input and adjusting first the zeroing potentiometer until oscillations just start and then the full-scale potentiometer so that the circuit's frequency output exactly doubles for each volt of input (1 v per octave for musical purposes). The comparator's output pulses while Q_5 amplifies the summing junction ramp for a sawtooth output.

The circuit in Fig. 3 will maintain the temperature of a quartz-crystal oven at 75°C. Five-volt single-supply operation permits the circuit to be powered directly from TTL-type rails.

A_1, operating at a gain of 100, determines the voltage difference between the temperature setpoint and the LM335 temperature sensor, which is located inside the oven. The temperature setpoint is established by the LM103 3.9-v reference and the 1-to-6.8-kilohm divider.

A_1's output biases the comparator, which functions as a pulse-width modulator and biases Q_1 to deliver switched-mode power to the heater. When power is applied, A_1's output goes high, causing the comparator's output to saturate low. Q_1 then comes on.

When the oven warms to the desired setpoint, A_1's output falls and the comparator begins to pulse-width–modulate the heater via the servo loop. In practice, the LM335 should be in good thermal contact with the heater to prevent oscillation in the servo loop. □

Antilog amplifiers improve biomedical signal's S/N ratio

by T. G. Barnett and D. L. Wingate
London Hospital Medical College, Department of Physiology, England

Low-voltage, biphasic signals recorded by instruments monitoring biomedical variables such as heart rate are often accompanied by high noise levels due to inadequate sensing, movement artefact, paging systems and power-line interference. Using paired antilogarithmic amplifiers, however, to provide the nonlinear amplification required, the level of the biphasic signals can be raised well above the amplitude of the interfering signals. The signal-to-noise ratio can thus be improved from 2:1 at the input to 10:1 at the output.

Such a scheme is superior to the use of paired logarithmic amps, which cannot handle biphasic signals at the zero-crossing points ($\log 0 = \infty$), and provides more sensitivity than conventional diode clippers, which introduce noise and cannot pass signals that drop below the circuit's 0.7 clipping threshold.

Input signals are amplified by A_1 and are separated into their positive and negative components by precision half-wave rectifier A_2 and inverter A_3. The corresponding outputs are then introduced into the AD759N and AD759P log/antilog amplifiers, which are wired to yield $e_o = E_{ref} 10^{-e_{in1}/K}$ for $-2 \leq e_{in1}/K \leq 2$, where E_{ref} is an internal reference voltage of approximately 0.1 volt and K is a multiplying constant that has been set at 1 as a consequence of utilizing input e_{in1}. The output voltages generated by A_4 and A_5 are of negative and positive polarity, respectively.

These components are then summed by A_6, whose output yields a bipolar, antilogged signal that can be introduced to appropriate trigger circuits. If desired, the original signal can be reconstructed by passing it through paired logarithmic amplifiers. □

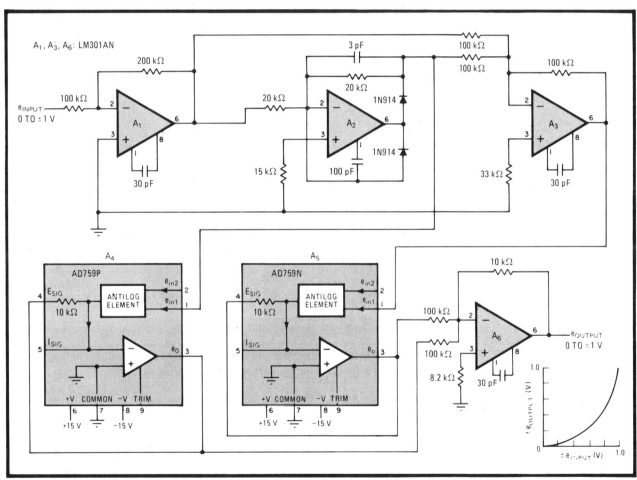

Biomedical booster. Paired AD759 antilog amps provide bipolar, nonlinear amplification, thus raising level of biphasic signals such as EKGs with respect to noise and so increasing S/N ratio. Circuit is superior to those using log amps, which cannot provide accurate output at zero crossings of signals, and is more sensitive than diode clippers, provides greater noise rejection than filters, and introduces no phase shift.

Differential amp cancels integrator's crosstalk

by Elzbieta Nowicka
Atomic Energy of Canada Ltd., Ottawa

The performance of a high-speed integrator can be improved considerably by utilizing the excellent common-mode rejection characteristics of an operational amplifier to reduce crosstalk caused by the switching of waveforms during the circuit's integrate-and-hold sequence. More specifically, operating the amp in its differential mode enables the integrator to virtually cancel the switching offsets that are generated by the almost ideal switch—the complementary-MOS analog gate, which is inexpensive and has low on-resistance but relatively high feedthrough. The supporting circuitry required (two extra gates) is minimal.

The technique generally used to start and terminate the three-step integration sequence is shown in (a), where S_1 and S_2 represent two electronic switches. These switches must be selected carefully, the main requirement being that the on-resistance of these devices be as low as possible, especially when the R_1C_1 time constant is small.

The C-MOS CD4066 transmission gate, with its nominal on-resistance of only 80 ohms for a supply voltage of 15 volts, would normally serve well in these applications, except that its crosstalk (the unwanted feedthrough of the gating signal) is high—typically, 50 millivolts for a 10-V square wave having $t_r = t_f = 20$ nanoseconds and a gate-output load of 1 kilohm. By introducing a pair of switches at each of both inputs of the LM218 op amp (b), however, the integrator will provide virtual elimination of the crosstalk by means of differential cancelation of gate feedthrough signals 1 and 2. There will be little error introduced by the differences between the on resistance of each individual gate in the 4066 package because of the balanced circuit arrangement. Typically, Δ_{on} will be less than 5 Ω for a 15-V supply.

Utilizing this scheme, the crosstalk will be reduced to less than 3 mV over a temperature range of 0° to 70°C. □

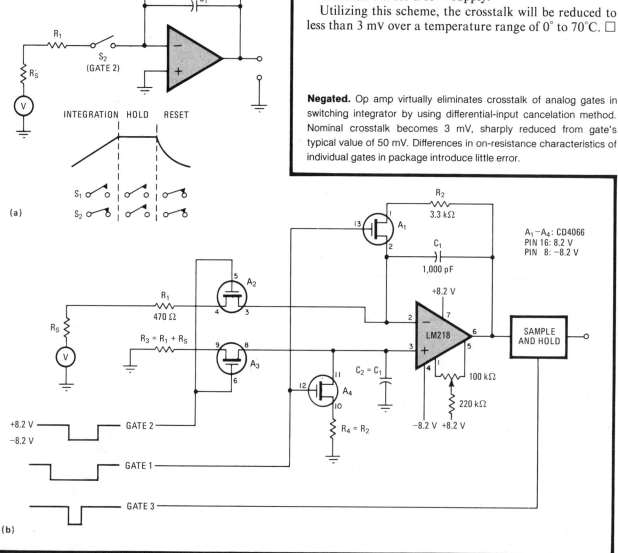

Negated. Op amp virtually eliminates crosstalk of analog gates in switching integrator by using differential-input cancelation method. Nominal crosstalk becomes 3 mV, sharply reduced from gate's typical value of 50 mV. Differences in on-resistance characteristics of individual gates in package introduce little error.

2. AUDIO CIRCUITS

OTAs and op amp form voltage-controlled equalizer

by Henrique S. Malvar
University of Brazilia, Brazil

An audio equalizer usually employs a manual control to regulate frequencies. However, with just two operational transconductance amplifiers, an op amp, and a constant-current source, a simple voltage-controlled audio-equalizer section can be made that, in effect, automatically controls the waveforms of a system. This section (a) can control a graphic equalizer of an audio system or, through a microprocessor, equalize the generated output from a speech or music synthesizer.

The transfer function of the circuit, H(s), is defined as $V_o(s)/V_{in}(s)$, or:

$$\frac{s^2 C_1 C_2 + s(C_2 G_3 + C_1 [G_4 + G_3 - (bg_{m1}/a)]) + G_3 G_4}{s^2 C_1 C_2 + s(C_2 G_3 + C_1 [G_4 + G_3 - (bg_{m2}/a)]) + G_3 G_4}$$

where $g_{m1} = I_{B1}/2V_t$ is the transconductance of OTA U_1, $g_{m2} = I_{B2}/2V_T$ is the transconductance of U_2, V_t is the volt equivalent of temperature (26 millivolts at 300 K) $G_3 = 1/R_3$, and $G_4 = 1/R_4$. Bias currents I_{B1} and I_{B2} alter only the first-order terms of H(s)—the requirement for a bump equalizer.

The gain at frequency ω_o, which is given by $1/(R_3 R_4 C_1 C_2)^{1/2}$, is flat when the externally applied bias currents I_{B1} and I_{B2} are equal. A boost in equalizer gain is attained for $I_{B2} > I_{B1}$ and vice versa. As a result, a positive value for control voltage V_c leads to a boost response, and a negative value for V_c results in a loss.

The gain versus frequency response (b) measured by the circuit corresponds to control voltages +4 volts, +2 v, 0 v, -2 v, and -4 v. The curve shows that the equalizer provides gain for positive voltages and attenuation for negative values. With the given values, voltage dividers R_2 and bR_2 (b is a constant) keep the signal levels at the inputs of U_1 and U_2 within their permitted linear range. The 400-microampere constant-current source can be implemented with a pnp transistor or a p-channel field-effect transistor. In addition, because the transconductance of an OTA decreases with temperature, the current source must have a positive temperature coefficient of about 0.3%/°C. □

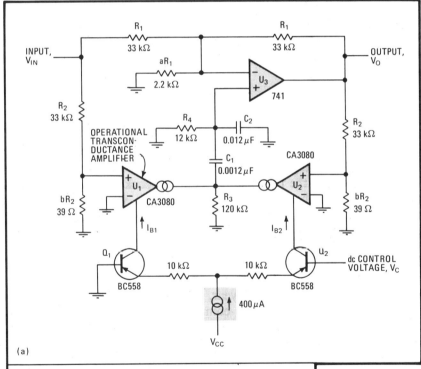

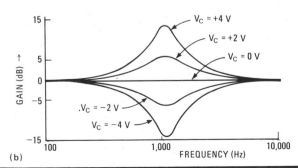

Equalizer. The circuit (a) uses two operational transconductance amplifiers CA3080, a 741 op amp, and a 400-microampere constant-current source to obtain a voltage-controlled audio-equalizer section. The parameters a and b are constants for voltage-dividing resistors R_1 and R_2. The measured frequency response (b) is attained for control voltages of +4 V, +2 V, 0 V, -2 V, and -4 V.

Digital processor improves speech-signal quality

by Brian Dance
North Worcestershire College, Bromsgrove, England

This microprocessor-controlled digital sampling technique improves the intelligibility of analog speech signals through substantially increasing the signal-to-noise ratio and therefore is particularly useful where noise becomes a difficult to solve problem. Its applications include public-address equipment, communication systems, and hearing aids.

In this system, the input signal is delayed by 8 milliseconds and fed to a divider whose gain is controlled by the feed-forward controller[1] (Fig. 1a). The input waveform is then sampled for every half a cycle and its maximum signal-amplitude value noted. This peak signal value thus determines the gain factor by which every sample taken during that half cycle is amplified. If the value of the input waveform exceeds the threshold value (Fig. 1b), the gain is automatically adjusted so that the instantaneous output voltage at that peak is increased to a specific value, V_{max}, which is the same for each half cycle.

Because gain changes occur only at the signal waveform's zero-crossing points, there is no sharp discontinuity in the output waveform that arises from gain changes. In addition, the waveform of the sampled and processed output is very similar to the input. Signals with an amplitude less than that of the threshold value are passed through the system with a gain of unity.

A detailed operation of the feed-forward level controller is illustrated in Fig. 2. Two random-access memories store the sample and gain values in the form of memory queues. The sample queue with a fixed length is accessed with a memory pointer and the gain-function queue is contained in a first-in, first-out memory, the contents of which depend on the number of peaks in the sample queue. The gain-function queue has input and output memory pointers. When a zero crossing is detected, the peak value from the previous half cycle is compared with the threshold value to select the gain, which is then placed in the gain-function queue.

The flow chart (see table), which is a software version

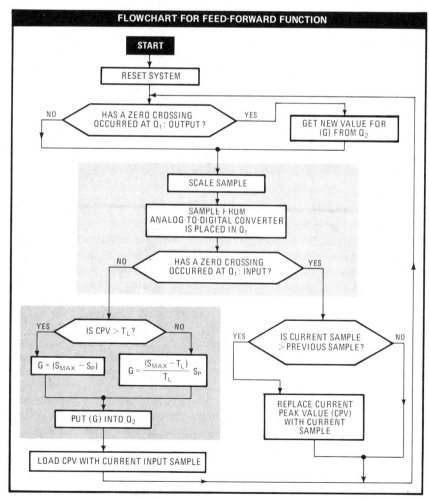

1. Processor. The feed-forward and delay technique (a) used in this processor system substantially improves the intelligibility of speech signals. The input signal, which is sampled at 16 kHz, is amplified with the appropriate gain factor to produce an output waveform that is similar to the input (b).

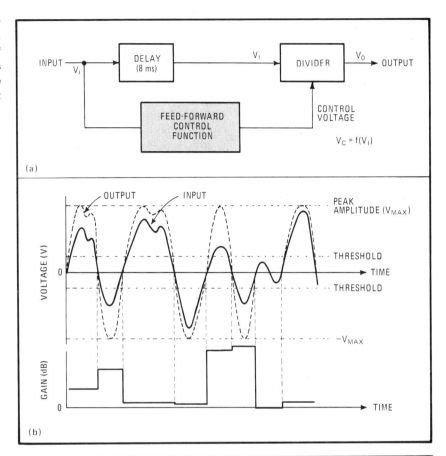

2. Controller. This circuit diagram illustrates the principle of operation of the feed-forward level controller. Sample and gain values are contained in the memory in form of queues. Peak value in each half cycle is compared with the threshold level to select the gain value placed in the gain-function queue.

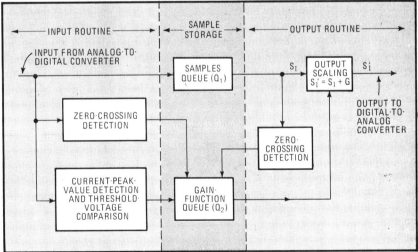

of the system's operation, indicates that the system relies on the correct start-up procedure for its continued operation. However, no noise problems were encountered. □

References
1. V. J. Phillips and L. D. Thomas, "A feed-forward level controller for speech signals," Report of the Department of Electrical and Electronics Engineering, University College of Swansea, University of Wales.

Electronic-music generator retriggers waveforms

by Thomas Henry
Electronic Music Studios, University of Iowa, Iowa City

Only three chips, one transistor, and a few passive components make up a circuit that produces the control signals needed by voltage-controlled oscillators and amplifiers in order to modulate musical parameters like loudness, timbre, and pitch in an electronic synthesizer. The circuit retains the simplicity and compactness of an exponential generator [*Electronics*, July 17, 1980, p. 123], and it also permits retriggerable operation.

This envelope generator controls the attack, decay and release times of a waveform and its sustained level. The circuit, however, also responds independently to the trigger and gate input signals of a keyboard, thereby creating a continuously repeating attack-decay portion of the waveform.

The synthesizer's keyboard initiates trigger and gate signals that control circuit timing. The gate pulse is produced if at least one key is pushed, and the trigger pulse is produced whenever the lowest note desired is changed by selecting a new key.

Thus a gate and a trigger pulse is generated when a key is pushed. Pressing a second key lower on the keyboard while holding the first key down generates a new trigger pulse, but there is no change in the gate pulse because it remains high.

In the circuit's quiescent state, the charge on timing capacitor C_3 is zero. When a single key is pressed, the 555 timer, which contains a comparator and R-S flip-flop, is enabled by the gate signal.

Following this action, the trigger signal is differentiated by the resistor-capacitor combination R_1C_1 and is applied to pin 2 of the timer to set its internal flip-flop and bring pin 3 high. Analog switch S_1 is thus turned on and current flows through resistor R_7 to charge C_3, initiating the attack portion of the waveform.

When the voltage across C_3 reaches 10 volts, the 555's flip-flop is reset. Pin 3 moves low and S_1 is turned off, thus terminating the cycle's attack portion.

At this time, the AND gate formed by D_1, D_2, and R_3 goes high, and analog switch S_3 is turned on. This switching permits C_3 to discharge through R_8, the decay-control potentiometer, to a voltage level specified by R_9, the sustain control. Resistor R_4 limits the range of sustain voltage from 0 to 10 V.

When the gate signal is removed, the AND gate goes low and S_3 is turned off. Transistor Q_1 is also turned off, which permits S_4 to turn on and C_3 to discharge through R_{10} for the release portion of the waveform.

If the generator is in the sustain portion of its output cycle, and a new trigger occurs, the flip-flop will be set again and the attack and decay portions of the cycle will recur. Coupling the gate signal to pin 4 of the 555 ensures retriggering even if the wave is not permitted to reach the conclusion of its attack cycle. □

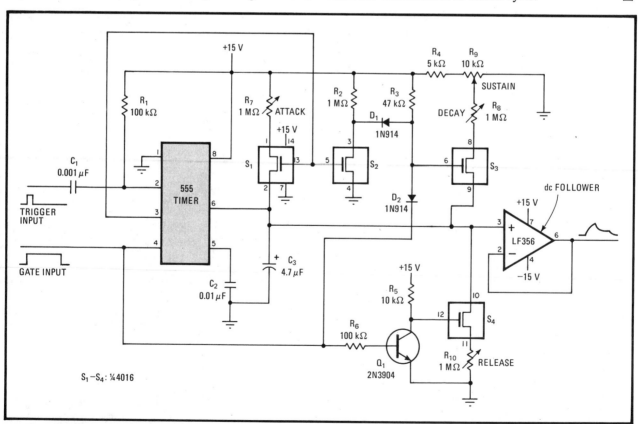

Repeating repertoire. The three-chip generator provides a four-step waveform, each cycle of which is adjustable, for modulating voltage-controlled oscillators in music synthesizers. Thus loudness, timbre, and pitch may be adjusted. The circuit costs under $8.

Agc prevents noise build-up in voice-operated mike

by Russell S. Thynes
Kirkland, Wash.

Hands-free operation of intercoms has several advantages over push-to-talk intercom systems. Constantly keyed "live" microphones, however, have the disadvantage of receiving undesirable environmental noise in the absence of speech. When such mikes are used in conjunction with intercoms having automatic gain control in the microphone mixing stages, this environmental noise will produce a swelling tide of sound each time normal communication is interrupted.

Shown here is an agc-VOX (voice-operated switch) scheme that allows constantly keyed microphones to be used in noisy environments without suffering from the effects of noise build-up.

Although the operation of the circuit is twofold, the function is primarily that of a gain-clamped agc circuit. Part (a) of the figure shows the transfer function of an agc with gain clamping and that of typical configuration. For input levels below those of normal speech, clamping the gain to a fixed value limits the area of the gain curve, reducing noise susceptibility—but without placing restrictions on the dynamic range of the agc itself.

The circuit is shown on the right (b). The agc section consists of operational amplifier A_1 and transistor Q_1, with diode D_1 and capacitor C_1 deriving the feedback control voltage. Q_1 is placed in a T configuration to achieve a wide control range and to ensure low levels of distortion. Distortion is further reduced by the gate-biasing resistors R_6 and R_7. As configured, this agc should provide 30 decibels of gain control with less than 0.5% distortion for most of the audio range.

A_3 is arranged as an adjustable noninverting amplifier, the gain of which can be varied from 20 to 40 dB. R_{12} (also in a T configuration) allows the user to set the VOX sensitivity to offset environmental noise conditions. A_4 simply compares the detected output of A_3 with a reference and switches to either a high or a low output limit depending on the VOX input level.

When input levels to the agc are below the VOX sensitivity setting, the output of A_4 will be at its lower limit, biasing Q_1 off and thus clamping the gain of A_1 to $(R_2+R_3)/R_1$.

When the VOX sensitivity level is exceeded, however, the comparator output swings to its upper limit and effectively disconnects the VOX from the agc feedback loop through blocking diode D_2. The gain is then expressed as:

$$-\frac{R_2+R_3}{R_1} \leq A_v \leq -\left(\frac{R_2+R_3}{R_1} - \frac{R_2R_3}{R_{on}R_1}\right)$$

where R_{on} is the practical on-resistance of the transistor.

This entire circuit can be configured using one quad op amp (such as an XR 4136) and requires no special considerations other than attention to the basic rules of grounding and supply bypassing. □

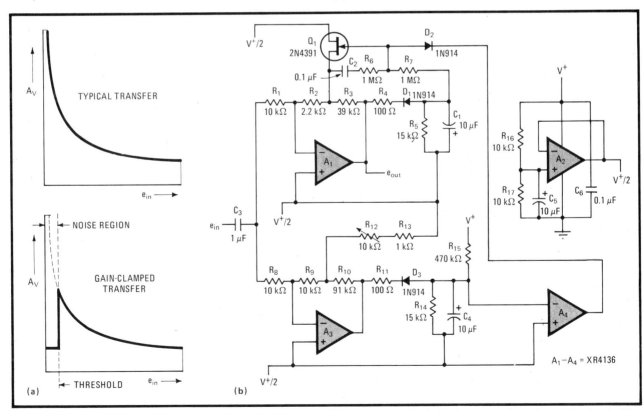

Hands off. If the gain is clamped to a minimum at input signal levels below the noise threshold, the surrounding noise is filtered out of the amplifier network (a), whereas speech kicks in the amplifier's automatic gain control. Both functions are performed by the circuit shown in (b).

Switching circuit fulfills three functions

by Charles Carson
WASH-FM, Metromedia Radio, Washington, D.C.

With just a single switching system, one audio processor may be shared by many different recording stations. The system can without noise or interference switch the processor to the station where it is needed, cancels its use at the locations where it is not being employed, and identifies its switching mode. As an example, the circuit is set up for three stations, but more functions can be added by using extra identical stages.

The output of the alternate-action switch consisting of inverters U_{1-a} and U_{1-b} changes state once S_1 is closed or optocoupler O_1 is turned on (Fig. 1). When the alternate-action switch's output at pin 4 is high, transistor Q_4 is turned on and relay 1 is energized. Also, this high output resets the other two alternate-action switches, turning them off. Closing S_1 again makes the output at pin 4 of U_{1-b} go low. As a result, relay 1 is disabled.

When station 1 is on, so is its corresponding lamp. The lamps for stations 2 and 3 flash to indicate that the system is in use elsewhere (Fig. 2). All the lamps are off when the system is not in use.

Inverters U_{2-a} and U_{2-b} form an astable oscillator operating at 1 hertz. Because transistor Q_7 is turned on when station 1 is on, the station's lamp burns continuously. The logic high is diode-coupled to AND gates U_{3-b} and U_{3-c} and now allows the output of the astable oscillator to be fed to transistors Q_8 and Q_9 through the gates and the diodes. This transistor output makes the lamps at the other two stations flash at 1 Hz. When the stations are idle, the control lines are in a low state, and consequently the indicator lamps remain off. □

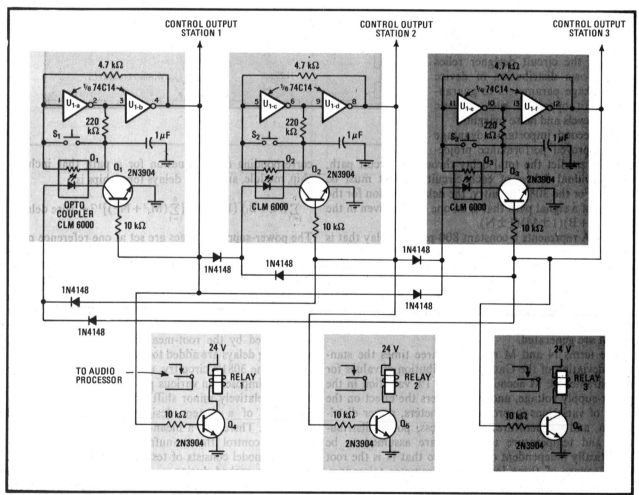

1. Switching. By means of this circuit, a single audio processor can switch between three recording stations. Inverters U_{1-a} and U_{1-b} form the alternate-action switch for station 1. Similarly inverters U_{1-c} through U_{1-f} simulate the switching action for stations 2 and 3. Relay 1 is energized when the output at pin 4 of U_{1-b} is high. In addition, this output resets the other two switches.

Parametric equalizer improves Baxandall tone control

by Henrique Sarmento Malvar
Department of Electrical Engineering, University of Brazilia, Brazil

Simple active filters are used here to build a continuously adjustable parametric equalizer having the same general response as the popular Baxandall circuit, which utilizes a switch-selectable scheme for bass and treble equalization. Center frequencies for both upper and lower bands, as well as their individual roll-off characteristics, may be independently controlled, and the depth of the equalization is also adjustable.

The circuit (see Fig. 1), an adaptation of an idea proposed by Thomas,[1] utilizes positive feedback and/or feed-forward principles to achieve the type and amount of equalization required. Five potentiometers set the aforementioned parameters, with the circuit operating on all simultaneously.

The center of the low-frequency passband is set by R_1. If the wiper of the bass control is moved towards the input operational amplifier, A_1, more of the low-frequency components of the input signal will pass through low-pass filter $C_1R_3R_8$ and appear at V_x, with potentiometer R_3 determining the roll-off. Because op amp A_2 inverts the signal, partial cancelation of the low-frequency components occurs and the total bass content is reduced at the output. As R_1 is moved in the opposite direction, a positive feedback loop around op amps A_2 and A_3 is formed, and the bass gain increases. In similar ST_0–ST_3, are set to LLLH during refresh. This output is decoded using the 74LS138 three-to-eight line decoder. Lines ST_1–ST_3 drive the decoder's select lines, A–C, and ST_0 drives the active-high enable input. The two active-low enable inputs are driven from the system clock and the address strobe, $\overline{AS}$, which goes low during each refresh cycle. When the rate counter times out, the address strobe moves low and the address of the row counter is placed on the address lines.

Thus, during a refresh cycle, the decoder's $\overline{Y}_0$ output will be driven low, synchronous with the $\overline{AS}$ and the system clock. This output is then fed to an active-low AND gate, along with the inverted AD_8 line, which only moves low when the row counter reaches a count of 128. Therefore the AND gate's output will go high only after the rate counter has timed out 128 times, assuming the refresh counter is initialized to zero.

Because the counter can be initialized to virtually any value, the refresh-interval time (which is the time between refresh cycles) is adjustable from 1 microsecond

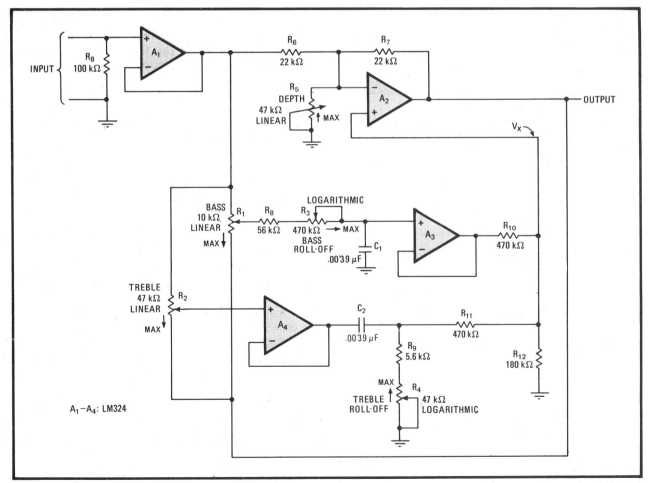

1. Trimming timbre. One-chip equalizer provides continuously variable control of bass and treble center frequencies, as well as individual roll-off characteristics. Depth of audio-band equalization is also adjustable. Unit costs little more than standard switch-selectable devices.

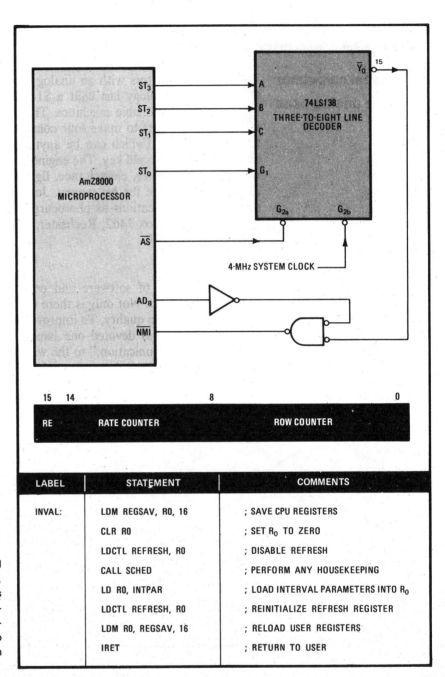

Internal interrupts. Software-based interval timer for refreshing dynamic memories, made possible by such versatile processors as the AmZ8000, has adjustable refresh-cycle and interval times. Only minimal hardware—a three-to-eight-line decoder and two gates for counting the number of refresh cycles per interval—is required.

to 8.194 milliseconds. The refresh-interval time is given by $T_0(128 - C_0)$, where T_0 is the time-out interval of the rate counter and C_0 is the initial value of the refresh counter.

The output from the AND gate, which indicates the end of the timing interval, is then introduced to the nonmaskable interrupt ($\overline{\text{NMI}}$) of the CPU. The CPU then enters the service routine, writing a 0 into bit 15 of the refresh register to disable the refresh operation so that it itself cannot be interrupted. Before returning control to the CPU, it reinitializes the refresh-row counter to the desired value and reenables refresh. This starts the timing of the next interval, at the end of which the CPU will be again interrupted. □

Audio-visual controller synchronizes museum display

by William S. Wagner
Northern Kentucky Unversity, Highland Heights, Ky.

A synchronized sound-and-transparency show that may be placed in any convenient area of a museum or science building can be created with this interface. Built entirely with off-the-shelf components, the cost of the circuit is below $20.

The interface controls a cassette player and a display having several illuminated panels. Each panel contains a color transparency and a source of light (in this case, a light bulb). The circuit causes these transparencies to light in sequence, while advancing the cassette tape, which contains a recorded message for each panel. Pairs of recorded audio tones control panel sequencing and thus synchronize the audio-visual display.

When the show ends, a second pair of audio tones shuts the entire display off. Because a continuous tape loop is used, the show may be restarted by pressing a start button. A pause button is included to extend the viewing period of any panel.

As for circuit operation, when the start button is pressed, the 4043 reset-set flip-flop sets, turning on the 2N2222 transistor and pulling in the double-pole, double-throw relay. Its normally open contact closes, turning on the cassette player's motor. At the same time, the relay's normally closed contact opens, allowing the audio signal to reach the LM324 amplifier. A 667- and a 1,200-hertz tone combine to form the initial sound heard and to activate their respective 567 phase-locked loops. This causes pin 3 of the first 4001 NOR gate to go high and advance the 4017 ring counter.

When pin 2 of the counter goes high, the 2N3904 and 2N3906 transistors turn on and the triac fires, causing the first light bulb to illuminate the first transparency. Then the recorded message corresponding to that transparency is played. At the end of the message a second pair of recorded tones (667 and 1,200 Hz) causes the 4017 counter to advance to pin 4, turning off the first light and turning on the second with its appropriate interfacing transistors and triac.

This process is repeated until all transparencies have been displayed and described. At the end of the show, recorded tones at 850 and 1,200 Hz activate their respec-

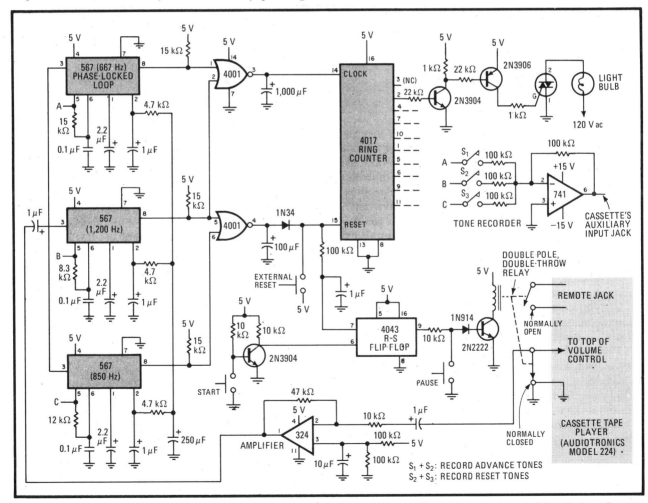

Show and tell. This interface synchronizes sound with illuminated panels and can be used in museum or science building displays. Cassette tapes hold recorded segments corresponding to information seen on illuminated display panels. When the circuit detects a chord preceding a given segment of text, the following panel is illuminated. Also, the interface has an automatic shut-down feature.

tive PLLs and the 4017 is reset so that all lights are turned off. The 4043 R-S flip-flop is also reset and turns off the 2N2222, which deactivates the relay and turns off the cassette player. □

Two-chip generator shapes synthesizer's sounds

by Jonathan Jacky
Seattle, Wash.

Generating the same adjustable modulating waveforms for a music synthesizer as the circuit proposed by Kirschman[1], but using only two integrated circuits, this generator also works from a single supply. It has, in addition, separate gate and trigger inputs for providing a more realistic keyboard response.

When gated or triggered, the generator, which is built around Intersil Inc.'s C-MOS 7555 timer, produces a waveform that passes through four states:
- An exponential attack.
- An initial decay, or fallback.
- A sustain, or steady dc level.
- A final decay, or release.

Each of these four parameters is continuously variable, so that waveforms having a wide variety of shapes can be generated.

The waveforms are generated by the sequential charging and discharging of capacitor C_1. Here, the 7555 controls the sequencing while diodes switch the currents, unlike Kirschman's circuit where comparators and flip-flops control the stepping and analog switches steer the currents. Furthermore, the 7555 is well suited for handling the two logic signals provided by most synthesizer keyboards—the gate, which is high as long as any key is depressed, and the trigger, which provides a negative pulse as each key is struck. The gate and trigger features eliminate the need to release each key before striking the next to initiate an attack phase.

In the dormant state (the gate input at pin 4 of the 7555 is low), capacitor C_2 is discharged. When the gate goes high and a trigger pulse appears at pin 2, the 7555 output (pin 3) goes high and charges C_1 through R_3, R_4, and D_1, producing the attack segment of the waveform. Note that diode D_2 is reverse-biased because pin 7 of the 7555 is high and that diode D_3 is back-biased by logic 1 signal applied to the gate input.

When the voltage across C_1 reaches 10 volts, pin 3 of the 7555 goes low and pin 7 is grounded, terminating the attack phase. D_1 and D_3 are now reverse-biased and C_1 discharges through D_2, R_5, and R_6 to produce the initial decay. The sustain level reached is determined by the voltage divider formed by resistor R_1 and potentiometer R_2. During this phase, a second attack can be obtained by striking another key (see timing diagram). When the last key is released, the gate goes low and C_1 will discharge through D_3, R_7, and R_8 to produce the final decay. The CA3130 operational amplifier serves as a buffer to protect C_1 from excessive loading. □

References
1. Randall K. Kirschman, "Adjustable e^x generator colors synthesizer's sounds," Electronics, July 17, 1980, p. 123.

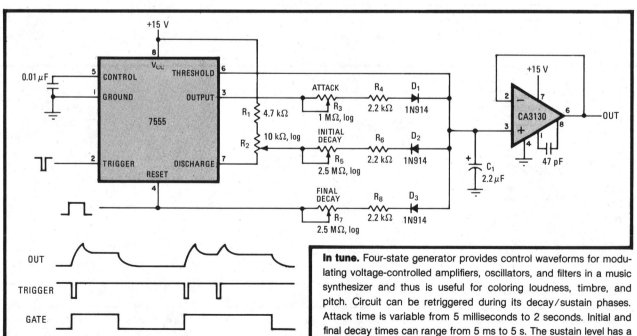

In tune. Four-state generator provides control waveforms for modulating voltage-controlled amplifiers, oscillators, and filters in a music synthesizer and thus is useful for coloring loudness, timbre, and pitch. Circuit can be retriggered during its decay/sustain phases. Attack time is variable from 5 milliseconds to 2 seconds. Initial and final decay times can range from 5 ms to 5 s. The sustain level has a dynamic range of 0 to 10 V.

On-chip transistors extend audio amp's design flexibility

by Jim Williams
National Semiconductor Corp., Santa Clara, Calif.

The availability of extremely low-cost audio-amplifier integrated circuits with on-chip transistor arrays, such as National's LM389, gives designers a great deal of flexibility in designing audio circuits. They make it much easier to develop low-cost versions of circuits unrelated to basic audio amplification, such as dc-dc converters, touch switches, stabilized frequency standards, scope calibrators, low-distortion oscillators, and logarithmic amplifiers. The designs of the often-needed converter, a bistable touch switch, and a tuning-fork frequency standard are discussed here in the first part of this two-part presentation.

The LM389 contains a 250-milliwatt audio amplifier and an array of three npn transistors, each of which is uncommitted. The amp has differential inputs and separate pins for setting its gain (from 20 to 200) via a resistor and runs off a single supply that may range from 4 to 15 volts. The three transistors have a minimum current-handling capability of 25 milliamperes and a minimum current gain of 100 for $V_{ce\,max} = 12$ v and for a wide range of collector currents. The chip is therefore ideal for general use.

One area in which the chip will be useful is in dc-dc switching conversion. The device in Fig. 1 is intended for use as a power supply in a digital system where it is necessary to supply ±15 v to a low-power load. As can be seen from the oscilloscope photograph, the LM389 switches at 20 kilohertz. That rate is determined by the triangular-wave feedback signal, whose time constant is set by $R_1 C_1 C_2$, and its square-wave output is applied to transistors Q_1 and Q_3. The series diodes ensure clean turn-off for Q_1 and Q_3.

Q_1's inverted output drives one half of the transformer primary through Q_2, while Q_3 drives the other half. The diodes across Q_2 and Q_3 suppress spikes. Thus there is an

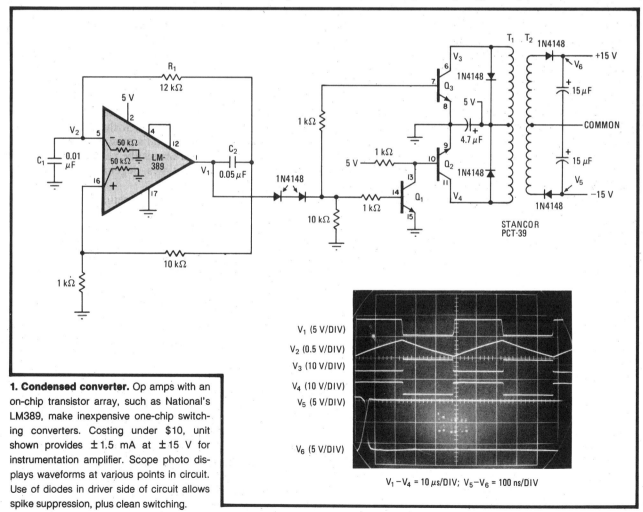

1. Condensed converter. Op amps with an on-chip transistor array, such as National's LM389, make inexpensive one-chip switching converters. Costing under $10, unit shown provides ±1.5 mA at ±15 V for instrumentation amplifier. Scope photo displays waveforms at various points in circuit. Use of diodes in driver side of circuit allows spike suppression, plus clean switching.

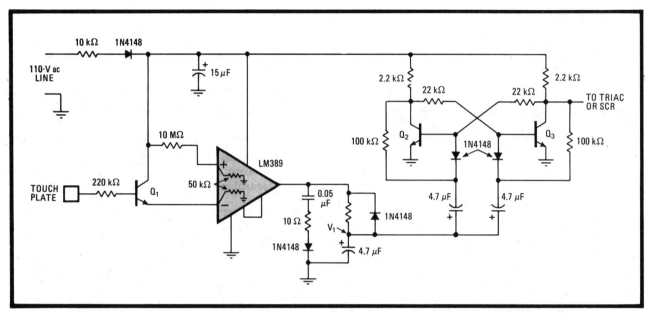

2. Simple switching. Simple bistable touch switch may be similarly constructed. Op amp works as comparator and trigger for flip-flop Q_2–Q_3, which changes state each time plate is contacted. Thus, SCR at output may be alternately fired and switched off on command.

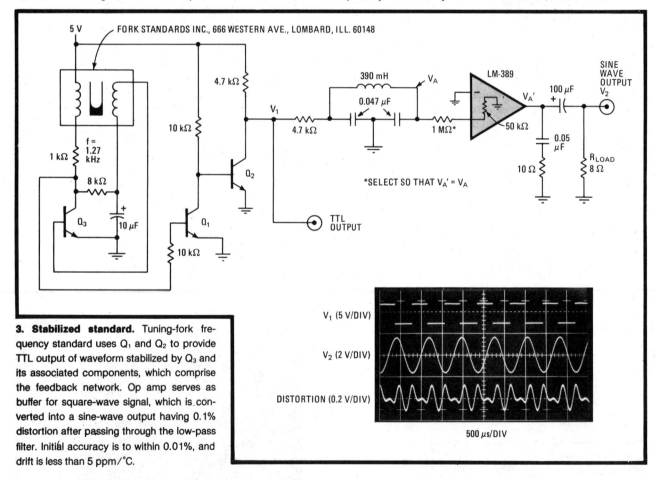

3. Stabilized standard. Tuning-fork frequency standard uses Q_1 and Q_2 to provide TTL output of waveform stabilized by Q_3 and its associated components, which comprise the feedback network. Op amp serves as buffer for square-wave signal, which is converted into a sine-wave output having 0.1% distortion after passing through the low-pass filter. Initial accuracy is to within 0.01%, and drift is less than 5 ppm/°C.

efficient step-up of voltage across T_2. This ac signal is rectified and filtered to produce complementary output voltages that may be used to power the desired linear components, in this case delivering ±1.5 mA, enough to power an operational or instrumentation amplifier.

The bistable touch switch (Fig. 2) allows a line-powered load to be controlled from a touch plate by means of a thyristor. Each time the plate is contacted, emitter-follower Q_1 conducts, permitting a fraction of the 60-hertz input signal to be applied to the inverting input of the amplifier. Consequently, the normally high output of the op amp follows the 60-Hz line input, causing V_1 to drop sharply.

This negative transition triggers a toggling flip-flop

formed by Q_2 and Q_3. In this manner, the output of the flip-flop changes state each time the touch plate is contacted, prompting the firing of the silicon controlled rectifier or triac that switches ac power to the load.

Figure 3 shows a tuning-fork frequency standard that is stabilized by appropriate feedback. Both sine-wave and TTL-compatible outputs are available. As the circuit needs only 5 V, it can run off a battery.

The tuning fork proper supplies a low-frequency output that is very stable (typically to within 5 ppm/°C) and has an initial accuracy of within 0.01%. Moreover, it will withstand vibration and shock that would fracture a quartz crystal. Here, Q_3 is set up in a feedback configuration that forces the fork to oscillate at its resonant frequency. Q_3's output is squared up by Q_1 and Q_2, which provide a TTL-compatible output. When passed through an LC filter and the op amp, which provides a low-impedance (8-ohm) output, the signal is converted into a sine wave having less than 1% distortion, as shown in the figure.

Several other useful circuits also can be built. The second part of this article will deal with the chip's use in a portable scope calibrator, a low-distortion oscillator, and as a logarithmic amplifier. □

Audio-range mixer calibrates intermod-distortion analyzers

by Michael Bozoian
Electric Consulting Service, Ann Arbor, Mich.

This low-cost system generates intermodulation distortion levels in the audio band from 0% to 100% and is thus a useful test and calibration instrument. Only two integrated circuits are required for it.

The LM3600N, an operational transconductance amplifier with linearizing diodes and a buffer, is wired as a wideband amplitude modulator (Fig. 1). The modulation level and phase is set by a 741 operational amplifier that works as a scaler and inverter. The modulation waveform generated by the circuit is symmetrical. As a result, the percent modulation m becomes $100(E_{crest} - E_{trough})/(E_{crest} + E_{trough})$ (Fig. 2). The calibrated circuit generates a value for m that becomes equal to the rms value of the modulating voltage, e_m.

The modulating frequencies applied range from 10 to 500 hertz, with 60 Hz most common. Thus, the general capability of the 741 as a modulation source is adequate, the only requirement being that its output offset be low.

Calibrating the circuit is simple and requires only an oscilloscope. With a carrier input of 7 kilohertz at 2 volts rms and a modulating input of 10 V rms at 60 Hz, the

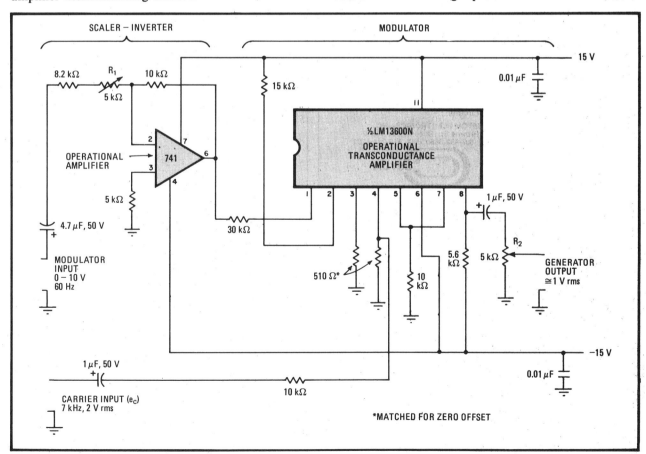

1. Nonlinear analysis. A two-chip audio and ultrasonic modulator generates preset levels of intermodulation distortion for a given carrier frequency suitable for testing and calibration purposes. The system is easily aligned, and the circuit's wideband modulator is flat to 120 kHz.

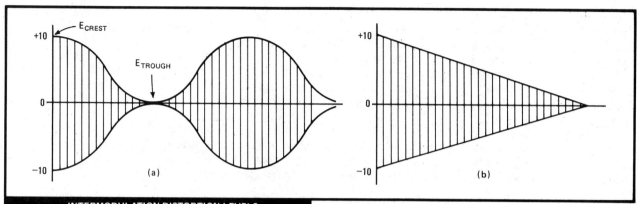

| INTERMODULATION DISTORTION LEVELS ||
Modulation voltage (rms)	Intermodulation distortion (%)
10.0	100
1.0	10
0.1	1
0.01	0.1

2. Response. This system can be calibrated at 100% intermodulation for the 7-kHz carrier frequency using a standard or trapezoidal pattern much as in any amplitude-modulated network. In (a), R_1 is adjusted for a merge of positive and negative troughs, holding the modulating level at 10 V rms. In (b), R_1 is adjusted until the modulation triangle just reaches its apex.

output should appear as shown.

When the vertical channel of the oscilloscope is driven by the output of the system and the horizontal time base is accordingly adjusted, the waveform of Fig. 2a is displayed. The triangle of Fig. 2b will be observed if the output of the system drives the vertical channel and the horizontal input receives the 60-Hz modulating signal. The inverting function of the 741 causes the apex of the modulation triangle to appear at the right of the trace.

The scaling function of the 741 serves to calibrate the system. For (a), trimmer potentiometer R_1 is adjusted until the positive and negative troughs of the curve merge for a 10-v rms input at 60 Hz. Expanding the horizontal sweep of the scope with the 5× or 10× magnification control will help to determine the merge point. In the alternative display of Fig. 2b, the gain of the 741 op amp must be adjusted until the modulation triangle just attains its apex, holding the modulating input at 10 V rms.

With the calibration complete, a direct relationship now exists between the modulation voltage level and the percent intermodulation distortion (see table).

The 10% point is used to calibrate the analyzer. Potentiometer R_2 provides level control and aids in determining the high-frequency–input sensitivity of the analyzer; a full-scale response of 25 mV is good. □

Angle-modulated signals suffer less a-m distortion

by Lowell S. Kongable
Motorola Semiconductor Products Sector, Phoenix, Ariz.

Frequency- or phase-modulated waveforms are subject to a carrier's amplitude modulation either through noise or through the system's modulating signal. This a-m distortion of the angle-modulated signal results in interruptions during demodulation, thereby distorting the received waveform. Limiters reduce the problem but cannot handle highly irregular signals. However, this circuit (a) substantially reduces distortion and may be

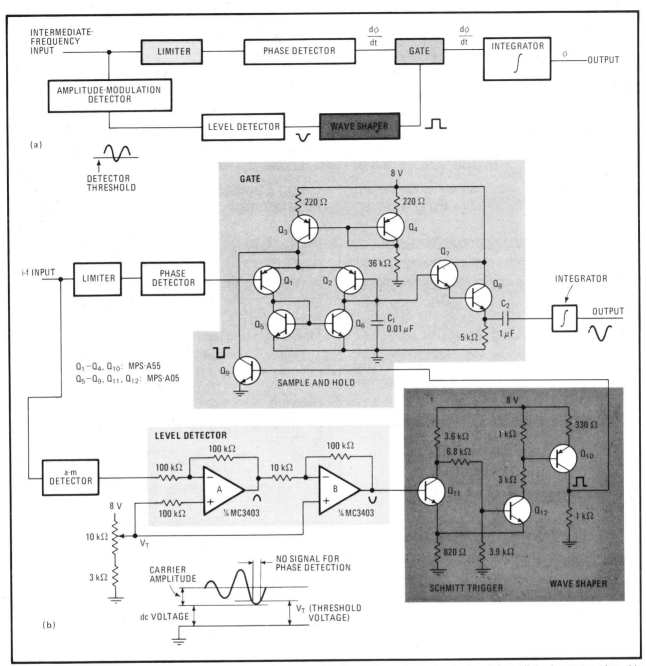

Uninterrupted. The circuit (a) provides a method to reduce a-m distortion of fm and pm signals. It switches off the integrator when this distortion is high (about 100% a-m) and obviates any interruptions in the system. The level detector, wave shaper, and gate are shown in (b).

incorporated into an integrated circuit. In addition, the circuit is automatic and independent of signal strength.

A high percentage of a-m waves allows insufficient signal for fm or pm detection. Instead, the deficiency is picked up by the phase detector as a large phase change and causes the integrator's output to emit an undesired noise. However, this circuit disconnects the integrator while the distortion is high and thus gets rid of the noise.

Comparator A (b) conducts at a level that is set by threshold voltage V_t. Below this level there is insufficient signal for phase detection. In addition, the direct-current voltage supplied by the a-m product detector is independent of the signal strength. As a result, the threshold voltage is independent of the signal.

The output of comparator A is amplified by B, which drives the Schmitt trigger. The trigger generates a control pulse for the sample-and-hold circuit consisting of transistors Q_1 through Q_6 and capacitor C_1. The output of the sample-and-hold circuit is fed into the integrator, which is interrupted only when the control pulse directs the current from Q_3 into Q_9 instead of Q_1 and Q_2. When Q_9 conducts, its collector is at a low voltage that reverse-biases the emitters of Q_1 and Q_2. This biasing retains the charge on C_1 until Q_1 and Q_2 conduct. With nearly 100% modulation, the voltages at the bases of Q_1 and Q_2 are large. The integrator "sees" only a small change at its input and, as a result, has no output. □

3. CLOCK CIRCUITS

Making a clock chip keep better time

by M. F. Smith
Department of Computer Science, University of Reading, England

Maintaining both the time and date functions in microprocessor applications became much easier when National's MM58167 and MM58174 microprocessor-compatible real-time clocks were introduced. The software approach that was used before their introduction simplified software and memory requirements, allowed increased flexibility of clock rates and selection of time resolution, and easily accommodated scheduling protocols. However, keeping time during a brown-out was still disastrous to system operation as was attempting to maintain the correct time despite the occasional timing difficulties that occur under software control.

Yet, occasional read errors and problems with spurious writing to the MM58174 when power is going down creates difficulties with the hardware-based system. These difficulties can be overcome with the software and hardware fixes prescribed here, which are intended for the MC6800 microcomputer system.

The problem with occasional read errors may be easily overcome by modifying software control to ensure that a valid binary-coded decimal number is read before the program continues and by ordering a rereading of the data if it has not been captured the first time (see printout of the partial listing, line 84). The cause of the read errors has never been definitely ascertained, although the problem has been encountered when other microprocessors have been used, such as National's 1NS8073. Thus, there may be a rare timing problem within the MM58174 itself, or the difficulty may occur between the microprocessor and the clock chip.

Trying to write to the clock chip when the power is going down will ordinarily cause a loss of timing information. A number of methods of preventing the loss were tried, and the one in the figure was the simplest and most successful.

Here, the CD4066 electronic switch will allow the chip to be selected only when the MC6800 clock enables

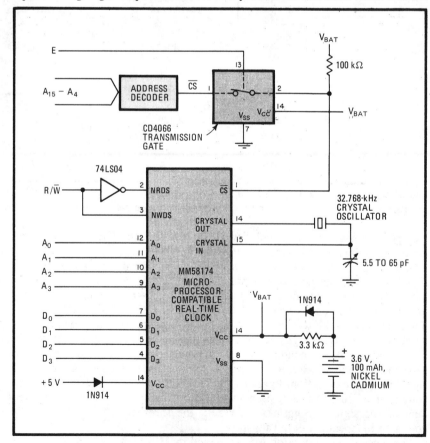

Glitch-free. The CD4066 transmission gate prevents loss of time-date information that is associated with an MM58174 hardware clock during a power-down condition. Read errors in time-date information may be eliminated in the software of a microprocessor system by writing a loop to ensure the program does not advance until the data has definitely been read correctly.

PARTIAL PROGRAM LISTING FOR CONTROL OF MM58167 MICROPROCESSOR-COMPATIBLE REAL-TIME CLOCK

Hexadecimal code	Location and label	Mnemonic	Operand	Comments
004E	82 HR_MIN_DISPLAY:			
004E B67807	83 H_T	LDAA	HOURS_TENS,E	;GET HOURS FROM CLOCK CHIP
0051 8D30	84	BSR	DIGIT_DISPLAY	;AND PUT IN BUFFER
0053 25F9	85	BCS	H_T	;RETRY IF CLOCK CHIP
0055 B67806	86 H_U	LDAA	HOURS_UNITS,E	;DOES NOT READ PROPERLY
0058 8D29	87	BSR	DIGIT_DISPLAY	
005A 25F9	88	BCS	H_U	;RETRY IF ERROR
	89			
005C 863A	90	LDAA	#":"	;DISPLAY COLON
005E BD0000	91	JSR	DISPLAY	
	92			
0061 B67805	93 M_T	LDAA	MINUTES_TENS,E	;DISPLAY MINUTES
0064 8D1D	94	BSR	DIGIT_DISPLAY	
0066 25F9	95	BCS	M_T	;RETRY IF ERROR
0068 B67804	96 M_U	LDAA	MINUTES_UNITS,E	
006B 8D16	97	BSR	DIGIT_DISPLAY	
006D 25F9	98	BCS	M_U	
	99			
006F 863A	100	LDAA	#":"	;DISPLAY COLON
0071 BD0000	101	JSR	DISPLAY	
	102			
0074 B67803	103 S_T	LDAA	SECONDS_TENS,E	;DISPLAY SECONDS
0077 8D0A	104	BSR	DIGIT_DISPLAY	
0079 25F9	105	BCS	S_T	;RETRY IF ERROR
007B B67802	106 S_U	LDAA	SECONDS_UNITS,E	
007E 8D03	107	BSR	DIGIT_DISPLAY	
0080 25F9	108	BCS	S_U	;RETRY IF ERROR
0082 39	109	RTS		
	110 ;			
0083	111 DIGIT_DISPLAY:			
0083 840F	112	ANDA	#0FH	;MASK OFF TOP PART
0085 8109	113	CMPA	#9	;MUST BE A NUMBER
0087 2207	114	BHI	A_S_ERROR	
0089 8B30	115	ADDA	#30H	;MAKE ASCII
008B BD0000	116	JSR	DISPLAY	;AND OUTPUT TO DISPLAY
008E 0C	117	CLC		;FLAG SUCCESSFUL
008F 39	118	RTS		
0090 0D	119 A_S_ERROR	SEC		;FLAG ERROR
0091 39	120	RTS		
	121 ;			
	122 ;			
0092	123 MO_DISPLAY:			
0092 B67809	124 D_T	LDAA	DAYS_TENS,E	;DISPLAY DAYS
0095 8DEC	125	BSR	DIGIT_DISPLAY	
0097 25F9	126	BCS	D_T	;RETRY IF CLOCK CHIP
0099 B67808	127	LDAA	DAYS_UNITS,E	;READ ERROR
009C 8DE5	128	BSR	DIGIT_DISPLAY	
	129			
009E BD0000	130	JSR	DISPLAY_SPACE	;SEPARATE
	131			
00A1 F6780C	132 MO_T	LDAB	MONTHS_TENS	;SHOW MONTHS
00A4 C40F	133	ANDB	#0FH	;MASK OFF TOP PART
00A6 C109	134	CMPB	#9	;MUST BE A NUMBER
00A8 22F7	135	BHI	MO_T	;RETRY IF ERROR
	136			
00AA 4F	137	CLRA		;CONVERT B INTO
00AB 5D	138 QUICK_CNV	TSTB		;BINARY OF DECIMAL
00AC 2705	139	BEQ	MO_U	;VALUE
00AE 5A	140	DECB		
00AF 8B0A	141	ADDA	#10	
00B1 20F8	142	BRA	QUICK_CNV	
	143			
00B3 F6780B	144 MO_U	LDAB	MONTHS_UNITS	
00B6 C40F	145	ANDB	#0FH	;MASK OFF TOP PART
00B8 C109	146	CMPB	#9	;MUST BE A NUMBER
00BA 22F7	147	BHI	MO_U	;RETRY IF ERROR
	148			

it. Once power to the microprocessor begins to fail, the signals from the E line cease and the MM58174 can no longer be selected. The 100-kilohm resistor is required to hold the chip-select line high when the electronic switch is open.

The software for using the hardware clock with the MC6800 is fairly straightforward. However, the program requires a few external routines for accessing the output port and a look-up routine for accessing the month, so that program size will be about the same as for a totally software-based clock. □

Recovering the clock pulse from NRZ-inverted data

by Doug Manchester
Halcyon, San Jose, Calif.

Although most data networks use nonreturn-to-zero–inverted encoding for the incoming data stream, few interface devices can recover a data-rate clock pulse from an NRZI source. However, this NRZI encoding and decoding circuit is capable of recovering not only that kind of clock pulse but also a data-rate clock pulse for synchronous data from an asynchronous modem.

An exclusive-NOR gate and a D-type flip-flop help turn serial binary data into NRZI data. The encoder truth table (Fig. 1a) shows that D_t is the binary data that is encoded at time t, Q_t is the state of the encoder output at time t, and Q_{t+1} is the state of the encoder output at the next bit time.

The NRZI data can be decoded by the circuit in Fig. 1b. This circuit performs the inverse of the encoding function and extracts the relevant information from the received encoded data. The clock frequency is synchronized with the clock pulse used for encoding the data.

The clock-pulse recovery circuit (c) uses a 4-bit counter u_1 to generate the bit-rate clock from the received data and the circuit clock that is 16 times $clock_1$. The counter is reset on every edge transition of data. The received data is sampled in the bit center for maximum distortion tolerance. Distortion as high as 2% per bit for 25 bits is tolerated without resynchronization. The master reset (MR) does not occur on a rising clock-pulse edge (d), for if it rose just prior to that rising edge, $clock_1$ would occur $1/16$ of a bit before the bit center. □

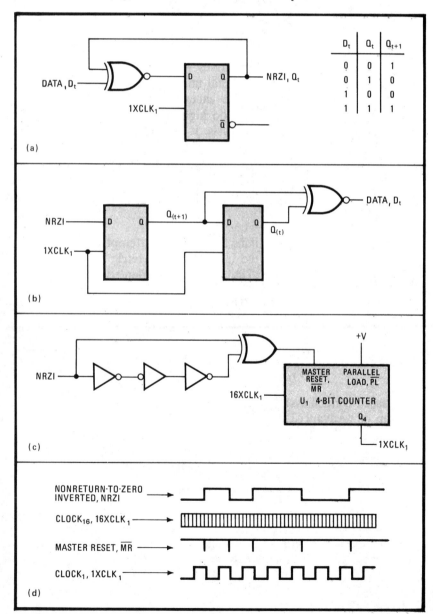

Recovery. Using the logic in (a), the binary data is encoded into NRZI data. This NRZI data can be decoded by performing the inverse of encoding function (b). The clock is recovered from the received data by the 4-bit counter U_1 that is clocked at 16 times the fundamental $clock_1$ (c). A typical timing diagram is depicted in (d).

Sync clock, counter improve programmable-width generator

by M. V. Subba Rao and V. L. Patil
Central Electronics Engineering Research Institute, Pilani, India

The typical programmable pulse-width generator either has limited programming capability or creates an initial timing error because of the asynchronism between an input trigger and the system's internal clock. Employing a binary-coded-decimal–programmable divide-by-n counter and a synchronous-start oscillator in a basic circuit overcomes these drawbacks, as shown here.

A trigger to the B_0 input of the 74121 monostable multivibrator generates a pulse that presets the 7476 flip-flop and also loads a preset number into the 74192 divide-by-n counter. The flip-flop then initiates pulse generation in the 74123 oscillator, whose output is counted down by the 74192 until zero is reached.

At this time, its borrow output, B_3, clears the flip-flop and the oscillator is disabled. The pulse width of the waveform at Q_3 is thus proportional to the number of clock pulses counted.

For the component values specified in the figure, the clock period is 1 microsecond. Thus, the circuit will generate pulses of from 1 to 9 µs, in steps of 1 µs. The 74192s can be cascaded to yield larger pulse widths. □

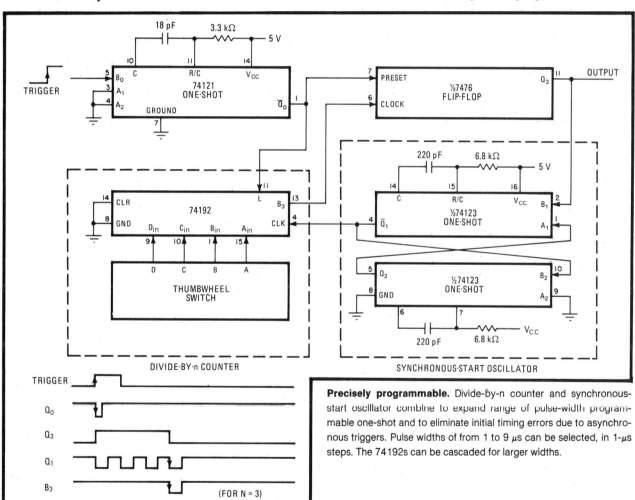

Precisely programmable. Divide-by-n counter and synchronous-start oscillator combine to expand range of pulse-width programmable one-shot and to eliminate initial timing errors due to asynchronous triggers. Pulse widths of from 1 to 9 µs can be selected, in 1-µs steps. The 74192s can be cascaded for larger widths.

Counter-timer resolves 1 µs over extremely wide range

by M. Antonescu
Ecole Polytechnique Fédérale de Lausanne, Lausanne, Switzerland

Providing much greater resolution and range than even the circuit proposed by Pinchak [*Electronics*, Nov. 6, 1980, p. 130], this timer-counter delivers programmable markers for periods of 1 to 120×10^6 seconds in 1-microsecond steps. The duty cycle is also selectable over a range of ratios from $1:10^6$ to $10^6:1$.

As seen in the figure, the 8640B time standard from Suwa Seiko, which has a built-in quartz-crystal oscillator, can be programmed by means of switches S_1 and S_2 for any of 64 output periods in the 1-µs-to-120-s range. When used in the timing mode, this oscillator drives the Toshiba TC 5070P counter, which is similar to Intersil's ICM8240 but has six digits and can drive a common-

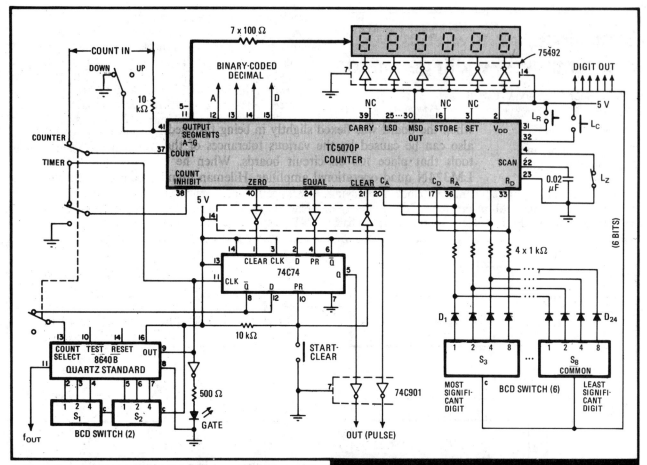

1. Show time. Start- and end-count numbers are loaded into a TC 5070P counter chip from switches S_3–S_8. For timing applications, another switch-selectable IC, the 8640B quartz standard, provides an accurate time base for marking precise intervals.

	setting	0	1	2	3	4	5	6	7
S_1	pin 2	L	H	L	H	L	H	L	H
	pin 3	L	L	H	H	L	L	H	H
	pin 4	L	L	L	L	H	H	H	H
	ratio	1/1	1/10	1/2	1/3	1/4	1/5	1/6	1/12
Multiplier	S_2	1	10	10^2	10^3	10^4	10^5	10^6	10^7

PROGRAMMING OF RATIO SWITCHES S_1 AND S_2

cathode light-emitting-diode display.

In the counter mode, switches S_3 through S_8 are used to set the start- and end-count numbers. Push-button switch L_C loads the counter with the start-count number, and switch L_R loads a register with the end-count number. When the latter is reached, the counter generates an "equal" signal that sets a flip-flop, and the setting is detected at the flip-flop's output through a buffer. A second signal clears the flip-flop when the count passes through zero, and this event is detected at the same output buffer. Both up and down counting is possible.

When timing, the 5070 counter chip counts the timing pulses from the 8640 time standard. Again, the equal and zero counts are used to toggle a flip-flop, but in this use the action occurs over a specified time span, rather than a specified number of counts. The time span marked by the counter is adjustable, from a period of one clock pulse up to that of a million clock pulses, yielding a maximum period on the order of years.

Both normal and complementary outputs are generated along with a visual indication of the gate period from the LED labeled "gate."

The timer has two modes, depending on the state of start/clear switch. Holding the switch on causes the counter to run continuously and reset itself every gate period, whereas a momentary depression of this switch cycles it through a single time period, stops the counter at the end, and displays the elapsed period.

For a gate time of 1 s, the time standard should be set for 2 s; that is, programming pins 3, 6, and 7 should be set high. The gate duration is the second half of the time period—in this example, 1 s—and is unaffected by the length of time the start/clear switch is closed.

The display output is multiplexed across the six seven-segment numerals. Switch settings are channeled to the display through a network of 24 diodes, providing a visual check of the setting.

Total current consumption with a fully lit display is about 40 milliamperes at 5 volts. With the display switched off, it falls to 7 or 8 mA. Overall accuracy is better than 10 parts per million.

Finally, an external clock can drive the time-standard chip so that special counting rates such as accelerated countdowns can be accommodated. □

4. COMPARATORS

Frequency comparator uses synchronous detection

by Israel Yuval
Video Logic Corp., Sunnyvale, Calif.

There are many different ways to monitor a frequency but most have the disadvantage that they also detect any noise inherent in the circuit along with the desired waveform. This circuit circumvents the noise problem through the use of a technique that samples the unknown frequency at the rate of the reference signal and then averages the measured waveform with a low-pass filter.

If the input frequency f_{in} is equal to the reference f_{ref}, a nonzero average results at the output. However, if $f_{in} \neq f_{ref}$, the average of the samples is zero. In addition, the circuit is immune to noise and works well between 100 hertz and 100 kilohertz.

Comparator U_2 (a) determines the sampling frequency that is derived from reference input signal V_{ref}. Resistor R_2 determines the duration of the sample pulse. The output of U_2 turns switch Q_1 on and off at the rate f_{ref}. The input signal, which is compared with the reference signal, is amplified by U_1 whose gain is set by R_1. Capacitor C_1 blocks any dc component of V_{in}. The cutoff frequency of the low-pass filter is determined by R_3 and C_3 and is given by the equation $f_c = 1/2\pi R_3 C_3$.

The output of the filter (b) keeps comparator U_3 in the high state as long as $|f_{in} - f_{ref}| < f_c$. The output of U_3 turns low otherwise. The response time of the circuit is determined by time constant $R_3 C_3$. To satisfy the response time and the frequency accuracy, the designer must appropriately select values for R_3 and C_3.

The harmonics of f_{ref} will also result in a nonzero average and therefore must be attenuated by adding a low-pass filter with a cutoff frequency less than $2f_{ref}$ at the input. Also, the phase delay between $V_{in}(t)$ and $V_{ref}(t)$ must not equal 90° because it will result in a zero average even when $f_{in} = f_{ref}$. Adding a phase delay to either signal will eliminate this problem. The circuit assumes that the phase difference between V_{in} and V_{ref} is less than ±90°. If no such certainty is guaranteed, a window comparator should replace U_3. □

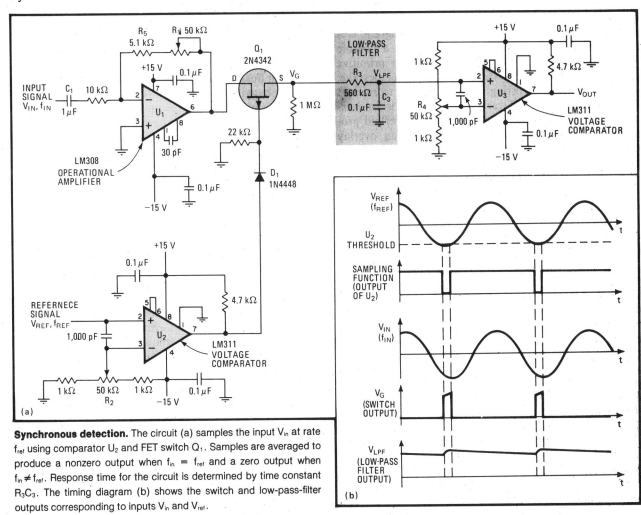

Synchronous detection. The circuit (a) samples the input V_{in} at rate f_{ref} using comparator U_2 and FET switch Q_1. Samples are averaged to produce a nonzero output when $f_{in} = f_{ref}$ and a zero output when $f_{in} \neq f_{ref}$. Response time for the circuit is determined by time constant $R_3 C_3$. The timing diagram (b) shows the switch and low-pass-filter outputs corresponding to inputs V_{in} and V_{ref}.

Comparator, one-shot produce pulse-width inversion for servos

by John Karasz
Norden Systems, Norwalk, Conn.

Frequencies usually trigger movements of servo-actuators within radio-controlled vehicles. However, these transmitted waves are difficult to produce and are noisy. This design overcomes both problems by employing pulse control with a comparator, two monostable multivibrators, and a pair of NOR gates. Although the circuit produces only three pulses, or three servo positions, more comparators and NOR gates may be added to create an n-position control.

The circuit compares the width of the input pulse with that of a reference. A reference pulse of 1.5 milliseconds is selected for the circuit through the RC network connected to one-shot U_{1a}. As a result, pulse durations of 1.0 ms and 2.0 ms produce servo-output positions of $+45°$ and $-45°$, respectively.

The reference pulse is triggered by the leading edge of the input pulse. The trailing edge of this pulse triggers U_{1b} to generate output Q_2, which is equal in duration to its input. When the width of the input pulse is equal to the reference pulse, output Q_2 passes through the NOR gates and is unchanged. Under this condition, no contributions are made by the comparator outputs.

However, when the width of the input pulse is greater than the reference pulse, the trailing edge of Q_1 triggers U_{1b} and comparator U_2 produces a high output at pin 5(A>B). This pulse starts at the trailing edge of the reference pulse and ends at the falling edge of the input pulse, thereby resulting in an output pulse of duration equal to the width of the reference minus the width of comparator output A>B.

Lastly, when the input-pulse width is shorter than the reference, the comparator generates a B>A output at pin 7. Its duration begins at the trailing edge of the input pulse and terminates at the falling edge of the reference pulse. The Q_2 pulse of U_{1b} and B>A output of the comparator are fed to gate U_{3a}, which generates an output of equal in duration to the width of the reference plus the width of comparator output B>A. Mirror-image pulse widths are produced about the reference pulse and the output is thus defined by $T_{out} = T_q \pm |T_s - T_{ref}|$, where T_{out} is the width of the output pulse, T_q is the duration of one-shot U_{1b}, T_s is the duration of the input, and T_{ref} is the width of the reference pulse. □

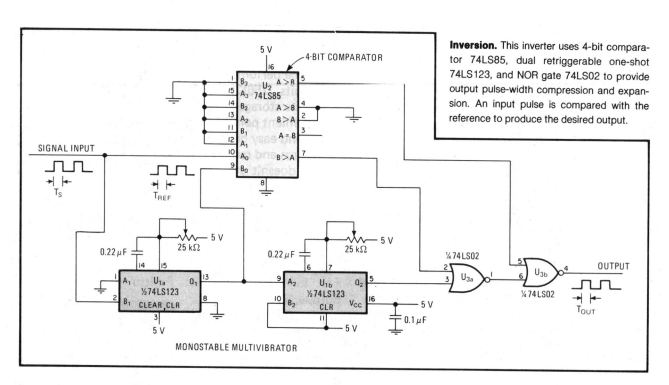

Inversion. This inverter uses 4-bit comparator 74LS85, dual retriggerable one-shot 74LS123, and NOR gate 74LS02 to provide output pulse-width compression and expansion. An input pulse is compared with the reference to produce the desired output.

Digital comparator minimizes serial decoding circuitry

by Harland Harrison
Memorex Inc., Communications Division, Cupertino, Calif.

Using significantly fewer chips than the comparator proposed by Patil and Varma[1], this two-word, 4-bit comparator offers other advantages as well—it accommodates any word length and is more easily modified to handle any bit width. The control signals needed to facilitate the comparison can also be more conveniently applied.

The circuit outputs are first cleared with a negative-going pulse from the start signal. This sets outputs O_{LESS} and $O_{GREATER}$ low. O_{EQUAL}, derived from O_{LESS} and $O_{GREATER}$, goes temporarily high. At this time, the numbers can be presented to data buses A and B for comparison, with the least significant bit pair introduced first. As a consequence of the configuration, any number of bit pairs per word can be compared without modifying the circuit at all. The data buses each accept up to 4 bits, but this number may be expanded simply by cascading 7485 comparators.

The result of each bit-pair comparison is then latched into the 7474 flip-flops by the D_i clock pulse, with the results of each bit-test being fed to the cascade inputs of the 7485. As a result, the comparator keeps track of the previous bit-pair check while continuing to update its results as each succeeding bit pair is introduced. Thus, the need for additional memory and logic elements is eliminated. The final result becomes valid after the D_m clock pulse, where m is the word length in bits, and remains valid until the next start pulse. □

References
1. V. L. Patil and R. Varma, "Digital comparator saves demultiplexing hardware," *Electronics*, Aug. 14, 1980, p. 129.

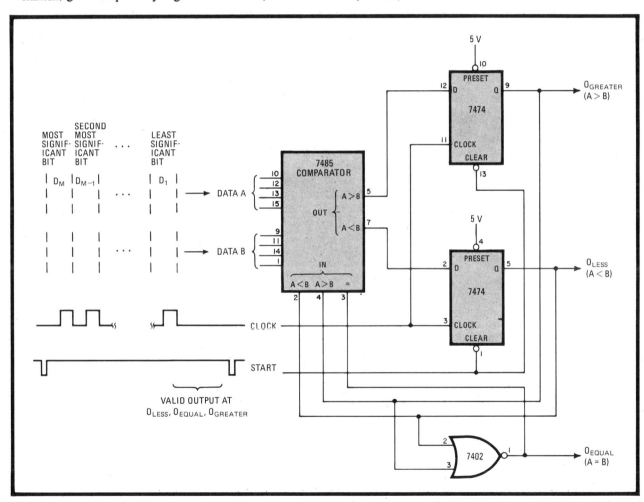

Less memory. Circuit performs 4-bit comparison of two numbers with minimal circuitry. 7485 comparator replaces large numbers of flip-flop–type memories and logic elements by keeping track of previous bit-pair checks in real time as each pair is introduced. Circuit accommodates any word length; bit width is expandable simply by cascading 7485 comparators.

5. CONTROL CIRCUITS

One-chip tachometer simplifies motor controller

by Henrique Sarmento Malvar
Department of Electrical Engineering, University of Brazilia, Brazil

Setting and stabilizing the angular velocity of a dc motor by means of a charge pump and a servoamplifier, one-chip tachometers such as National Semiconductor's LM2917 serve well as a simple but elegant motor-speed controller. Such an arrangement is preferable to the widely used scheme in which both positive and negative feedback is utilized to keep the motor's back electromotive force, and thus its speed, constant by generating a voltage that is proportional to a given load.

As shown in the figure, a magnetic pickup coil detects the angular velocity of a motor-driven flywheel and feeds the low-amplitude pulses, whose frequency is proportional to the motor speed, to the LM2917's charge pump. As a result, the pump generates a current, I_1, whose average value is directly proportional to the input frequency.

The operational amplifier that follows compares this voltage to a user-set reference and, through power transistor Q_1, generates a voltage for the motor's armature of $V_A = (R_2/R_1)(V_{ref} - I_1 R_1)$. Thus, potentiometer R_6 sets the motor's speed, for when $V_1 > V_{ref}$, voltage V_A decreases, and vice versa.

In this application, the gain of the operational amplifier, determined by resistors R_1 and R_2, has been set at approximately 150. The greater the gain, the lower the variation of motor speed with changes in load resistance. However, the setting of very high gains should be avoided, because there will be a reduction in the gain and phase margins—that is to say, a loss of stability in the feedback control loop.

As for the selection of other components to meet any particular application, note that capacitor C_1 serves a double purpose: it integrates pulsed current I_1, thereby performing a smoothing function, and it sets a low-frequency pole for the amplifier, thereby ensuring stability. C_2 sets the conversion factor of the tachometer and should be increased for low-speed motors. R_3 minimizes the offset due to the amplifier's bias currents at pins 4 and 10. Finally, C_3 functions as a noise filter.

As seen, the LM2917's tachometer conversion factor will be almost independent of its supply voltage, as a consequence of the zener diode connected at the device's supply port. The supply voltage should not fall outside the range of 10 to 15 volts, however. □

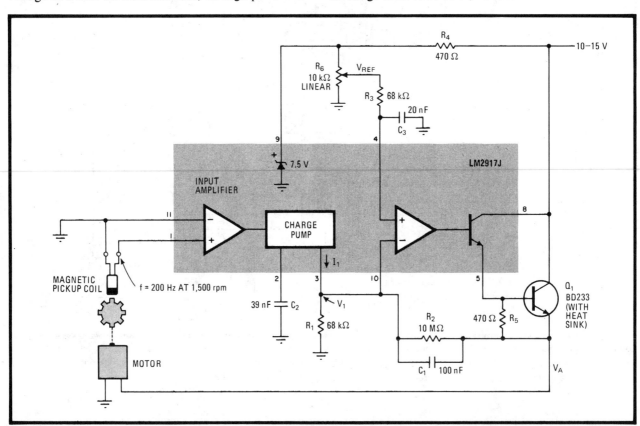

Setting speed. LM2917J tachometer, which is basically a frequency-to-voltage converter, sets and stabilizes motor speed. Few RC components are required, thereby simplifying circuitry. Power transistor Q_1 is the only external active element needed.

Thermocouple simplifies temperature controller

by V. J. H. Chin
National Research Council, Division of Chemistry, Ottawa, Canada

Virtually all the designs for low-noise, high-temperature controllers of the type that use zero-crossing switches have thermistors in the sensing circuit—an impractical configuration in many cases because of the size requirements and availability of the thermistor itself. To overcome these inconveniences, this simple circuit substitutes an ordinary thermocouple for the thermistor, yet works as well as a thermistor-based one—for instance, it controls the environmental (furnace) temperature from room temperature to 1,100°C to within ±2%.

In general operation, the furnace is heated from the ac line through a triac, triac driver A_1, and the CA3079 zero-voltage switch A_2. The CA3079, in turn, switches on when the output differential from the thermocouple drops below a value corresponding to a given furnace temperature. Switching occurs because the amplified thermocouple voltage, V_T, at the output of A_3 falls below the user-preset reference potential, V_R, at the input to A_4. Note that in addition, the CA3079 must be biased so that potential V_{13} is initially less than the comparator's output, V_9, in order that the circuit containing R_L of the furnace will be completed and current will flow when the furnace is cold.

The thermocouple voltage is linearly proportional over its entire range to the temperature in the furnace; consequently, if potentiometer R_1 is linear, it can be directly calibrated in terms of temperature. For optimum switching, the voltage V_{13} should be set at half of the LM324's supply voltage.

As for cost, the prototype circuit was built for an outlay of less than $20.

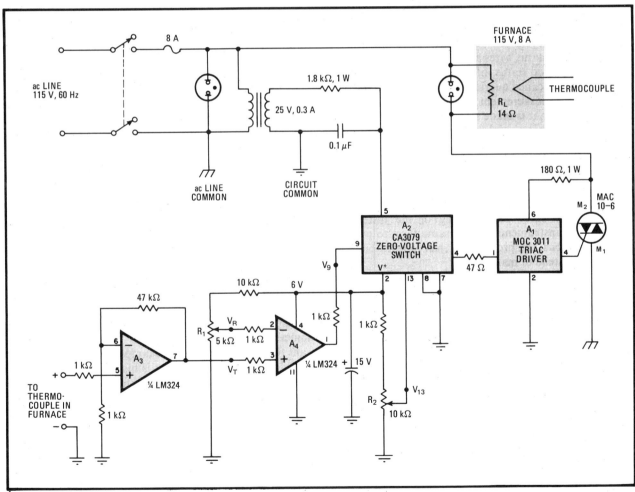

Hot-wired heater. Chromel-Alumel thermocouple eases design of zero-crossing–switched temperature controllers. Potentiometers R_1–R_2 set the reference voltage for switching on furnace from CA3079 switch without the need for setting up a complicated reference scheme. Linear response makes it easy to calibrate R_1 as a direct function of desired temperature. Circuit works to 1,100°C, is accurate to within ±2%.

Stepper resolves motor's angular position to 0.1°

by Jaykumar Sethuram
Electronic Associates Inc., West Long Branch, N. J.

One-chip digital comparators and counters simplify the design of this controller, which resolves the position of a stepping motor to 0.1°. Using complementary-MOS circuitry, the unit is low in cost and power consumption is minimal.

The set of four binary-coded decimal numbers, D_1 to D_4, introduced to the cascaded 40085 4-bit comparators, are the command signals that order the motor to the desired bearing expressed in hundreds, tens, units, or tenths of a degree, respectively. The range of the input set is thus 0000 to 3600. With the aid of the comparators and sequential logic, the 40192 up/down counters track the position of the stepper at every instant, updating its count and thus rotating the motor until its contents match the setting of D_1–D_4.

As can be seen, the sequential logic circuitry determines the direction of rotation of the stepper and counter by monitoring the A = B, A > B, and A < B outputs of the output comparator. The logic is designed to rotate the stepper from its current position to the desired position in the minimum number of steps. Thus, if the motor's present position is at 5° and the intended position is 300°, the stepper will automatically be rotated in a counterclockwise direction.

The circuit can be easily modified for applications where the input data is available for only a very short time. In such cases, it is only necessary to add input latches to capture the data. □

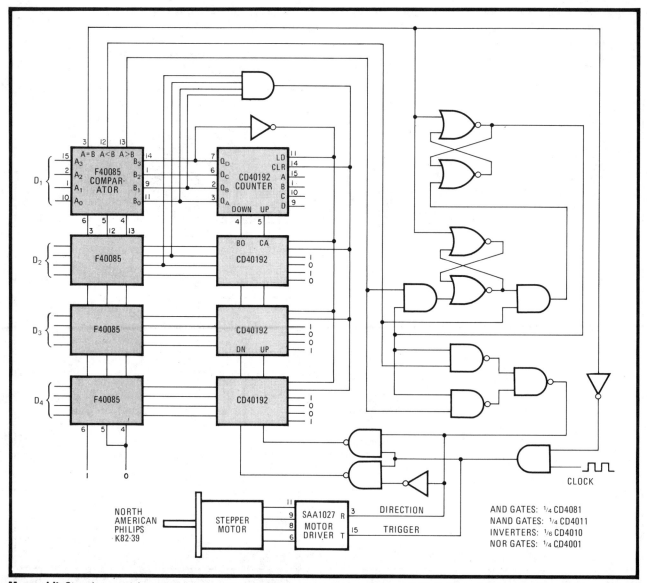

Move a bit. Stepping motor is rotated into desired position with comparators and sequential logic that minimizes the difference between the 4-bit command set D_1–D_4, and the output of the up/down position-tracking counter. Angular position is resolved to 0.1°.

Current-biased transducer linearizes its response

by Jerald Graeme
Burr-Brown Research Corp., Tucson, Ariz.

Transducers that work on the principle of variable resistance produce a nonlinear response when voltage-biased, as in common bridge configurations. However, a single operational amplifier configured to provide current biasing for the transducer eliminates this difficulty and allows output offset voltages to be controlled or removed.

As a voltage-biased transducer's resistance varies, so does the current through it. Thus, any signal voltage taken from the transducer will be a function of both current and voltage variations and will be a nonlinear function of the transducer's resistance.

Current biasing rather than voltage biasing avoids this nonlinearity, but reference current sources are not as readily available as voltage ones. Fortunately, an op amp can convert a reference voltage for this purpose and in addition provide other benefits.

In (a), a voltage-to-current converter circuit is adapted for voltage control of the supply current and of the output offset voltage. As laid out, the transducer's bias current is $I_x = (V_2 - V_1)/R_1$, where V_1 and V_2 are the externally applied control voltages. The current polarity can be set at either + or −, allowing an inverted or noninverted voltage response to variations in transducer resistance.

The resulting current flow in the nominal transducer resistance, R, produces an offset voltage at the circuit's output with a counteracting voltage developed by V_1: $V_{oso} = (1 + (R_2/R_1))I_xR - (R_2/R_1)V_1$. Thus, through the proper selection of V_1, the output offset may be nulled to zero or set to either a positive or negative polarity.

Signal variations about that level result from a change in transducer resistance. The net voltage output then becomes $e_o = [1 + (R_2/R_1)]I_x\Delta R + V_{oso}$. This response is linearly related to ΔR. Also the signal is amplified.

To set the gain, R_2 should be adjusted after R_1, which sets the level of I_x, is selected. Because the output signal is taken directly from the amplifier's output, the transducer is buffered against loading effects.

Deviations in the described performance result primarily from voltage and resistor tolerances and from resistor-ratio error. Mismatch of the $R_2:R_1$ ratio is particularly serious, as this will make I_x somewhat a function of ΔR, thereby reintroducing nonlinearities in the circuit. Such a mismatch will alter the term in the denominator of the I_x formula to read $((R+\Delta R)/R_2) \times ((R_2/R_1) - (R_2'/R_1'))$, where R_1' and R_2' are the mismatched counterparts of R_1 and R_2, respectively. Further errors will result from the dc input-error signals of the op amp. As a result, the deviation in e_o will be:

$$\Delta e_o = \left[1 + \frac{R_x}{R_2}\left(1 + \frac{R_2}{R_1}\right)\right]\left[\left(1 + \frac{R_2}{R_1}\right)V_{os}\right] + \left(1 + \frac{R_2}{R_1}\right)R_xI_{os} + R_2I_{B-}$$

where V_{os}, I_{os}, and I_{B-} are the input offset voltage, input offset current, and the inverting input-bias current of the op amp, respectively. Making the R_2/R_x ratio large reduces this error.

In practice, it is either inconvenient or sometimes undesirable to supply two voltage references to the circuit. A single reference voltage may be applied, as shown in (b), which is the Thévenin equivalent of the circuit in (a). For the specific component values given, the amplifier will deliver a 1.0-volt full-scale output signal with a zero offset in response to a transducer resistance that ranges from 300 to 350 ohms. □

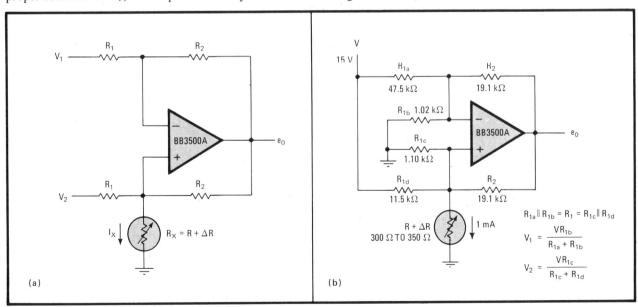

Holding constant. A voltage-controlled current source drives transducer (a) so that its output voltage is a function of its resistance change only, thereby reducing the circuit's nonlinear response. The output offset can also be virtually eliminated. The single-reference current source circuit (b), which is the Thévenin equivalent of (a), may be more attractive to implement in some cases.

Pulse generator has independent phase control

by Roberto Tovar Medina
Institute of Applied Mathematics, University of Mexico, Mexico

Many phase-locked–loop applications need a circuit to generate signals whose phase can be controlled independent of their other characteristics. Using a 555 timer and a few discrete components, this design provides a pulser with independent phase control between 0° to 180°. In addition, the phase is continuously adjustable.

Timer U_1 (a) together with transistor Q_1 and capacitor C_1 generates a sawtooth signal whose amplitude is between $V_{cc}/2$ and $2V_{cc}/3$ (b). For every cycle of this sawtooth wave, a short pulse is produced at the output of U_1. This pulse clocks flip-flop U_{3-a} to generate reference signal Q_a. By comparing the sawtooth signal with a reference voltage provided by potentiometer R_4, the comparator output clocks flip-flop U_{3-b} to generate pulse Q_b that is phase-shifted with respect to the reference.

Because this phase difference bears a linear relationship to the reference voltage at the noninverting terminal of U_2, R_4 is calibrated in terms of the phase control, with $V_{cc}/3$ corresponding to 0° and $2V_{cc}/3$ to 180°. Since both Q_b and $\overline{Q}_b$ are available from the output flip-flop, the circuit provides both phase-advance and phase-lag versions of the reference signal. □

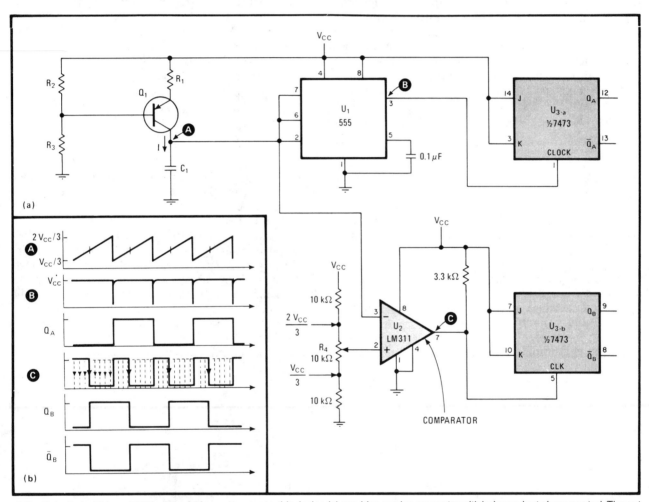

Adjusting phase. Using a 555 timer and a few components, this design (a) provides a pulse generator with independent phase control. The output can be either delayed or advanced with respect to the reference at Q_a. R_4 is calibrated in terms of the phase difference, with $V_{cc}/3$ corresponding to 0° and $2V_{cc}/3$ corresponding to 180°. The timing diagram (b) depicts the phase relationship between the reference and the outputs.

C-MOS circuit controls stepper motor

by Otto Neumann
Square D Canada, Waterloo, Ont., Canada

Of the many circuits around that can drive stepper motors, this stepper controller is one of the simplest to construct. It uses just a few popular complementary-MOS integrated circuits and Darlington power transistors, its cost is low, its power consumption is minimal, and it may be controlled remotely.

Motorola's binary up-down counter MC14516 is connected to binary decoder MC14555 and quadruple OR

TRUTH TABLE FOR CLOCKWISE ROTATION									
Counter output		Decoder output				Motor terminal			
B	A	Q_3	Q_2	Q_1	Q_0	5	1	4	3
0	0	0	0	0	1	1	1	0	0
0	1	0	0	1	0	0	1	1	0
1	0	0	1	0	0	0	0	1	1
1	1	1	0	0	0	1	0	0	1

Driver. This stepper-motor driver uses complementary-MOS integrated circuits and Darlington power transistors to control the speed and direction of the stepper's rotation. Up-down counter MC14516 is linked with decoder MC14555 and quad OR gate MC14071 to generate the pulses for driving the motor through transistors Q_1–Q_4. The unit can be controlled remotely.

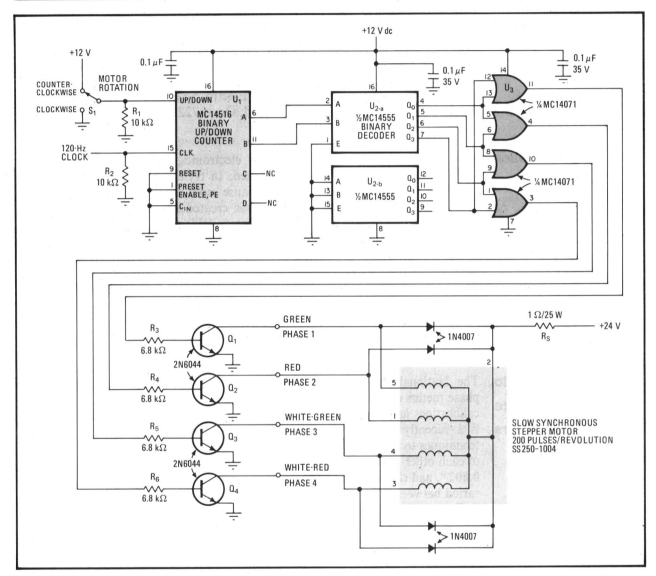

gate MC14071 to generate the pulses that will drive a four-phase stepping motor (see figure). The clock input to counter U_1 runs at 120 hertz and is derived from a 555 timer. The stepper is driven through transistors Q_1–Q_4. In addition, each output of decoder $U_{2\text{-}a}$ drives the base of each transistor through resistors R_3–R_6. This setup reduces the load on the output of $U_{2\text{-}a}$ to less than 1.3 milliamperes, which is enough current to drive the Darlington power device.

The motor's rotation is controlled by switch S_1. A +12-volt dc source is supplied to pin 10 of U_1 for counterclockwise rotation; grounding pin 10 gives clockwise rotation. Unused section $U_{2\text{-}b}$ of the decoder is grounded to shield the IC from transients. Pull-down resistors are recommended when the clock input and the clockwise-counterclockwise selection switch are located more than 1 foot from the driver board.

With a high-torque motor, such as an SS250-1004 with a torque of 225 ounce-inch, sensing resistor R_s must be employed. This resistor protects the stepper motor. With a low-torque motor, about 50 oz-in., it is not necessary to use R_s, and the motor supply should be reduced to +14 v dc. □

Low-cost controller stabilizes heater-type cryostats

by S. K. Paghdar, K. J. Menon, R. Nagarajan, and J. Srivastava, *Tata Institute of Fundamental Research, Bombay, India*

Controllers for maintaining small objects at a low temperature use either complicated gas-flow techniques, that regulate gas pressure or costly commercial units that are built for general applications and must be modified to suit a particular need. Finger-type cryostats, in contrast, which control a heater on the principle of error-sensing and feedback techniques, are extremely simple, easy to implement, and inexpensive. As for stability, this circuit holds the temperature to ±0.2°K over 24 hours (short-term variation is ±0.01 K) in the range of 80 to 200 K.

As shown in the figure, the circuit may be built around an OC23 or other power transistor, that has a fairly linear base-current-to-collector-current characteristic, eliminating the need for the additional amplification usual in circuits of this type. Error signals from a Wheatstone bridge are applied to the 741 operational amplifier act as the temperature control. The signal is then amplified by transistor Q_1 and passes to Q_2, which drives a resistive load and associated container that, placed inside the liquid-nitrogen–filled cryostat, heats the environment.

Thermal changes are sensed by a copper resistance thermometer, also located inside the cryostat. This element, placed in one arm of the Wheatstone bridge, reflects changes in the equilibrium temperature between the container and environment and in effect acts to cancel the original error signal applied to the op amp.

This part of the circuit serves only to correct temperature variations and does not set the temperature of the cryostat. Potentiometer R_1 fulfills this task by setting the bias current at the output without affecting the base-driving circuit of the power transistor.

To calibrate the system, switch S_1 is placed in the manual position and R_1 adjusted to obtain the desired cryostat temperature. Next, the Wheatstone bridge should be balanced by adjusting the standard resistor, whose value will be determined by the center value of the circuit's copper resistance thermometer.

The gain of the amplifier stage, R_3/R_2, should be matched to the response of the cryostat. A nominal value might be 1,200, as in this circuit.

The system can handle heater currents of up to 850 milliamperes. Above this value, however, the current gain of the power transistor begins to fall, so that the circuit will not function effectively in the specified temperature range. □

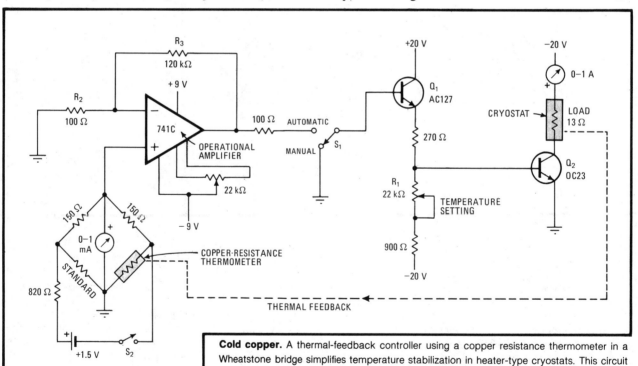

Cold copper. A thermal-feedback controller using a copper resistance thermometer in a Wheatstone bridge simplifies temperature stabilization in heater-type cryostats. This circuit holds temperatures to within ±0.2 K over 24 hours over the range of 80 to 220 K.

Resistor-based servo replaces mechanical governors in motors

by John Gaewsky and Herbert Hardy
Polaroid Corp., Cambridge, Mass.

Mechanical governors in motors having two speed settings may now be replaced by an equally efficient feedback circuit that uses a resistor instead of a bulky and expensive transducer to sense motor speed. The circuit shrinks the motor's size without impairing its efficiency and also lengthens motor life, partially because it eliminates regulator-contact arcing. Only one integrated circuit and some discrete components are used. A light-emitting diode indicates an out-of-regulation state and the accompanying acceleration.

The servo system (a) measures the back electromotive force generated by the motor and uses it to generate an electrical signal proportional to the motor's speed. This signal is then filtered by a four-pole low-pass active filter whose cutoff frequency is determined by the fundamental frequency of the brush noise at the motor's lowest controllable speed. The four-pole filter is obtained by cascading two second-order low-pass filters.

The filter output is compared with an adjustable reference, which is set by zener diode D_1 and adjusted using potentiometer R_3. When the speed signal is lower than the reference, power is applied to the motor, causing it to accelerate until the signal is greater than the reference.

Servo control. The servo system in (a) allows a continuous speed adjustment of a electrically governed motor while retaining the efficiency of a mechanical governor. The circuitry in (b) uses only one IC (LM324) and few discrete components to emulate the mechanical governor. The circuit measures the back emf generated by the motor and uses it to generate an electrical signal pro-proportional to the motor's speed. It also indicates out-of-regulation and acceleration.

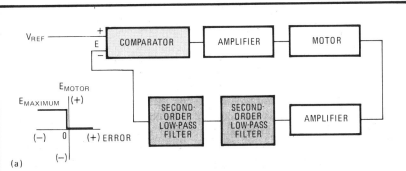

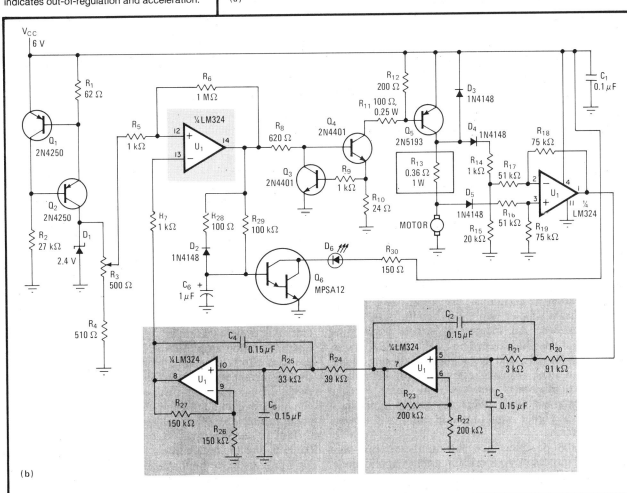

Because of the lag generated by the motor's time constant, the motor's speed response to an applied voltage is not immediate. The power that is switched to the motor is controlled by the comparator output. To limit the driver current, a low-power switching circuit is connected as a constant current source.

Sense resistor R_{13} is placed between V_{CC} and the armature resistance (R_a) of the motor in the speed-sensing circuit (b) and is used to form a bridge circuit satisfying the relationship $R_{14}/R_{15} = R_{13}/R_a$. This condition together with an appropriate gain of the amplifier, given by R_{18}/R_{17}, results in a differential voltage output that is proportional to the motor's speed. Losses in the sense resistor are minimized by selecting a value of $0.1R_a$ for it. In addition, diodes D_4 and D_5 prevent the input to the differential amplifier from exceeding the maximum specification.

When the motor is accelerating or the load-voltage variation requires a continuous power application (100% duty cycle), the out-of-regulation indicator lights. A duty cycle of less than 100% causes C_6 to discharge through R_{28}, which prevents Q_6 from turning on. □

Interface lets microprocessor control stepper motor

by V. L. Patil
Central Electronics Engineering Research Institute, Pilani, India

With the aid of the 8255 programmable peripheral interface and a few discrete components, a microprocessor can control the speed of a stepper motor and its direction of rotation by regulating the number and frequencies of the driving pulses the stepper receives. Such a setup is needed and widely used in process- and numerical-control instrumentation.

Port A of programmable peripheral interface U_1, used

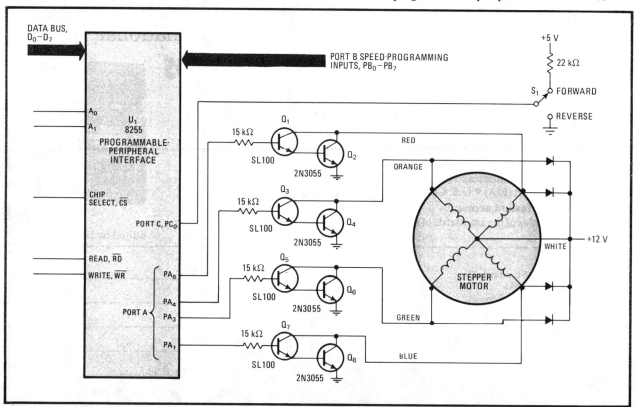

Stepper interface. The speed and rotational direction of the stepper motor can be controlled by a microprocessor interfaced with the motor through programmable peripheral interface device U_1 and transistors Q_1–Q_8. The transistor outputs provide the driving pulses for the motor.

STEPPER-MOTOR CONTROL PROGRAM

Location		Instruction code			Label	Mnemonic			Comments
85	00	3E	8B		STEP PGM	MVI A	8B		DATA 8B LOADED IN ACCUMULATOR
85	02	D3	03			OUT	03		8255 PROGRAMMED
85	04	3E	CC			MVI A	CC		DATA CC LOADED IN ACCUMULATOR
85	06	D3	00			OUT	00		DATA AT PORT A IS CC
85	08	4F				MOV	C	A	ACCUMULATOR DATA SAVED
85	09	CD	20	85		CALL	85	20	DELAY CALLED
85	0C	DB	02			IN	02		PORT C DATA TO ACCUMULATOR
85	0E	E6	10			ANI	10		ACCUMULATOR BIT TESTING
85	10	C2	30	85		JNZ	85	30	CONDITIONAL JUMP
85	13	79				MOV	A	C	DATA FROM C TO ACCUMULATOR
85	14	07				RLC			LEFT SHIFT
85	15	D3	00			OUT	00		DATA TO PORT A
85	17	4F				MOV	C	A	ACCUMULATOR SAVED
85	18	C3	09	85		JMP	85	09	UNCONDITIONAL JUMP
85	20	DB	01		DELAY	IN	01		PORT B DATA TO ACCUMULATOR
85	22	47				MOV	B	A	PORT B DATA TO REGISTER B
85	23	3E	FF			MVI A	FF		DATA FF TO ACCUMULATOR
85	25	3D				DCR A			ACCUMULATOR DECREMENTED
85	26	C2	25	85		JNZ	85	25	CONDITIONAL JUMP
85	29	05				DCR B			REGISTER B DECREMENTED
85	2A	C2	23	85		JNZ	85	23	CONDITIONAL JUMP
85	2D	C9				RET			RETURN
85	30	79			SUB PGM	MOV	A	C	DATA FROM REGISTER C TO A
85	31	0F				RRC			RIGHT SHIFT
85	32	D3	00			OUT	00		DATA TO PORT A
85	34	4F				MOV	C	A	ACCUMULATOR SAVED
85	35	C3	09	85		JMP	85	09	UNCONDITIONAL JUMP

in the basic input/output mode 0, drives the stepper via transistors Q_1 through Q_8 (see figure). The transistor outputs provide driving pulses with an amplitude of 12 volts at 1.5 amperes. Port B sets the speed of the motor and is programmed by the user. Bit PC_0 from port C controls the direction of rotation of the motor and is set either by switch S_1 or an external source.

When the accumulator is loaded with the hexadecimal number CC and rotated to either left or right, depending on the direction of rotation desired, a proper sequence of logic 1s and 0s is obtained at four of the port A bits—PA_1, PA_3, PA_4, and PA_6. The time gap between two consecutive sequences may be programmed through port B. The logic level at pin PC_0 of port C determines the direction of rotation. The presence of a 1 results in a forward direction, while a 0 gives the reverse direction. The table is a control program for the motor. □

Sequencer plus PROM step three motors independently

by Vikram Karmarkar
Hindustan Computers Ltd., New Delhi, India

Using a 4-bit bipolar microprogram sequencer and a 32-word-by-8-bit programmable read-only memory as a storage element, this processor easily meets the frequent requirement for multiple, independent stepping of up to three motors in either open-loop or servo-type control systems. The software approach makes possible the building of a compact and extremely flexible position controller.

The Am2911 program sequencer provides dynamic control for each motor, which in turn is dependent upon the information residing in the 6311 PROM. Here, forward- and reverse-step sequences are stored (that is, the sequence to be taken for a specified direction), as well as the individual motor selection/disable codes and a 3-bit-wide instruction field for the sequencer. Note that each motor has separate step-sequence codes in PROM so that it is not necessary for all three stepping motors to be of the same type.

Information for setting the actual direction of each motor is introduced to the program sequencer through lines FWD1–FWD3 of the 74LS153 multiplexer, with FWD = 1 specifying the forward condition. The active state of each motor is determined by the $\overline{MOT1}$–$\overline{MOT3}$ lines, with $\overline{MOT}$ = 0 denoting the on condition. The system is clocked at a rate dependent upon system requirements and the capability of the stepping motors.

As for the software (see table), the main control loop, which occupies locations 00 to 04, handles all vectored subroutine calls. After each individual step sequence, program control returns to this main loop. Three no-operation instructions separate the control loop from the body of the program that samples each motor's status. Sampling of the $\overline{MOT}$ line of each motor determines whether the program control will branch or bypass a particular motor's subroutine. It may be observed from the notational shorthand that BR specifies an unconditional jump to the address presented to the Am2911 and that BRC is a branch to a given subroutine if its $\overline{MOT}$ line is at logic 0. PC specifies a step to the next address, and RTN commands return to the main program.

A typical sequence map for three motors is shown, where the forward sequence for any motor is 011, 101, 110 (3, 5, 6 in decimal). The reverse-direction modes are 110, 101, and 011 (6, 5, 3 in decimal). □

Address	Motor code P_7, P_6 (decimal equivalent)	Step output P_5, P_4, P_3 (decimal equivalent)	Next address instruction P_2, P_1, P_0	Hex code
00	0	0	PC	07
01	1	0	BRC	40
02	2	0	BRC	80
03	3	0	BRC	C0
04	0	0	BR	04
05	X	X	X	X
06	X	X	X	X
07	X	X	X	X
08	1	6	PC	77
09	1	5	PC	6F
0A	1	3	PC	5F
0B	0	0	RTN	05
0C	1	3	PC	5F
0D	1	5	PC	6F
0E	1	6	PC	77
0F	0	0	RTN	05
10	2	6	PC	B7
11	2	5	PC	AF
12	2	3	PC	9F
13	0	0	RTN	05
14	2	3	PC	9F
15	2	5	PC	AF
16	2	6	PC	B7
17	0	0	RTN	05
18	3	6	PC	F7
19	3	5	PC	EF
1A	3	3	PC	DF
1B	0	0	RTN	05
1C	3	3	PC	DF
1D	3	5	PC	EF
1E	3	6	PC	F7
1F	0	0	RTN	05

BR: unconditional jump to address at D_2 of Am2911
BRC: branch to subroutine if $\overline{MOT}$ = 0
PC: continue to next address
RTN: return from subroutine

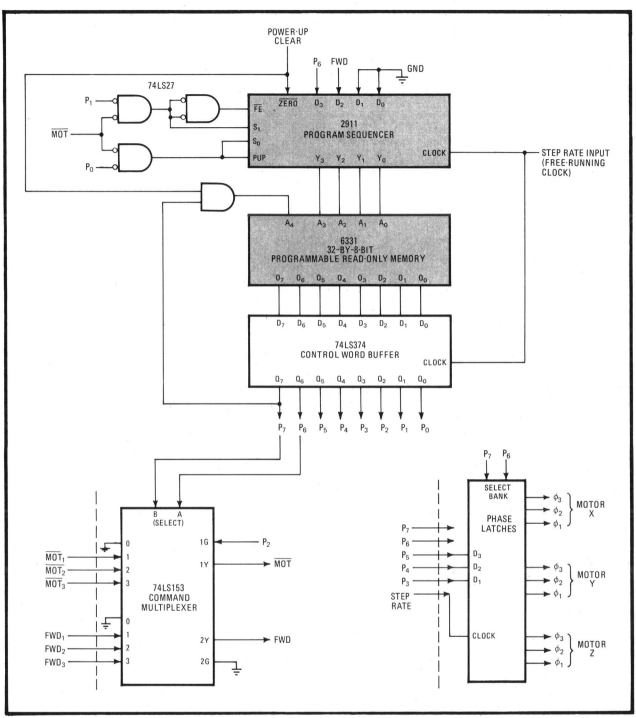

Simultaneous stepping. A microprogram sequencer and 32-word-by-8-bit PROM control the individual directions and stepping sequences of up to three servo motors. Its software approach gives the position controller great flexibility, as stepping sequences are ordered in PROM to suit particular applications. The servo-control system provides direction information through FWD1–FWD3 lines.

Thermistor probe helps regulate liquid levels

by Leonard Sherman
National Semiconductor Corp., Santa Clara, Calif.

There are many techniques for sensing and controlling the level of a liquid in a container. However, this design uses a simple circuit to regulate liquid levels and a thermistor probe to measure the liquid's thermal conductivity relative to the surrounding air or gas. By removing heat from the sensor, the controller determines the level of the fluid. The circuit works well even under harsh environments and may be an economical solution to level-control problems.

Dual timer U_2 provides a delayed, adjustable drive pulse for the solid-state relay comprising optically isolated triac driver U_3 and triac Q_1. This pulse is generated only after comparator U_1 is driven high. When the comparator's output goes low, delay-timing capacitor C_1 is discharged. As a result, the astable section of timer U_2 produces a trigger pulse for the monostable section whose output drives the solid-state relay circuitry. The turn-on delay time and the drive pulse width are adjusted by potentiometers R_1 and R_2, respectively.

The thermal reference point of comparator U_1, with reference to the ambient temperature, is set by resistor R_3. The ideal value, for maximum noise immunity, is half way between the wet and dry voltage on the thermistor sensor. □

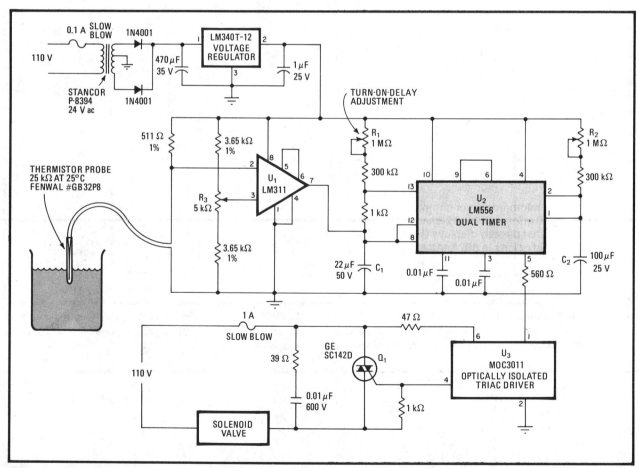

Sensing. The thermistor probe senses the level of the liquid and triggers the controller circuit composed of comparator U_1 and dual timer U_2. The output of the timer drives the solid-state relay comprising driver U_3 and triac Q_1. Resistor R_3 sets the thermal reference point for comparator U_1. The probe measures the liquid's thermal conductivity relative to the surrounding air or gas.

Octal register links dissimilar a-d converters

by K. Russell Peterman
Radian Corp., Austin, Texas

Many microprocessor-based systems use more than one type of analog-to-digital converter and thus require a circuit to compensate for the differences in output coding. Two of the more popular varieties—complementary-2's complement for successive-approximation converters and sign-magnitude binary for integrating converters—may use this design (see figure). The circuit matches outputs by converting sign-magnitude–binary data into 2's complement with only one multifunction octal register chip and can be applied in all sign-magnitude–binary converters.

When the sign bit is 1, the converter inverts all bits except the sign bit, and when the sign bit is 0, no bits are inverted (see table). The binary code contains two 0s (+ or −). However, the 2's-complement code does have an asymmetry about 0, thus eliminating the double zero. As a result, there is a 1-bit discrepancy between the dc voltages represented by the two codes for input values of 0 or negative. Higher resolution converters may be obtained by adding a second 74LS380 chip to the circuit, and sign-magnitude, binary-coded–decimal converters require BCD-to-binary conversion preceeding the multifunction register. □

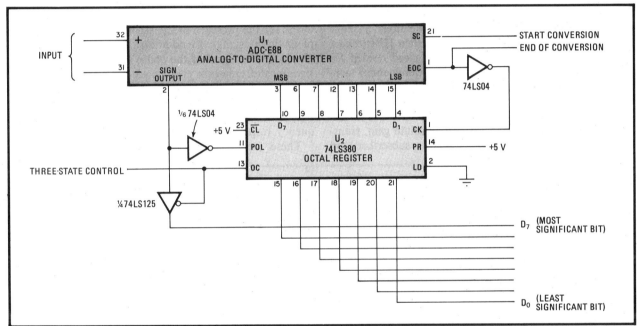

Conversion. Using just multifunction octal register U_2, this circuit converts sign-magnitude binary to complementary-2's complement. The circuit inverts all bits except the sign bit when the sign bit is 1 and inverts none when the sign bit is 0.

CONVERTING SIGN-MAGNITUDE BINARY TO 2's COMPLEMENT															
	Sign-magnitude binary									2's complement						
Scale	Sign	Most significant bit						Least significant bit	Scale	MSB						LSB
+FS−1 LSB	1	1	1	1	1	1	1	1	+FS−1 LSB	1	1	0	0	0	0	0	0
+1 LSB	1	0	0	0	0	0	0	1	+1 LSB	1	1	1	1	1	1	1	0
+0	1	0	0	0	0	0	0	0	0	1	1	1	1	1	1	1	1
−0	0	0	0	0	0	0	0	0	−1 LSB	0	0	0	0	0	0	0	0
−1 LSB	0	0	0	0	0	0	0	1	−2 LSB	0	0	0	0	0	0	0	1
−FS+1 LSB	0	1	1	1	1	1	1	1	−FS	0	0	1	1	1	1	1	1

Conductive foam forms reliable pressure sensor

by Thomas Henry
Transonic Laboratories, Mankato, Minn.

Pressure-sensitive resistors made with conductive foam usually suffer from mechanical and electrical reliability problems — the electrodes of the unit are prone to short, and its sensor rarely returns to its initial value once the pressure is released. However, this circuit, which uses a low-cost electronic pressure sensor, overcomes these problems and provides additional control voltages.

The electronic pressure sensor (a) comprises a conductive foam that is sandwiched between two copper-clad boards that act as electrodes. This configuration creates a pressure-sensitive resistor that has a high resistance in an uncompressed state. Its value drops considerably under pressure; when compressed, the sensor's high resistance value of 10 to 50 kilohms drops to several hundred ohms.

Sensor. This pressure-sensitive resistor (a) together with the circuit (b) provides a reliable electronic pressure sensor. Conductive foam that is sandwiched between two electrodes forms the pressure sensor. Ordinary insulating foam rubber, surrounding the conductive layer, is placed within the sandwich to prevent the boards from shorting. The output from the sensor is sensed by op amp A_1.

The insulating foam rubber placed in the sandwich prevents the electrodes from shorting and also evens out the action of the sensor. Common integrated-circuit packaging foam is used as the conductive material.

Operational amplifier A_1 senses the output generated by voltage divider R_6 and the pressure sensor (b). Any noise in the system is grounded by capacitor C_1. In addition, C_3 functions as a low-pass filter to provide a smooth voltage at the output of A_1. When the pressure sensor has a nominal uncompressed resistance of 10 kΩ, the voltage at pin 3 of A_1 is about -5 volts when the sensor is uncompressed and -15 V when compressed. This voltage swing is offset by a fixed value of $+7.5$ v,

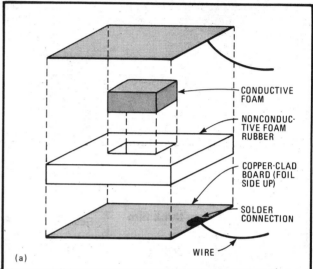

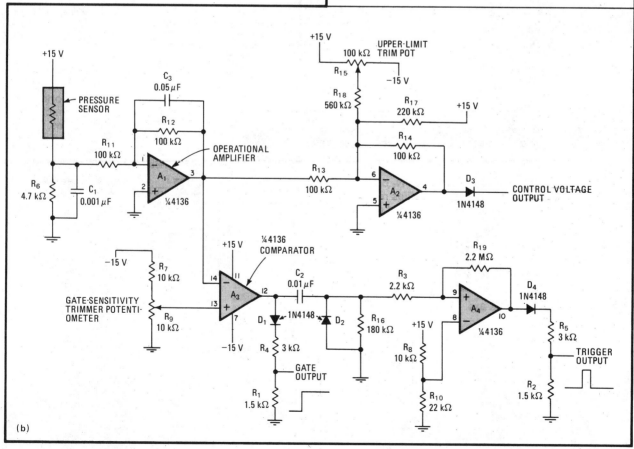

produced by R_{17}, and the sum is inverted by A_2 whose output then swings from -2.5 v to $+7.5$ v. In addition, this output is further truncated by diode D_3 to provide a range of 0 to $+7$ v. As a result, the sensor will always indicate a return to a constant value.

Comparator A_3 generates a gate output that is based on the amount of pressure exerted on the pad, which is set by trimmer potentiometer R_9. The comparator output is differentiated by A_4 to provide a 1-millisecond, 5-v trigger pulse.

This circuit is designed to control an electronic-music synthesizer. The control-voltage output of the circuit controls the voltage-controlled oscillator while the gate and trigger pulses fire the envelope generator of the synthesizer. Thus one transducer is used to control several parameters of a design simultaneously.

Though the circuit provides a reliable uncompressed and compressed voltage output, there is no guarantee that the voltages between these two extremes follow a linear progression. The plot of the voltages depends both on the physics of the sensor and the voltage drop across diode D_3. □

Positive pulse triggers 555 integrated-circuit timer

by Rudy Stefenel
San Jose, Calif.

Many applications require a circuit capable of producing timing intervals, and the most popular monolithic timer is the 555. Though versatile, this timer is limited by a negative-going trigger input. However, a careful study of the functional block diagram shows that pin 5, which is connected to the noninverting input of comparator 2 through a resistor, can be treated as a positive-going trigger input point. Thus pin 5 can now serve both as the control voltage input for which it was originally intended by the 555 designers and as the positive trigger input.

Because the trigger pulse disappears by the time the timing capacitor is charged to the control voltage, the trigger input at pin 5 does not affect the control voltage. The sensitivity of pin 5 to trigger input is controlled by the voltage difference between it and pin 2. This is done by connecting pin 2 to a voltage divider network.

As shown in figure, the monostable multivibrator comprising the 555 timer is driven by the rising edge of the positive-going input trigger pulses. Pin 2 is connected to the center of the resistor network between supply and ground. In addition, a bypass capacitor is connected at pin 2 to make it insensitive to stray pulses coupled from nearby circuits. □

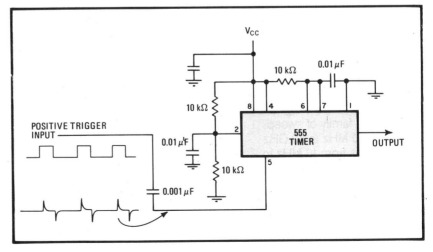

Trigger. The internal block diagram of IC timer 555 shows that pin 5 is connected to the noninverting input of the comparator 2 through a resistor and therefore can be used as a positive–trigger-input terminal. The monostable multivibrator consisting of timer 555 and its associated circuitry is driven by positive input pulses.

Speeding floppy data transfer under program control

by S. Shankar
Sangamo Weston/Schlumberger, Energy Management Division, Atlanta, Ga.

The transfer of data between a floppy-disk drive and memory is normally carried out through a microprocessor's direct-memory-access (DMA) port—a bit of overkill, considering that the disk drive's data rate is relatively slow (1 byte every 32 microseconds) while the DMA's write time is less than 1 μs. Still, using DMA represents one of the best ways to process the data because other methods, such as input/output or interrupt control, are just not fast enough to accommodate the floppy. But data can be transferred under program control, while avoiding the time-consuming test-flag and jump steps that are so much a part of the aforementioned methods, if the processor's read and write operations are synchronized with the rate at which the floppy handles data.

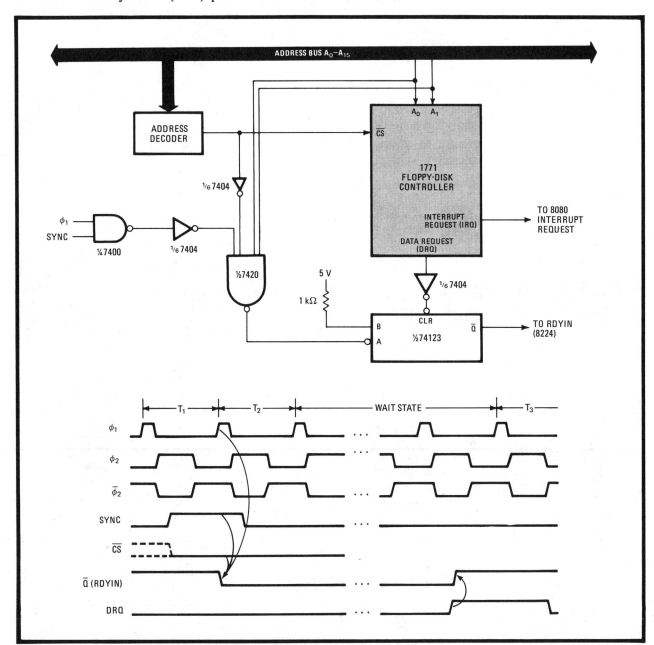

Synchronous. Software-based controller minimizes processor-to-floppy-disk data-transfer time by ensuring that processor's read and write operations are synchronized with floppy's data rate. Scheme works with any processor having the ready and wait facilities.

	8080 PROGRAM FOR DATA TRANSFER BETWEEN MICROPROCESSOR AND FLOPPY DISK	
Label	Source statement	Comments
	LXI H, FDCRGG	Set up (H,L) to point to FDC data register
	LXI B, BUFFER	Set up (B,C) to point to buffer area where data is to be stored
	― ― ― ― ―	
	― ― ― ― ―	
	― ― ― ― ―	Set up 1771 for Read
	― ― ― ― ―	
	― ― ― ― ―	
	JMP READ	Go to READ
	.	
	.	
READ	MOV A, M	Move byte from FDC Register to 8080 accumulator. The 8080 goes into WAIT until data becomes available.
	STAX B	Store contents of accumulator in buffer area.
	INX B	Increment (B,C)
	JMP READ	Go back for more
	.	
	.	
RSTX	― ― ― ― ―	Interrupt Service Routine for IRQ. Routine is executed on completion of READ
	― ― ― ― ―	
	― ― ― ― ―	

The implementation is shown for the Western Digital 1771 floppy-disk controller working with an 8080 microprocessor. As seen with the aid of the program, the processor initially generates an output command to the floppy-disk controller to start the data-read operation. It then proceeds to the read routine and attempts to read a data byte from the floppy's data register.

The data register is accessed when $A_0 = A_1 = 1$ is placed on the data bus and $CS = 0$. Then the 74123 monostable (one-shot) multivibrator is triggered on the occurrence of the synchronizing signal, forcing the processor into the wait state.

When the floppy disk is ready to generate a data byte, line DRQ goes high and clears the one-shot. The processor comes out of the wait state and reads the byte. It then loops back and repeats the operation.

The read loop could go on indefinitely. To afford some flexibility, however, provision is made for interrupt control. Because the IRQ command is always generated on completion of a read-command execution, the processor can never hang up.

The same technique is used to write data into the floppy. The only difference is that the processor writes data into the disk each time a DRQ signal is generated.

The total time required to transfer one byte of data with this method is 29 machine cycles, or 14.5 μs using a 2-megahertz clock. The method should work just as well with double-density floppy disks where the rates are twice as fast. □

Variable–duty-cycle circuit controls valve mixer

by Gary S. Kath
Merck Institute for Therapeutic Research, Rahway, N. J.

High-pressure liquid-chromatography analysis usually requires the disproportionate mixing of two solvents to form a buffer. A three-way solenoid-valve mixer under the control of a variable–duty-cycle circuit is one way to combine the two substances accurately. A programmable timer selects the mix in 1% increments by setting the flow time of each fluid as a percentage of the circuit's total cycle period.

The solenoid-valve–control circuit (see figure) uses programmable timer U_5 to select the percentage mix, from 1% to 99%, for one fluid. The other substance flows for the remaining percentage of the cycle. Selecting a valve-cycle period of 1, 5 or 10 seconds is accomplished through switch S_1. In addition, a miniature three-way solenoid valve diverts the two fluids to the outlet.

Timer U_1 functions as an astable oscillator whose frequency is set by S_1. Integrated circuits U_2, U_3, and U_4 divide this output by 100 and generate a trigger pulse of one clock cycle duration for every 100 clock cycles generated by U_1. In addition, the output of U_1 is fed to monostable timer U_5.

Upon receiving the positive edge of the trigger pulse, the output of U_5 goes low. This low signal cuts off the Darlington power amplifier, consisting of transistors Q_1 and Q_2, that switches the solenoid valve to port A. The thumbwheel switch setting determines the duration of the clock cycle for which the output of U_5 remains low. When the output of U_5 goes high, the amplifier is driven into saturation and the solenoid valve switches to port B. This cycle repeats every 1, 5, or 10 s depending on the setting of S_1. □

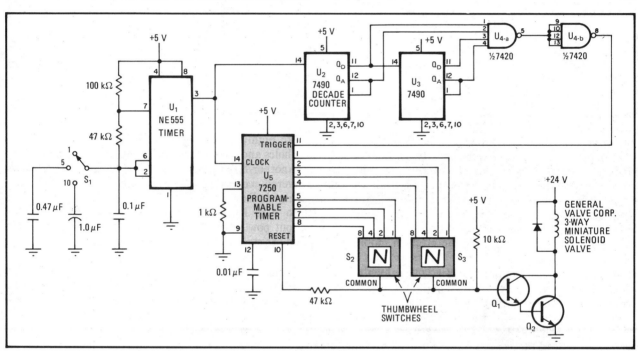

Right chemistry. A miniature three-way solenoid valve is controlled by this variable–duty-cycle oscillator circuit. S_1 sets the total valve operation period, while S_2 and S_3 set the percentage of the cycle operation period for port A. Port B is open for the remaining time of the cycle.

Appliance-controller interface aids the paralyzed

by David Rye
BSR (USA) Ltd., Blauvelt, N.Y.

The popular BSR System X-10 controller, which serves so well in home-security systems, can be easily adapted for use by the paralyzed with this interface, whose puff-and-sip switches enable the handicapped to turn on appliances. Here, the switches are used to activate the desired output device via a stepped counter, instead of the push-button console normally required. Light-emitting diodes provide the visual feedback to the user.

The heart of the X-10 controller is BSR's 542C chip, which transmits a common-carrier code on the ac power line corresponding to the user-defined location of the desired appliance in the system, when it is appropriately addressed. The control signal is then decoded by plug-in modules connected to the individual appliances. This standard scheme is modified by adding circuitry to accommodate the sip switch, which is used to select the desired appliance, and the puff switch, which is used to transmit the corresponding code onto the power line.

The modification is made with the aid of a 74C161 4-bit counter for stepping and a 4051 demultiplexer for code selection, as shown. In this configuration, the address lines of the 4051 are controlled by the divide-by-16 counter, whose outputs step through a count of 0 to 7 twice in each 16-count cycle when advanced by the sip switch, which serves as a clock (see truth table). The eight K inputs of the 542C will therefore be selected in sequence, twice for each complete cycle of the counter. The circuit is wired so that numbers 1 to 8 corresponding to the numbered wall module are transmitted during the first eight clocked states of the counter. The D output of the counter is low at this time, and odd numbers are sent when $\bar{S}_1$ is at logic 1. Even numbers occur during the time that $\bar{S}_3 = 1$, meaning that $\bar{S}_1$ or $\bar{S}_3$ can be selected by the A output of the counter, which is high on odd counts and low on even counts.

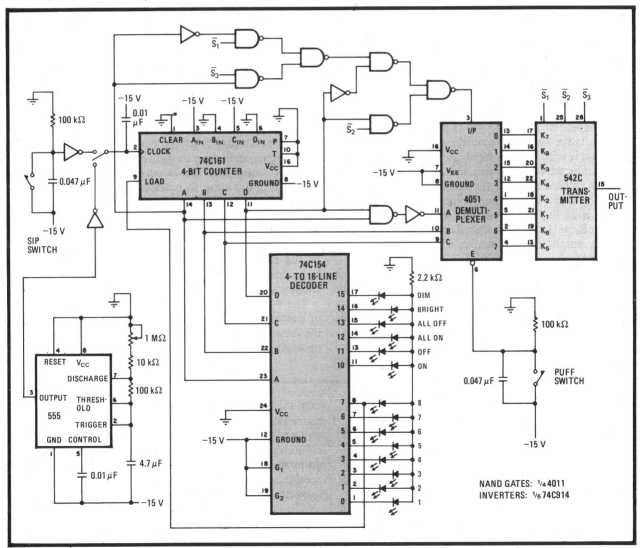

Handy. BSR X-10 controller, the heart of the Sears Home Control System, is easily adapted for the physically handicapped with interface that uses tongue-depress, head-activated, or sip and puff switches. Eight appliances can be activated. Eight additional commands provide all-on and all-off functions, light-dimming duties, and so on. Light-emitting diodes supply user with required visual feedback of controller's state.

Alternatively, the 555 timer can be used for the clock to yield automatic code selection. This eliminates the need for the sip switch, making the controller easier to operate. An external potentiometer is included to set the scanning speed to the user's preference and ability.

In either case, outputs 0 to 7 of the 4051 are enabled in sequence. Note that each is connected to the K inputs of the 542C in such a manner as to ensure that the transmission of numbers is in the correct order. It is seen that the proper sequence can be transmitted by using the counter's A output to select $\overline{S}_1$ or $\overline{S}_3$, using the D output to disable the A input of the 4051 so that only even-number outputs on the multiplexer are enabled, and connecting the four even-numbered outputs to the appropriate K inputs of the 542C.

Eight additional functions can be sent on the second eight counts, including the off, on, all-off, all-on, bright, and dim commands. The D output of the counter is set high during this time. $\overline{S}_2$ is selected when $D = 1$. Under this condition, the input to the 4051 is a binary number that steps from 0 to 7, and all its outputs will be enabled in sequence.

Note that during counts 8 and 9, the 542C transmits invalid codes. It is therefore suggested that these states be avoided. To achieve this, it is necessary to preset the counter to 10 after count 7 is decoded.

The 4-to-16-line decoder (74C154) monitors the state of the counter. Fourteen light-emitting diodes indicate the function addressed, thus providing the user with the necessary feedback information. □

	TRUTH TABLE: CONTROLLER FOR USE BY THE MOTION-HANDICAPPED										
Count	Counter				4051			4051 O/P	K	S	Code sent
	A	B	C	D	A	B	C				
0	0	0	0	0	0	0	0	0	K_7	$\overline{S}_1$	1
1	1	0	0	0	0	0	0	0	K_7	$\overline{S}_3$	2
2	0	1	0	0	0	1	0	2	K_3	$\overline{S}_1$	3
3	1	1	0	0	0	1	0	2	K_3	$\overline{S}_3$	4
4	0	0	1	0	0	0	1	4	K_2	$\overline{S}_1$	5
5	1	0	1	0	0	0	1	4	K_2	$\overline{S}_3$	6
6	0	1	1	0	0	1	1	6	K_6	$\overline{S}_1$	7
7	1	1	1	0	0	1	1	6	K_6	$\overline{S}_3$	8
8	0	0	0	1	0	0	0	0	K_7	$\overline{S}_2$	*
9	1	0	0	1	1	0	0	1	K_8	$\overline{S}_2$	*
10	0	1	0	1	0	1	0	2	K_3	$\overline{S}_2$	on
11	1	1	0	1	1	1	0	3	K_4	$\overline{S}_2$	off
12	0	0	1	1	0	0	1	4	K_2	$\overline{S}_2$	all on
13	1	0	1	1	1	0	1	5	K_1	$\overline{S}_2$	all off
14	0	1	1	1	0	1	1	6	K_6	$\overline{S}_2$	bright
15	1	1	1	1	1	1	1	7	K_5	$\overline{S}_2$	dim

PROM forms flexible stepper-motor controller

by Gregory C. Jewell
Caldisk, Provo, Utah

With just a 256-by-4-bit programmable read-only memory, a decoder, and a driver, this design provides a programmable stepper-motor controller. The example given here is a four-phase stepping motor incorporated in a floppy-disk drive. However, the idea is easily adapted to other applications.

The controller uses four of the PROM's eight input-address bits. They are connected to the outputs to obtain a stable state only when the input address—consisting of four direct and four feedback inputs—produces an output equal to the feedback address. Therefore, when one of the direct inputs changes state, the resulting output changes in sequence until a stable state is attained.

A working version of the design (see figure) uses 1-K PROM 74S287, two-to-four-line decoder 25LS2539, and drivers Q_1 through Q_4. The four direct inputs are initialize ($\overline{INT}$), step ($\overline{ST}$), direction (DI) and track 00 (TRK 00). The feedback inputs are NSF, NDF, A_1F, and A_0F. In addition, the step input is a pulse generated by the rising edge of the step signal that is sent by the disk controller.

The direction of motor rotation is controlled by input DI, which determines the order of activation for the motor phases. TRK 00, which is provided by a sensor circuit, is active when the read/write head is positioned above it. In addition, the PROM does not allow the motor controller to step lower than TRK 00.

Two of the four output signals, NS and ND, indicate internal states and are also fed back to the input. Outputs A_1 and A_0 are decoded by U_{2-A}. This decoder output is used to control one of the four motor phases when the motor-enable input is active. The relationship between PROM outputs A_1, A_0, and motor phase is

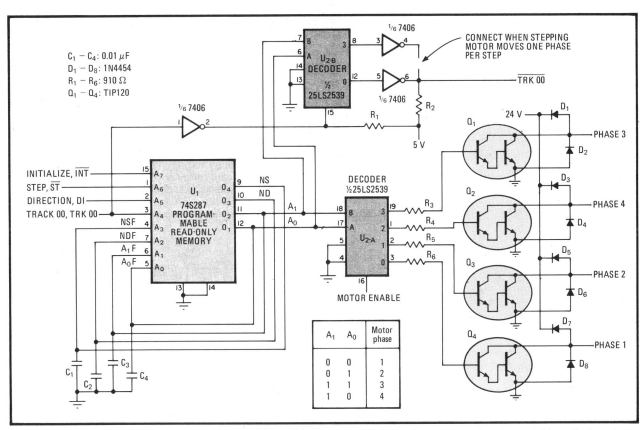

Controller. The stepper-motor controller employs a 256-by-4-bit programmable read-only memory, a decoder, and drivers Q_1 through Q_4. Decoder U_{2-B} is provided for track 00 detection. The chart illustrates the relationship between A_1, A_0, and the motor phases.

| TABLE 1: STEPPER-MOTOR TRUTH TABLE TO MOVE TWO PHASES PER STEP ||||||||||||
| Inputs |||||||| Outputs |||| Comments |
$\overline{INT}$	DI	$\overline{ST}$	TRK 00	NSF	NDF	A_1F	A_0F	NS	ND	A_1	A_0	
0	X	X	X	X	X	X	X	0	0	0	0	initialization
1	X	1	X	0	0	0	0	0	0	0	0	stable at even track
1	0	0	1	0	0	0	0	0	0	0	0	cannot step lower than track 00
1	0	0	0	0	0	0	0	0	1	0	0	step to lower intermediate phase state
1	X	0	X	0	1	0	0	1	1	0	0	
1	X	0	X	1	1	0	0	1	1	0	1	
1	X	0	X	1	1	0	1	1	1	0	1	stable at intermediate phase state
1	X	1	X	1	1	0	1	0	1	0	1	step to lower (odd) track
1	X	X	X	0	1	0	1	0	0	0	1	
1	X	X	X	0	0	0	1	0	0	1	1	
1	X	1	X	0	0	1	1	0	0	1	1	stable at odd track
1	0	0	X	0	0	1	1	0	1	1	1	step to lower intermediate phase state
1	X	0	X	0	1	1	1	1	1	1	1	
1	X	0	X	1	1	1	1	1	1	1	0	
1	X	0	X	1	1	1	0	1	1	1	0	stable at intermediate phase state
1	X	1	X	1	1	1	0	0	1	1	0	step to lower (even) track
1	X	X	X	0	1	1	0	0	0	1	0	
1	X	X	X	0	0	1	0	0	0	0	0	
1	X	1	X	0	0	0	0	0	0	0	0	stable at even track
1	1	0	X	0	0	0	0	1	0	0	0	step to higher intermediate phase state
1	X	0	X	1	0	0	0	1	0	1	0	
1	X	0	X	1	0	1	0	1	0	1	0	stable at intermediate phase state
1	X	1	X	1	0	1	0	1	0	1	1	step to higher (odd) track
1	X	1	X	1	0	1	1	0	0	1	1	
1	X	1	X	0	0	1	1	0	0	1	1	stable at odd track
1	1	0	X	0	0	1	1	1	0	1	1	step to higher intermediate phase state
1	X	0	X	1	0	1	1	1	0	0	1	
1	X	0	X	1	0	0	1	1	0	0	1	stable at intermediate phase state
1	X	1	X	1	0	0	1	1	0	0	0	step to higher (even) track
1	X	1	X	1	0	0	0	0	0	0	0	
1	X	1	X	0	0	0	0	0	0	0	0	stable at even track
1	X	1	X	0	1	0	0	0	0	0	0	define invalid states
1	X	1	X	0	1	1	1	0	0	1	1	
1	X	1	X	1	1	0	0	1	0	0	0	
1	X	1	X	1	1	1	1	1	0	1	1	

1 = HIGH   0 = LOW   X = DON'T CARE

| TABLE 2: STEPPER-MOTOR TRUTH TABLE TO MOVE ONE PHASE PER STEP ||||||||||||
| Inputs ||||||||| Outputs |||| Comments |
$\overline{INT}$	DI	$\overline{ST}$	TRK 00	NSF	NDF	A_1F	A_0F	NS	ND	A_1	A_0	
0	X	X	X	X	X	X	X	1	1	0	1	initialization at odd track
1	X	1	X	1	1	0	1	1	1	0	1	
1	X	0	X	1	1	0	1	1	1	0	0	step to higher (even) track
1	X	0	X	1	1	0	0	0	1	0	0	
1	X	0	X	0	1	0	0	0	1	0	0	stable at even track (see state 5)
1	X	1	X	0	0	0	0	0	0	0	0	state 1: stable at even track
1	0	0	1	0	0	0	0	0	0	0	0	cannot step lower than track 00
1	0	0	0	0	0	0	0	1	0	0	0	step to lower (odd) track
1	0	0	X	1	0	0	0	1	0	0	1	
1	X	0	X	1	0	0	1	1	0	0	1	state 2: stable at odd track
1	X	1	X	1	0	0	1	0	0	0	1	enable next step sequence
1	X	1	X	0	0	0	1	0	0	0	1	stable at odd track
1	0	0	X	0	0	0	1	0	1	0	1	step to lower (even) track
1	0	0	X	0	1	0	1	0	1	1	1	
1	X	0	X	0	1	1	1	0	1	1	1	state 3: stable at even track
1	X	1	X	1	1	1	1	0	0	1	1	enable next step sequence
1	X	1	X	0	0	1	1	0	0	1	1	stable at even track
1	0	0	1	0	0	1	1	0	0	1	1	cannot step lower than track 00
1	0	0	0	0	0	1	1	1	0	1	1	step to lower (odd) track
1	0	0	X	1	0	1	1	1	0	1	0	
1	X	0	X	1	0	1	0	1	0	1	0	state 4: stable at odd track
1	X	1	X	1	0	1	0	0	0	1	0	enable next step sequence
1	X	1	X	0	0	1	0	0	0	1	0	stable at odd track
1	0	0	X	0	0	1	0	0	1	1	0	step to lower (even) track
1	0	0	X	0	1	1	0	0	1	0	0	
1	X	0	X	0	1	0	0	0	1	0	0	state 5: stable at even track
1	X	1	X	0	1	0	0	0	0	0	0	enable next step sequence
1	X	1	X	0	0	0	0	0	0	0	0	stable at even track (see state 1)
1	1	0	X	0	0	0	0	1	0	0	0	step to higher (odd) track
1	1	0	X	1	0	0	0	1	0	1	0	
1	X	0	X	1	0	1	0	1	0	1	0	stable at odd track (see state 4)
1	1	0	X	0	0	1	0	0	1	1	0	step to higher (even) track
1	1	0	X	0	1	1	0	0	1	1	1	
1	X	0	X	1	1	1	1	0	1	1	1	stable at even track (see state 3)
1	1	0	X	0	0	1	1	1	0	1	1	step to higher (odd) track
1	1	0	X	1	0	1	1	1	0	0	1	
1	X	0	X	1	0	0	1	1	0	0	1	stable at odd track (see state 2)
1	1	0	X	0	0	0	1	0	1	0	1	step to higher (even) track
1	1	0	X	0	1	0	1	0	1	0	0	
1	X	0	X	0	1	0	0	0	1	0	0	stable at even track (see state 5)
1	0	1	X	0	1	0	1	0	0	0	1	define invalid states
1	1	1	X	0	1	0	1	0	0	0	1	
1	0	1	X	0	1	1	0	0	0	1	0	
1	1	1	X	0	1	1	0	0	0	1	0	
1	0	1	X	1	0	0	0	0	0	0	0	
1	1	1	X	1	0	0	0	0	0	0	0	
1	0	1	X	1	0	1	1	0	0	1	1	
1	1	1	X	1	0	1	1	0	0	1	1	
1	X	1	X	1	1	0	0	0	1	0	0	
1	X	X	X	1	1	1	0	1	0	1	0	
1	X	X	X	1	1	1	1	0	1	1	1	

1 = HIGH   0 = LOW   X = DON'T CARE

shown in the figure. The decoder $U_{2\text{-}B}$ generates $\overline{TRK\ 00}$ to enable the drive interface only when the proper motor phase exists.

Tables 1 and 2 show the functioning of the circuit in mode 1 and mode 2, respectively. Mode 1 moves the motor two phases per step. In this mode, the motor is adjusted to position the read/write head above the even tracks when phase 1 is active and above the odd tracks when phase 3 is active. In mode 2, which moves the motor one phase a step, the motor is adjusted to position the read/write head above even tracks when phase 1 or phase 3 is active. An active phase 2 or phase 4 indicates odd-numbered tracks. A power-on problem is overcome by initializing the stepper motor at an odd track and automatically stepping to the next higher even track. □

6. CONVERTERS

D-a converter also performs a-d translation

by Michael Parsin
Precision Monolithics Inc., Santa Clara, Calif.

Complex microprocessor-based systems that require both analog-to-digital and digital-to-analog conversion, such as gas-flow and motor-speed units and antenna positioners, will be simplified by using multiplexing one-chip converters such as PMI's DAC-76. One of the advantages gained is elimination of an a-d converter and the inherent interfacing problems between two converters that would normally exist. System accuracy is also maintained, because the a-d and d-a conversions are processed and controlled via the same chip.

As shown in Fig. 1, the DAC-76 logarithmic (companding) d-a converter is alternately switched from its encoding or measurement (a-d) mode to its decoding or control (d-a) mode by placing the appropriate signals at its E/D port. An 8085 microprocessor and an 8155 input/output port provide the digital data (see program table). Operational amplifier A_1, comparator A_2, voltage reference A_3, and sample-and-hold amplifier A_4 come into use during the a-d portion of the cycle.

The measurement is initiated when the E/D line goes high, and the bipolar 0- to 5-volt analog input signal is sampled. The 8085 then commands the d-a device to generate a ramp, which appears at the I_{oe} (+) port. When the ramp voltage equals the analog input, A_2 fires, generating an end-of-conversion signal. The processor's internal counter, which had initiated counting at the beginning of the cycle, is then frozen and its binary contents stored in a 1-byte memory location. Note that the DAC-76's companding feature makes it possible to compress 12-bit accuracy into a 7-bit format. This format, combined with a sign bit, makes it an ideal interface for the typical 8-bit microprocessor.

During the decode cycle, the data previously measured is compared with the control variable, which is stored away in a separate register, and the system acts to minimize the difference. For example, if the set point in a temperature controller has been adjusted at 100°, and the measured value is 110°, a correcting voltage that corresponds to the 10° offset must be generated for

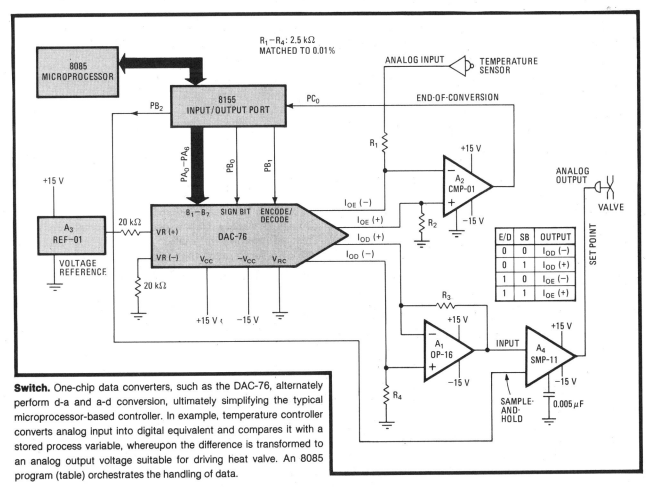

Switch. One-chip data converters, such as the DAC-76, alternately perform d-a and a-d conversion, ultimately simplifying the typical microprocessor-based controller. In example, temperature controller converts analog input into digital equivalent and compares it with a stored process variable, whereupon the difference is transformed to an analog output voltage suitable for driving heat valve. An 8085 program (table) orchestrates the handling of data.

\multicolumn{4}{c}{8085 PROGRAM FOR DUAL DATA CONVERSION}			
Location	Label	Source statement	Comments
2000		ORG 2000H	
2000	BEGIN:	LXI SP,20C2H	; INITIALIZE STACK POINTER
2003		MVI A,03H	; SET POINTS A AND B TO OUT AND C TO IN
2005		OUT 20H	; THIS SETS 8155 CSR
2007	ENCOD:	MVI A,07H	; START ENCODE
2009		NOP	
200A		OUT 22H	; SET ENCODE/DECODE (E/D) TO ENCODE AND SIGN BIT TO (−)
200C	LOOP:	INR B	; INCREMENT UP COUNTER
200D		NOP	
200E		MOV A,B	; COUNT TO ACCUMULATOR
200F		OUT 21H	; OUT COUNT TO DAC-76 DIGITAL IN
2011		IN 23H	; MONITOR END OF CONVERSION
2013		ANI 01M	; MASK PC_0
2015		JNZ LOOP	; LOOP WHEN END-OF-CONVERSION IS HIGH
2018		MOV A,B	; MOVE RESULT TO ACCUMUALTOR
2019		STA 203CH	; STORE RESULT IN MEMORY 203C
201C		MVI B,00	; CLEAR B REGISTER
201E	DECOD:	MVI A,7FH	; START DECODE
2020		OUT 21H	; OUT TO DIGITAL-TO-ANALOG CONVERTER ALL 1's +5 V
2022		MVI A,01H	; SET E/D TO DECODE AND SHIFT BIT TO (+)
2024		OUT 22H	; OUTPUT TO SAMPLE-AND-HOLD (S/H) AND DAC-76
2026		MVI A,05H	; S/H TO HOLD ANALOG OUT TO +5 V
2028		OUT 22H	; OUT TO S/H
202A		MVI A,00H	; ANALOG OUT TO NEGATIVE
202C		NOP	
202D		NOP	
202E		OUT 22H	; OUT TO S/H AND DAC-76
2030		MVI A,04H	; S/H TO HOLD ANALOG OUT TO −5 V
2032		OUT 22H	; OUT TO S/H AND d-a CONVERTER
2034		NOP	
2035		NOP	
2036		JMP ENCOD	; START ENCODE AGAIN
		END	

controlling the heat valve. The processor does this by addressing a lookup table to determine the required change in valve position that is proportional to the offset. The E/D line is then brought low, and a constant current is steered by processor control through the d-a converter's I_{od} (+) port and thus to the inverting input of the operational amplifier, A_1.

Thus capacitor C_H of the sample-and-hold device charges linearly as long as digital signal PB_2, which has been set low by the 8155 output port initiating data proportional to the offset voltage, remains low. When A_4 is placed in the hold state, the sampling period ends, and the voltage across the capacitor is transferred to the analog output line in order to drive the valve and thus set its position.

The aforementioned stand-alone configuration represents one good example of a direct digital control system, where a single host computer can be placed at the hub of a multitude of remote controllers. It should be kept in mind, however, that this system could easily be made part of a distributed digital control network, where the host computer has many remote minicomputers that have a hand in presiding over the major control loops. □

5-V converter powers EE-PROMs, RS-232-C drivers

by Richard A. McGrath
Studio 7 Technical Documentation, San Carlos, Calif.

Many of today's electrically erasable programmable and electrically alterable read-only memories require different voltages from the RS-232-C drivers needed to interface them with microprocessors. This dc-to-dc converter changes 5 volts into ±11 v for these drivers and 21 v for programming or reading EE-PROMs. Because the RS-

	\multicolumn{4}{c}{TYPICAL TEST RESULTS FOR A 5-V SUPPLY}			
Test point voltage (V)	\multicolumn{4}{c}{Logic state of IH5143 analog switch, pin 15}			
	\multicolumn{2}{c}{Switch S_2 open}	\multicolumn{2}{c}{Switch S_2 closed (grounded)}
	Low	High	Low	High
MC1488* pin 1	−0.2	−11.8	−4.3	−13.5
MC1488* pin 14	+21.7	+10.6	+8.7	+7.5
2817 pin 1 1.5-kΩ load	+20.7	+10.7	+3.5	+3.0
+V_{output}	+22.4	+11.3	+9.4	+8.2
−V_{output}	−0.9	−12.4	−4.9	−14.0

*300-Ω load on output pin 3, logic high on pin 2

232-C drivers and EE-PROMs operate at different time intervals and the circuit's switching time is much shorter than that of serial communication, power requirement conflicts do not occur.

Semiconductor Circuits' power converter U_1 can generate ± 12 V or $+24$ V with the addition of single-pole double-throw switch S_1, diode D_1, and capacitor C_1(a). If S_1 is in position 1, ± 12 V are produced. In position 2, this switch gives $+24$ V. In addition, D_1 ensures that a positive voltage is generated from pin 3 during switching, and surge voltages are suppressed by C_1.

The dc-to-dc converter (b) uses Intersil's analog switch S_1 (IH5143) to switch the converter more quickly from ± 11 to $+21$ V. The switching is controlled by logic, and many voltages can be selected from the circuit with switch S_2 and the logic at pin 15 of S_1 (see table).

Any negative voltage across the drain-to-body junction of the body-puller field-effect transistor of S_1 is stopped by diode D_2. Zener diode D_5 and resistor R_2 reduce $+24$ to $+21$ V when EE-PROM U_2 is programmed or read. In addition, this combination reduces ± 12 to ± 11 V, which powers RS-232-C driver U_3. Capacitor C_2 prevents voltage override when the switching frequency is about 5 kilohertz. Grounding resistor R_1 together with S_2 provides correct power switching for EE-PROMs when the circuit is turned on or off. The 5-V supply also is directly tapped in order to power one of the inputs of the EE-PROM.

Tradeoffs in memory organization (2-K by 8 bits), programming convenience, product availability, and power requirements led to the choice of Intel's 2817 EE-PROM for U_2. Many other voltages can be selected from this circuit, and as a result, any EE-PROM may be used in it. This cost-effective design eliminates bulky transformers and voltage regulators and thus requires only a few external components. It finds applications in lightweight airborne or robotic systems. □

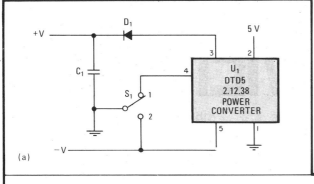

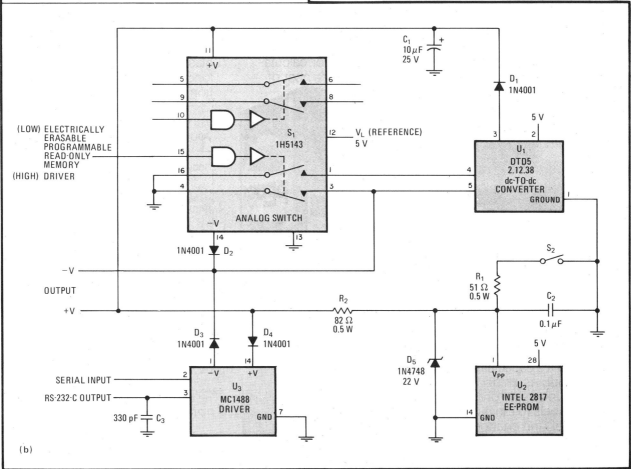

Converter. The dc-to-dc converter (a) uses minimum components to generate ± 12 or $+24$ volts. To provide faster switching, mechanical switch S_1 is replaced with the Intersil's analog switch IH5143 (b). The switching of this converter is controlled by logic and used to program or read 2817 EE-PROMs or to power RS-232-C drivers.

Interfacing 10-bit a-d converter with a 16-bit microprocessor

by Sorin Zarnescu
Westwood, Calif.

Eight-, 10-, and 12-bit analog-to-digital converters cannot easily impart their knowledge to 16-bit microprocessors, like Intel's 8086, because the received output is not in 2's complement form and the sign bit has to be extended. However, a match can be made with this design, which can offer programmable interrupt control. In addition, only one input instruction is needed to read the contents of the a-d converter. As an example, Analog

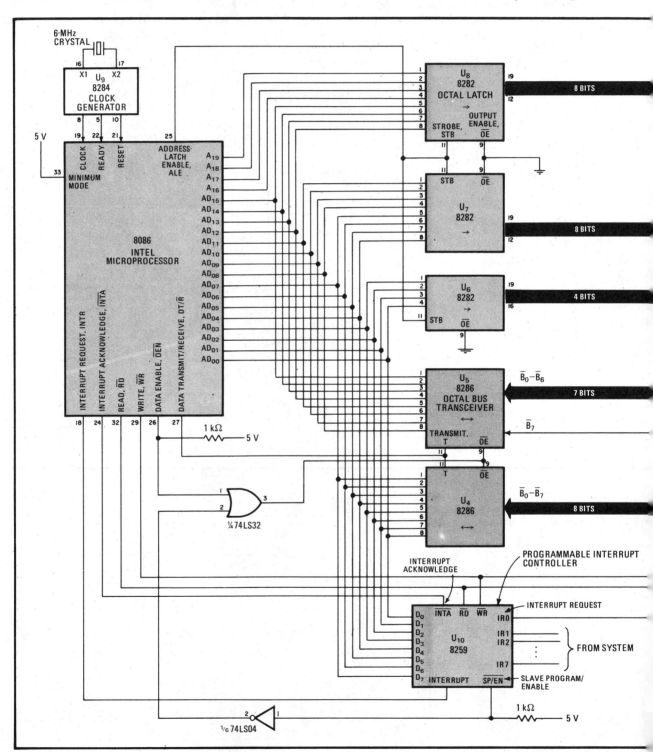

Devices' 10-bit a-d converter, AD571, is used.

Because the a-d converter's output code is offset binary, the most significant bit is inverted and used to control octal line driver-receiver U_3. When the input is positive, the MSB is zero and the octal driver is open. As a result, seven MSBs are 0. On the other hand, when the input is negative, the octal driver is in its tristate mode and the MSB is 1, resulting in seven MSBs also being 1.

After receiving the data-ready interrupt through programmable interrupt controller U_{10}, the microprocessor

Interface. The circuit provides an easy and fast way to interface Analog Devices' AD571 10-bit a-d converter with Intel's 16-bit 8086. The decode logic provides the a-d converter with signals for reading the data and starting the conversion.

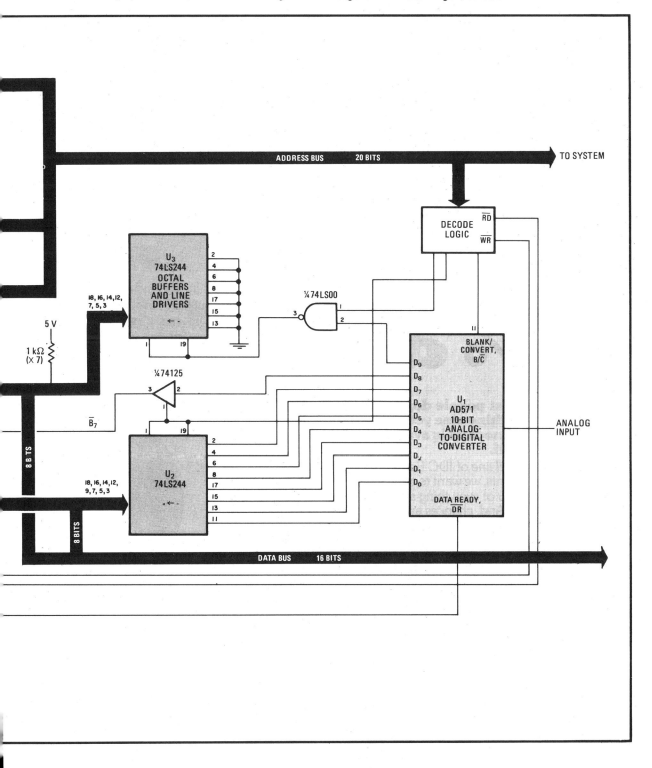

A SEQUENCE FOR THE INTERRUPT ROUTINE		
Mnemonics	Symbolic address	Comments
INP	CONV	; READ DATA
OUT	CONV1	; B/$\overline{\text{C}}$ LINE BROUGHT HIGH
⋮	PROCESSING	
OUT	CONV2	; B/$\overline{\text{C}}$ LINE BROUGHT LOW (START CONVERSION)
IRET		; RETURN

reads the data through an input command also supplied by U_{10}. Since the 8086 has a multiplexed address and data bus, U_6, U_7, and U_8 are used as latches for the address. In addition, the address-latch–enable output of the processor clocks U_6, U_7, and U_8. Octal bus transceivers U_4 and U_5 provide buffers for the data bus.

The direction of data on this bus is controlled by the data transmit-receive pulse that is generated by the processor. In addition, the data-ready output from the converter is used as an interrupt-request input for U_{10}. The decode logic provides the signals for reading the data and starting the conversion (see table). □

Access control logic improves serial memory systems

by Robert G. Cantarella
Burroughs Corp., Paoli, Pa.

File storage systems used in large scientific processors or computers need serially organized dynamic random-access-memory systems to transfer blocks of data from one memory unit to another in a serial order. Dynamic RAM devices used in such secondary stores need to be refreshed in a fixed cyclic order to keep the cost and latency low. This refresh logic loop further reduces latency, while retaining the cost advantage of a serial organization, and refreshes the memory at twice the minimum rate. In addition, zero latency is guaranteed with block transfers of at least L milliseconds.

When an initial transfer request XFERRQST occurs, the requester's address RQSTADDR is loaded into the refresh counter, which resets the refresh loop. This setting initiates the transfer (XFERSTART) with zero latency. The refresh address is then held in the refresh counter. Because the refresh loop is cycling at twice the required frequency, all bits are properly refreshed within L ms. The reset is then enabled only after a full cycle has completed when the time-out down counter is loaded with L ms. However, this action only occurs when the time-out down counter is zero.

After the transfer request is initiated, access is granted either when the requester's address is identical to the refresh address or when the time-out down counter is zero. Thus latency is a function of the requester's address and the time since the last reset. As a result, a block transfer of L ms guarantees that the next transfer request will be granted immediate access.

The system performs like a random-access memory but retains the cost advantages of a serial-memory system. The constant refresh rate results in a steady current drain, thereby reducing the cost of the power-supply system and storage cards. □

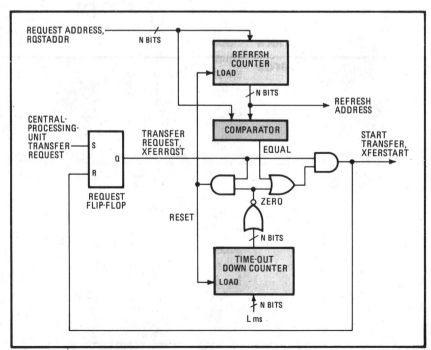

Access. The access control logic refreshes the memory at twice the requisite frequency to keep the latency low. Access is granted when the requester's address is identical to the refresh address or the time-out down counter is zero, whichever occurs first. Block transfer of L ms guarantees zero latency.

Voltage-controlled resistance switches over preset limits

by Chris Tocci
Halifax, Mass.

Using two field-effect transistors as switches, this voltage-controlled resistor network can order up any value of resistance between two preselected limits. It is unlike other circuits in that it does not employ the drain-to-source resistance of matched FETs, whose R_{ds} characteristics are usually proportional to a control voltage. As for circuit linearity, it will far exceed that of conventional networks using a single FET in various feedback configurations.[1]

In operation, oscillator A_1–A_2 generates a 0-to-10-volt triangle wave at 100 kilohertz, which is then compared with the control signal, V_c, at A_3. During the time that the control exceeds switching voltage V_T, FET Q_1 is turned on, and resistor R_1 is placed across resistance R_{out} (disregarding the R_{ds} of Q_1). At all other times, FET Q_2 is on and resistor R_2 is placed across R_{out}. Thus R_{out} is equal to an average value proportional to the time each resistor is placed across the output terminals, with the actual resistance given by $R_{out} = (R_1 - R_2)V_c/10 + R_2$, for $R_1 > R_2$. This relationship will hold provided any potential applied to the R_{out} port from an external device is less in magnitude than the supply voltages; that any signal processing at R_{out} be done at a frequency at least one decade below the 100-kHz switching frequency; and that the upper and lower resistance limits, R_1 and R_2, are much greater than the on-resistance of Q_1 and Q_2, respectively.

Potentiometer R_{11} adjusts the baseline of V_T to zero so that with $V_c = 0$, $R_{out}n = R_2$, where n is a constant. Further calibration can be carried out by trimming R_1 and R_2 to precise values.

This circuit is readily adapted to many applications, such as a one-quadrant multiplier. This is achieved by connecting a voltage-controlled current source into the R_{out} port to build a dc-shift amplitude modulator whose carrier frequency is the switching frequency. The audio information or data is taken from V_c, but with the signal offset by 5 volts. Thus the dynamic range of the circuit will be 10 v. □

References
1. Thomas L. Clarke, "FET pair and op amp linearize voltage-controlled resistor," *Electronics*, April 28, 1977, p. 111.

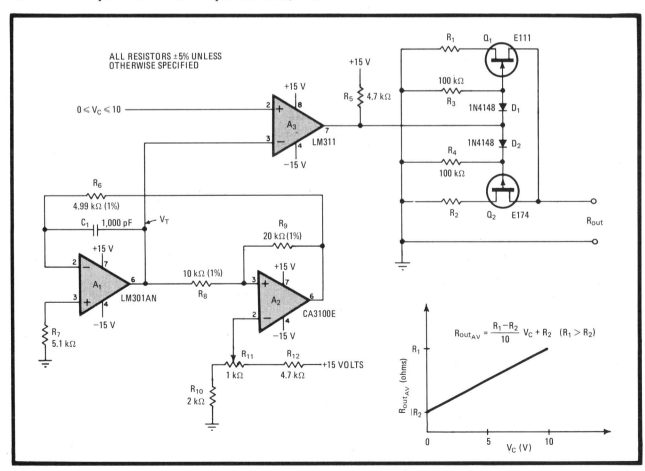

Ohmic linearization. FET switches in voltage-controlled resistor network place maximum-minimum resistors R_1 and R_2 across R_{out} so that the resistance is proportional to the average time each is across output port. Switching technique ensures piecewise-linear operation. This circuit lends itself to many applications, such as a-m modulator, by placing voltage-controlled current source across R_{out}.

Current-biased transducer linearizes its response

by Jerald Graeme
Burr-Brown Research Corp., Tucson, Ariz.

Transducers that work on the principle of variable resistance produce a nonlinear response when voltage-biased, as in common bridge configurations. However, a single operational amplifier configured to provide current biasing for the transducer eliminates this difficulty and allows output offset voltages to be controlled or removed.

As a voltage-biased transducer's resistance varies, so does the current through it. Thus, any signal voltage taken from the transducer will be a function of both current and voltage variations and will be a nonlinear function of the transducer's resistance.

Current biasing rather than voltage biasing avoids this nonlinearity, but reference current sources are not as readily available as voltage ones. Fortunately, an op amp can convert a reference voltage for this purpose and in addition provide other benefits.

In (a), a voltage-to-current converter circuit is adapted for voltage control of the supply current and of the output offset voltage. As laid out, the transducer's bias current is $I_x = (V_2 - V_1)/R_1$, where V_1 and V_2 are the externally applied control voltages. The current polarity can be set at either + or −, allowing an inverted or noninverted voltage response to variations in transducer resistance.

The resulting current flow in the nominal transducer resistance, R, produces an offset voltage at the circuit's output with a counteracting voltage developed by V_1: $V_{oso} = (1+(R_2/R_1))I_xR - (R_2/R_1)V_1$. Thus, through the proper selection of V_1, the output offset may be nulled to zero or set to either a positive or negative polarity.

Signal variations about that level result from a change in transducer resistance. The net voltage output then becomes $e_o = [1+(R_2/R_1)]I_x\Delta R + V_{oso}$. This response is linearly related to ΔR. Also the signal is amplified.

To set the gain, R_2 should be adjusted after R_1, which sets the level of I_x, is selected. Because the output signal is taken directly from the amplifier's output, the transducer is buffered against loading effects.

Deviations in the described performance result primarily from voltage and resistor tolerances and from resistor-ratio error. Mismatch of the $R_2:R_1$ ratio is particularly serious, as this will make I_x somewhat a function of ΔR, thereby reintroducing nonlinearities in the circuit. Such a mismatch will alter the term in the denominator of the I_x formula to read $((R+\Delta R)/R_2) \times ((R_2/R_1) - (R_2'/R_1'))$, where R_1' and R_2' are the mismatched counterparts of R_1 and R_2, respectively. Further errors will result from the dc input-error signals of the op amp. As a result, the deviation in e_o will be:

$$\Delta e_o = \left[1 + \frac{R_x}{R_2}\left(1 + \frac{R_2}{R_1}\right)\right]\left[\left(1 + \frac{R_2}{R_1}\right)V_{os}\right] + \left(1 + \frac{R_2}{R_1}\right)R_xI_{os} + R_2I_{B-}$$

where V_{os}, I_{os}, and I_{B-} are the input offset voltage, input offset current, and the inverting input-bias current of the op amp, respectively. Making the R_2/R_x ratio large reduces this error.

In practice, it is either inconvenient or sometimes undesirable to supply two voltage references to the circuit. A single reference voltage may be applied, as shown in (b), which is the Thévenin equivalent of the circuit in (a). For the specific component values given, the amplifier will deliver a 1.0-volt full-scale output signal with a zero offset in response to a transducer resistance that ranges from 300 to 350 ohms. □

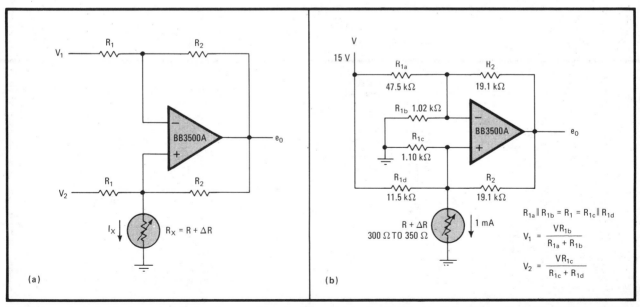

Holding constant. A voltage-controlled current source drives transducer (a) so that its output voltage is a function of its resistance change only, thereby reducing the circuit's nonlinear response. The output offset can also be virtually eliminated. The single-reference current source circuit (b), which is the Thévenin equivalent of (a), may be more attractive to implement in some cases.

Low-power f-V converter turns portable tachometer

by Dan Watson
Intersil Inc., Cupertino, Calif.

Placing a frequency-to-voltage converter in the form of a complementary-MOS timer and operational amplifier at the front end of an analog-to-digital converter reduces the power, wiring complexity, and costs associated with designing an efficient digital tachometer or anemometer for field use. When combined with the multifunctional capability of such a-d converters as Intersil's ICL7106, a direct or scaled reading of the input parameter expressed in revolutions per minute can be readily determined with few additional parts.

The ICL7106 contains not only all of the required clock and display-driving circuitry, but also a reference, so that the external reference voltage normally required in circuits of this type can be omitted.

Signals applied by a magnetic or optical transducer to the input of the ICM7555 timer (powered by the converter's reference voltage) are converted into fixed-width pulses of corresponding frequencies. The ICL7611 micropower op amp integrates these pulses, and consequently the smoothed signal introduced to the a-d converter is a direct function of the input frequency. Thus, the signal will have an amplitude of

$$V_{in} = (RPM/60)(t_{pw})(V_r)(E)(R_4/R_3)$$

where
t_{pw} = pulse width of timer = $1.1\,R_2C_2$
V_r = reference voltage of ICL7106 = 2.8 volts
E = number of events per revolution from the magnetic or optical sensor, the number of fan or propeller blades, or the number of point closures per revolution in an automotive application.

The converter's full-scale output is 200 millivolts. Note that the V⁻ timer port is powered by the internal low-reference voltage of the converter, precluding the need for a second reference because of the rail-to-rail output swing of the ICM7555.

The output of the converter is given by $n = (V_{in}/V_r)(R_7+R_8)/R_{8a}$. The a-d converter contains on-chip display circuitry for driving a liquid-crystal display. If a light-emitting-diode display is desired, the a-d converter can be replaced with its sister unit, the ICL7107. □

Restless wind. Complementary-MOS chips combine to make a simple, cost-effective flea-powered digital tachometer or anemometer for field use. Input signals of corresponding frequency from magnetic or optical transducers are converted into voltages by A_1 and A_2, and then into the equivalent digital output by A_3. R_3–R_4, and R_7–R_8 set the scaling multipliers.

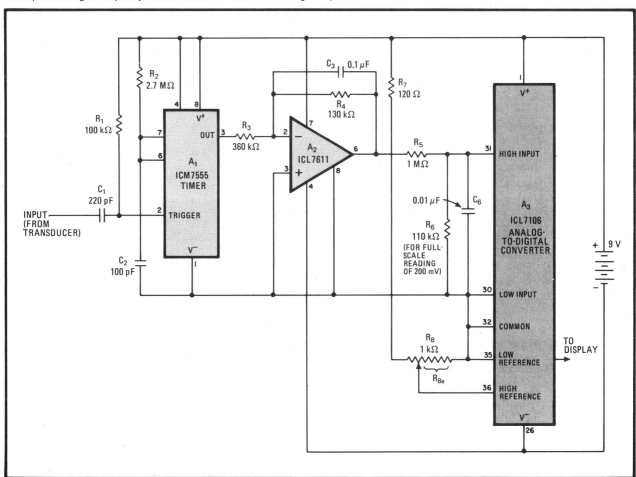

Frequency comparator uses synchronous detection

by Israel Yuval
Video Logic Corp., Sunnyvale, Calif.

There are many different ways to monitor a frequency but most have the disadvantage that they also detect any noise inherent in the circuit along with the desired waveform. This circuit circumvents the noise problem through the use of a technique that samples the unknown frequency at the rate of the reference signal and then averages the measured waveform with a low-pass filter.

If the input frequency f_{in} is equal to the reference f_{ref}, a nonzero average results at the output. However, if $f_{in} \neq f_{ref}$, the average of the samples is zero. In addition, the circuit is immune to noise and works well between 100 hertz and 100 kilohertz.

Comparator U_2 (a) determines the sampling frequency that is derived from reference input signal V_{ref}. Resistor R_2 determines the duration of the sample pulse. The output of U_2 turns switch Q_1 on and off at the rate f_{ref}. The input signal, which is compared with the reference signal, is amplified by U_1 whose gain is set by R_1. Capacitor C_1 blocks any dc component of V_{in}. The cutoff frequency of the low-pass filter is determined by R_3 and C_3 and is given by the equation $f_c = 1/2\pi R_3 C_3$.

The output of the filter (b) keeps comparator U_3 in the high state as long as $|f_{in} - f_{ref}| < f_c$. The output of U_3 turns low otherwise. The response time of the circuit is determined by time constant $R_3 C_3$. To satisfy the response time and the frequency accuracy, the designer must appropriately select values for R_3 and C_3.

The harmonics of f_{ref} will also result in a nonzero average and therefore must be attenuated by adding a low-pass filter with a cutoff frequency less than $2f_{ref}$ at the input. Also, the phase delay between $V_{in}(t)$ and $V_{ref}(t)$ must not equal 90° because it will result in a zero average even when $f_{in} = f_{ref}$. Adding a phase delay to either signal will eliminate this problem. The circuit assumes that the phase difference between V_{in} and V_{ref} is less than ±90°. If no such certainty is guaranteed, a window comparator should replace U_3. □

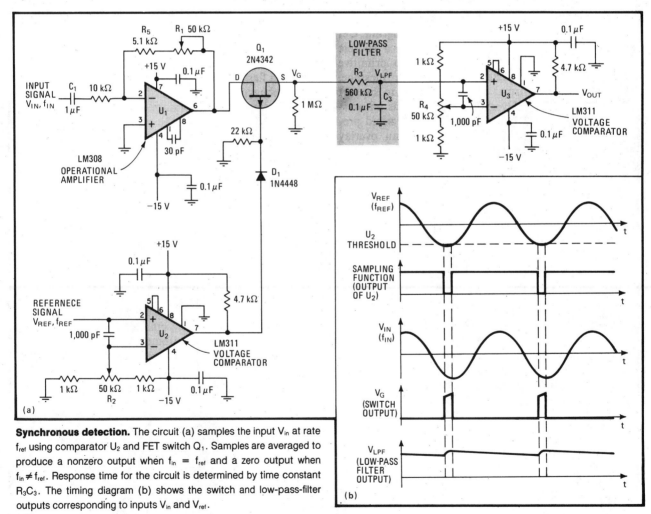

Synchronous detection. The circuit (a) samples the input V_{in} at rate f_{ref} using comparator U_2 and FET switch Q_1. Samples are averaged to produce a nonzero output when $f_{in} = f_{ref}$ and a zero output when $f_{in} \neq f_{ref}$. Response time for the circuit is determined by time constant $R_3 C_3$. The timing diagram (b) shows the switch and low-pass-filter outputs corresponding to inputs V_{in} and V_{ref}.

Voltage-detector chip simplifies V-f converter

by Lloyd Powell
David Taylor Naval Ship Research and Development Center, Annapolis, Md.

One-chip voltage detectors such as Intersil's ICL8212 contain all the necessary reference, discharge, and hysteresis circuitry needed to build a simple voltage-to-frequency converter. Providing a 0-to-1-kilohertz output for a 0-to-2-volt input in its basic range, the low-cost circuit requires only a few additional passive components, including an operational amplifier.

As shown, A_1 integrates incoming signals until its output voltage, and hence also the voltage across capacitor C_1, becomes $V'_{ref} = V_{ref}(R_2+R_4)/R_2$, where V_{ref} is the internal reference voltage of the ICL8212. At that instant, A_2's comparator goes high and switches its output transistor. While transistor Q_1 provides a positive-going output pulse, C_2 holds A_2's comparator high, so that C_1 is discharged quickly, in just 5 microseconds. Thus the process of integration and discharge occurs at a rate given by output frequency $f = V'_{in}/V_{ref}R_1C_1$.

Linearity and offset are better than 0.2% over the 0-to-1 kHz voltage-to-frequency range, assuming the operational amplifier used has low offset voltage and bias current. Suggested capacitance values for other frequency segments are given in the box in the figure. The input-voltage range can also be selected by means of R_1. For greatest accuracy, capacitors having the highest stability should be selected, and thus those of the silver-mica and polycarbonate type should be used where practical. For less stringent applications, those of the ceramic and disk type will suffice.

Supply voltages may range from ±2 to ±18 v without loss of circuit linearity. □

Cost-effective conversion. Micropower voltage detector chip and op amp form integrating V-f converter. Simple, low-cost circuit has 0.2% linearity over 0-to-2-volt input range. Frequency range, nominally 0 to 1 kilohertz, is chosen by appropriate selection of C_1 and C_2 (see box).

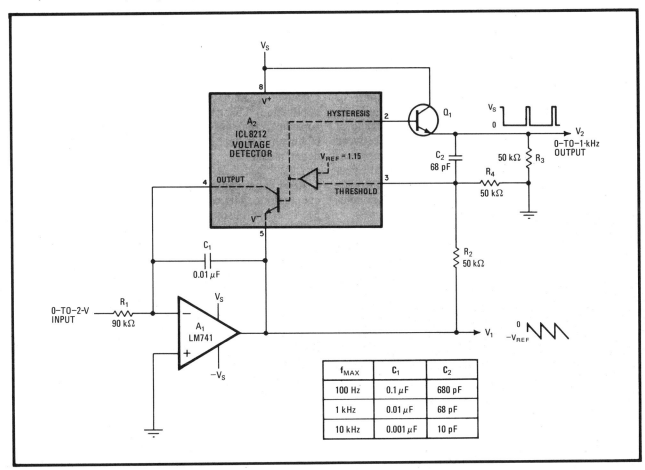

Photocurrent-to-frequency converter notes light levels

by Robert Nowotny, *Institute for the Study of Radium and Nuclear Physics, University of Vienna, Austria*

Illumination levels are usually measured in voltage for photometric and photographic applications. However, this circuit detects them in terms of hertz with a converter that uses the current-source characteristics of photodiodes, under reverse bias, to transform light levels into frequency.

The current, I_{ph}, obtained from the photodiode charges integrating capacitor C_1 to a voltage that is slightly greater than the threshold level of the ICM7555 timer (a). As a result of the charging, the timer turns on and C_1 discharges down to the lower trigger level. The timer is shut off once this level is hit, and a new charging and discharging cycle is started.

T_{ch} is the charging time and is given by $\Delta V \times C_1 / SE$, where ΔV is the voltage excursion between two threshold levels, E is the illuminance, and S is the diode response, which is a function of wavelength. At a wavelength of 800 nanometers, the response for p-i-n photodiode BPX66 is typically 0.5 microampere per microwatt irradiance. The sensitivity for an unfiltered tungsten light source (2,856 K) and a diode having an area of 1 millimeter square is 10 nA per lux illuminance.

Since ΔV is adjustable at pin 5 of the timer, the converter is calibrated with a voltage divider. C_1 is chosen through $C_1 = SE_{max}/\Delta V f_{max}$, where f_{max} is the maximum frequency that may be produced by the circuit. The actual frequency that is generated by the circuit is a function of the charge and the discharge times. The discharge time (T_d) depends on C_1 and the discharge current.

This circuit provides 10 pulses per lux with T_d equaling 450 nanoseconds. The frequency offset at a zero light level is determned by the sum of the input current of the timer (typically 20 picoamperes at 5 volts) and the dark current of the photodiode (≈ 1 nA). Because of low dark current and capacitance, p-i-n photodiodes are most suitable. A low-frequency offset (less than 1 Hz) enables the circuit to cover a frequency range of five decades. □

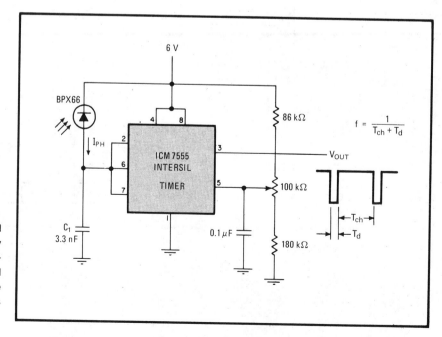

Illumination level. Photodiode BPX66 and timer ICM7555 in the astable mode directly convert photocurrent to frequency. Capacitor C_1 is charged by the photocurrent and discharged by the internal switch of the timer. A voltage divider at pin 5 calibrates the converter to measure illumination levels.

Delay circuit produces precision time intervals

by Jozef Kalisz
Warsaw, Poland

Using only four emitter-coupled-logic integrated circuits, this simple low-cost time-delay generator produces precise time intervals between start and stop pulses and creates a pulse whose width is controlled by the preset delay. The output delay can be adjusted in clock-period increments.

Initially, two out of six D-type flip-flops contained in U_2 are set by the synchronization pulse. Also, the clock pulse sets the $\overline{Q}$ output at pin 15 of U_2 to a low level, which is inverted by gates G_1 and G_2. This inverted output presets flip-flops $U_{1\text{-}a}$ and $U_{1\text{-}b}$ and in addition provides the starting pulse. The rising edge of the next clock pulse terminates the starting pulse and shifts the low state to pin 11 of U_2. Subsequently, the successive clock pulses shift the low state in sequence from one to another. These $\overline{Q}$ outputs are inverted by gates G_3 and G_4 to provide the stop pulse. Since pins 13 and 15 are set simultaneously, the output at pin 13 produces zero delay.

The leading edges of the start and stop pulses are directly related to the falling edges of the appropriate clock pulses. The duration of the output pulses is determined by the low state of the clock pulse, and the time interval between the start and stop pulses is selected by means of switch S_1. The range attained with this circuit is 0 to 10τ in increments of τ, where τ is the period of the clock input.

The pulse generated at the width output has a duration equal to the preset delay. The three outputs, whose repetition frequency is determined by the frequency of the sync input signal, can directly drive 50-ohm loads. The time-delay accuracy is better than 100 picoseconds and depends mainly on the propagation delay introduced by the gates. Also, the circuit requires that the external clock and sync source must have ECL outputs. □

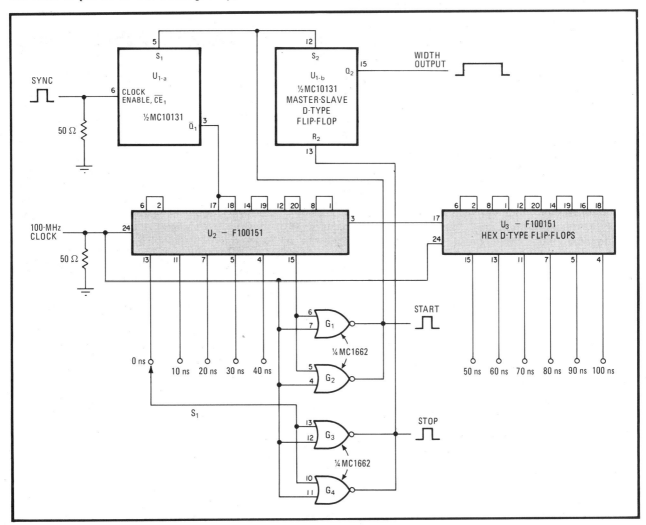

Selectable. This pulse-delay generator provides selectable time intervals, in steps of 10 nanoseconds, between the start and stop pulses. For a clock input of 100 megahertz, the range covered by the circuit is 0 to 100 ns with an accuracy of better than 100 picoseconds. The pulse generated at the width output has a duration equal to the preset delay. The external clock and sync source must have ECL outputs.

OTA multiplier converts to two-quadrant divider

by Henrique S. Malvar
Department of Electrical Engineering, University of Brazilia, Brazil

An inexpensive two-quadrant divider that is useful in tunable and tracking filters and special-purpose modulators and demodulators may be built using an operational transconductance amplifier (see figure). This circuit is based upon a multiplier circuit by W. G. Jung[1].

The desired circuit response is achieved by placing the CA3080 multiplier within the feedback loop of the 308 comparator. This method implements the divider function more easily than—and just as accurately as—the logarithmic and antilog converters often employed.

The transfer function of the circuit is given by:

$$V_{out}/V_{in} = -(1+k)/kR_2 g_m = -[2(1+k)V_T/kR_2 I_B]$$

where k = the resistance scaling ratio and $g_m = I_B/2V_T$ for the CA3080. V_T is the thermal voltage (26 millivolts at 23°C). The divider gain is thus inversely proportional to the input bias current, I_B. The plot in the figure shows the divider's nearly ideal response. Gain measurements were made with a selective voltmeter having a bandwidth of 10 hertz (HP 3581C) to eliminate noise effects—the circuit's linearity extends over five decades.

The offset of the 308 is also amplified as the circuit's bias current is reduced. Thus, I_B's lower limit should be around 20 nanoamperes. Also, the compensating capacitor must be at least equal to $A_{max} \times 30$ picofarads, where A_{max} is the maximum attenuation given by the divider circuit. The circuit uses a 300-pF capacitor because the maxiumum attenuation is 20 decibels.

The circuit's response to temperature variations is minimal, but in critical applications compensation is necessary for V_T. A temperature-compensating resistor having a thermal coefficient of 0.33% per °C for the resistor value kR_1 is used in this instance. □

References
1. W. G. Jung, "Get gain control of 80 to 100 dB," Electronic Design, June 21, 1974, pp. 94–99.

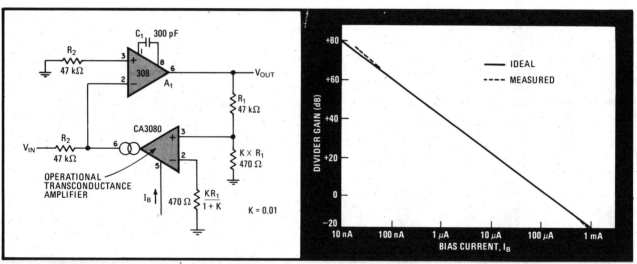

Inverse. A basic two-chip multiplier is easily transformed into a two-quadrant divider by the appropriate feedback. This scheme is simplier, no less accurate, and less costly than those using log and antilog converters. Circuit linearity extends to five decades of control current, I_B. Except for the most critical applications, temperature compensation is not required.

Current mirror linearizes remote-controlled timer

by George Hughes and S. A. Hawley
Eye Research Institute, Boston, Mass.

Although setting the pulse duration of timers of the 555 variety by remote means is most conveniently done with a single control such as a potentiometer, often there is an undesired nonlinear relationship between wiper-arm setting and output width because of the simple methods employed to achieve control. Adding a current mirror and feedback loop to the basic circuit solves the problem of linearity with little additional complexity or cost.

In general, any current passing through the pot's wiper should be minimized and the pot placed as close as possible to the circuit's interfacing operational amplifier, especially in remote-control applications, where the effects of stray coupling from various processing circuits can be considerable. A typical configuration is shown in (a). In this type of circuit, however, difficulties arise because the op amp's output voltage supplies charging currents to the one-shot's timing capacitor through a fixed resistor. As a result, the pulse capacitor width will be inversely proportional to the current driving C_T and will not be a linear function of the wiper-arm position.

Adding the current mirror and the feedback loop to the circuit, as shown in (b), overcomes this drawback. Here, the mirror's charging current is made a constant whose magnitude is proportional to only the voltage at the amp's noninverting input, V_i, and hence to the potentiometer's setting. In the feedback loop, the average value of the timer's output is compared with a voltage that represents the wiper-arm position, where current injected into timing capacitor C_T is such that the difference is kept small by the virtual-ground properties of the op amp. The average value of the timer's output is itself a linear representation of pulse duration, so that overall linear control is maintained.

This circuit will function with any TTL timer. Parts values are not critical and can be varied to suit a wide range of triggering rates and pulse durations. Substitution of matched dual transistors or packaged current mirrors is recommended to improve the circuit's temperature stability. □

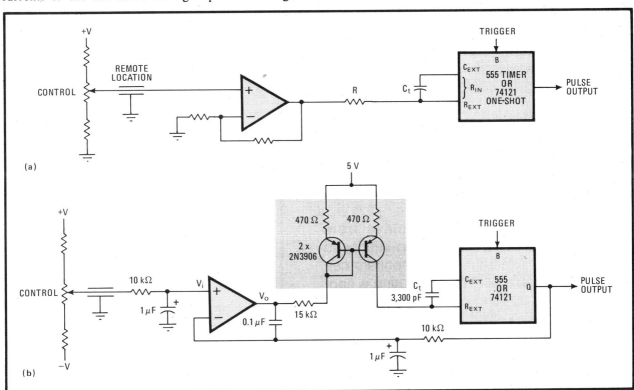

Current correspondence. A typical single-control pot arrangement for setting the on-time of a one-shot (a) has a nonlinear relationship of duration to wiper position because the current-driving capacitor, C_T, is proportional to $V_i(R+R_{IN})$. Adding a current mirror and feedback loop to the circuit (b) linearizes the relationship by generating a constant current set by V_i. The values shown are for $8 < T_{out} < 50\ \mu s$.

Analog multiplexer and op amp unite for precise d-a converter

by Dil Sukh Jain
National Remote Sensing Agency, Hyderabad, India

This simple low-cost digital-to-analog converter uses just an operational amplifier and an analog multiplexer to convert a 4-bit digital input into an analog output. Cascading an additional 16-channel analog multiplexer will extend the input digital word length of the d-a converter to 8 bits. The accuracy and the stability of the converter depend mainly on the accuracy of the resistors and stability of the reference voltage.

Operational amplifier U_1 operates as an inverting amplifier with a weighted-resistor switching network connected in its feedback path. The 16-channel analog multiplexer, U_2, functions as the resistor switching network that is controlled by the four binary inputs A_0–A_3. A 4-bit input, whose decimal equivalent is N, switches multiplexer channel Y_n – Z on and provides a feedback resistance of R_f = NR. Thus U_2 sets as the gain of the amplifier a value that corresponds to the equivalent digital input.

The analog output of the d-a converter is $V_O = -(R_f/R_i)V_R = -(NR/R)V_R = -NV_R$, where V_R is a stable reference voltage used in the circuit. The above relationship shows that the analog output V_O is proportional by a factor of V_R to the digital input:

As an example, consider a 4-bit input 0101, whose decimal equivalent is N = 5. Using, for simplicity, a reference voltage of $V_R = -1$ volt, the circuit produces an analog output of $V_O = -5(-1)v = +5$ v.

Op amp NE531 offers a high slew rate for high-speed operation. The circuit may be used as a programmable-gain-control amplifier whose desired gain can be set by thumbwheel switches. In addition, by interchanging input resistor R and multiplexer U_2, the circuit can serve as a programmable attenuator. □

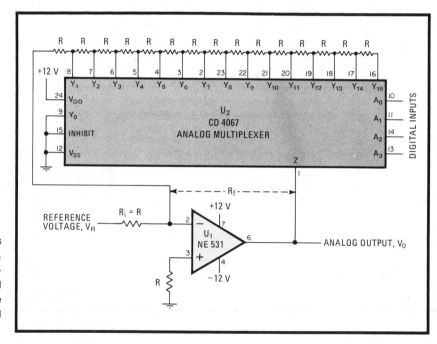

D-a converter. Analog multiplexer U_2 places resistors in the feedback path of amplifier U_1 to control the latter's gain and thereby produce an output proportional to the digital input. The analog output $V_O = -NV_R$, where N is the decimal equivalent of the digital input and V_R is the reference voltage.

A-d converter linearizes 100-ohm temperature detector

by Anthony Parise
Neutronics Inc., King of Prussia, Pa.

Conventional methods of linearizing resistance temperature detectors use a diode-resistor ladder circuit but do not produce a true straight line. However, this circuit (see figure) linearizes a 100-ohm platinum resistance temperature detector and uses few parts. With an analog-to-digital converter, such as ICL7106, a digital thermometer can easily be built.

For this detector, temperature is:

$$T = \frac{1{,}000(R_T - k_0)}{k_2 - k_1(R_T - k_0)}$$

where R_T is the resistance of the detector in ohms at temperature T and k_0, k_1, and k_2 are constants that are listed in the figure. Current I develops a voltage across R_0 that compensates for the 0° reading (in either Celsius or Fahrenheit) and the detector's lead resistance by raising input low V_0 volts above reference point A.

The input to the converter is now $V_{input} = V_T - V_0$. Voltage V_2, across R_2, corresponds to reference constant k_2, and the ratio R_a/R_b is selected so that it equals k_1. As a result, U_1's reference input voltage $= V_2 - k_1 V_{input} = V_2 - k_1(V_T - V_0)$.

The value displayed with the 7106 is equal to 1,000 ($V_{input}/V_{reference}$). Substituting for V_{input} and $V_{reference}$, the equation for display reduces to:

$$\frac{1{,}000 I (R_T - R_0)}{IR_2 - k_1 I(R_T - R_0)}$$

The expression for T may be reduced by canceling I in the equation and substituting k_0 for R_0 and k_2 for R_2. R_0 and R_2 must be temperature-stable and adjustable to provide easy calibration and have similar temperature coefficients. If a binary output is needed for a microprocessor interface, 12-bit binary a-d converter ICL7109 is recommended for U_1. □

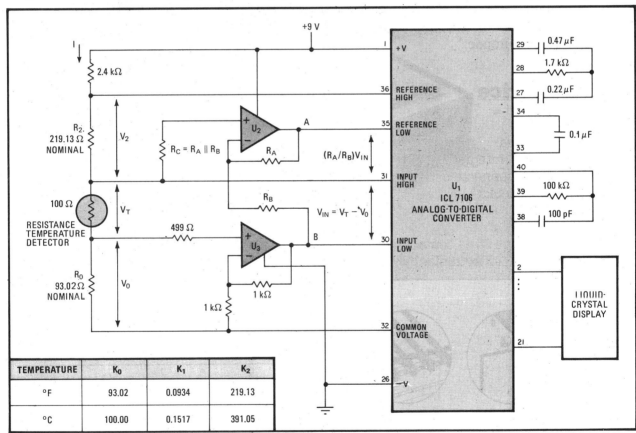

Linearize. The general linear equation relating resistance to temperature is implemented with two op amps and a-d converter ICL7106. The display corresponds to an output given by 1,000 ($V_{input}/V_{reference}$), which when simplified is identical to the equation for temperature T.

TEMPERATURE	K_0	K_1	K_2
°F	93.02	0.0934	219.13
°C	100.00	0.1517	391.05

Low-cost coordinate converter rotates vectors easily

by Arthur Mayer
Sperry Systems Management, Great Neck, N. Y.

Especially useful for graphics display applications, this simple $15 vector rotator, which takes coordinates in the x-y cartesian system and adds an angle of rotation to produce new coordinates x', y', is faster and cheaper than others currently available.

As shown in the schematic, the analog voltage pair (x,y) represents the vector $r\angle\theta$, where $r^2 = x^2+y^2$ and tangent $\theta = y/x$. The two inputs x_{IN}, y_{IN}—together with $-x_{IN}$, $-y_{IN}$ obtained from inverting amplifiers A_1 and A_2—are applied to the CD4052 dual analog multiplexer, which is controlled by the two most significant bits of the binary-coded rotation angle Φ. Each dual multiplexer output signal passes through a unity-gain amplifier, A_3 or A_4, and then through a tandem of inverting amplifiers (A_5, A_7 or A_6, A_8) to the final output.

Each tandem of inverting amplifiers is coupled with an AD7533 multiplying digital-to-analog converter to make a four-quadrant multiplier: A_5 and A_7 are coupled with M_1, and A_6 and A_8 are coupled with M_2. The digital input to both converters is provided by the remaining bits of Φ.

The analog input to M_1 is the average of the signals from A_4 and A_8, and the analog input to M_2 is the average of the signals from A_3 and A_7. The output currents from the cross-fed d-a converters feed the summing junctions of A_5 through A_8, where they add to the inputs that have been selected by the multiplexer, thus producing the output voltages x_{OUT}, y_{OUT}.

All resistances in the circuit are 30 kilohms so it is convenient to use dual in-line packages, like Beckman's 698-3, with eight resistors per DIP. Another DIP, Bourns's 7102, could replace the two 15-kΩ trimmers needed to raise the effective input impedance of each AD7533 to $15(2)^{1/2}$ kΩ, the value required in this design.

Regardless of the value of Φ, $x^2_{OUT}+y^2_{OUT} = x^2_{IN}+y^2_{IN}$. In other words, the output vector's magnitude is always equal to that of the input vector. However, the relationship between the input and output vectors is given by $\theta_{OUT} = \theta_{IN}+\Phi'$, where $\tan(\Phi'/2)$ is equal to $(2^{1/2}-1)(\Phi-45°)/45°$ and Φ is between 0° and 90°. The difference between Φ' and $\Phi-45°$ vanishes for $\Phi = 0°$, 45°, and 90° and is always less than 1° for other values of Φ in the first quadrant. Note that the error and its variation with angle recur in the other three quadrants.

The 45° offset in Φ' is due to the bipolar operation of the AD7533 converter. The offset may be corrected by simply adding 45° to the digital equivalent number at the Φ input lines. The remaining error will be small enough to go unnoticed on most graphical displays.

To calibrate the vector rotator, x_{IN} is set to some constant voltage and set $y_{IN} = 0$. Then the trimmers are adjusted to make $x_{OUT}+y_{OUT} = 0$ when $\Phi = 0°$ and $x_{OUT}-y_{OUT} = 0$ when $\Phi = 90°$.

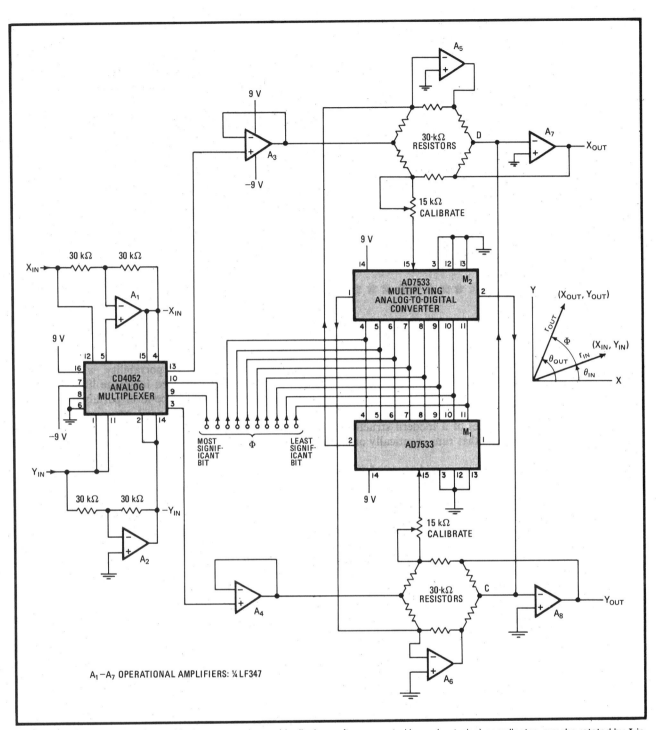

Transformation. Input data for positioning a cursor in graphic displays, often presented in x-y (cartesian) coordinates, may be rotated by Φ in steps of 0.35° to new location x′, y′ One analog multiplexer, two multiplying d-a converters and eight op amps in a unique cross-fed summing network perform the operation. Alternatively, the circuit will transform a vector from polar (r,θ) into rectangular (x,y) form.

With the addition of a clock and a counter to make $\Phi = \omega t$, the vector rotator becomes a sine-cosine generator. For example, for a 5-volt root-mean-square output, x_{IN} and y_{IN} is set to 5 v dc; then $x_{OUT} = 5(2)^{1/2} \cos \omega t$ and $y_{OUT} = 5(2)^{1/2} \sin \omega t$.

Because of the functional error in the angle as given by the formula for $\tan(\Phi'/2)$, either output will contain third and fifth harmonics each having a magnitude 0.8% that of the fundamental. Total harmonic distortion, therefore, is 1.1%. □

7. ENCODERS/DECODERS

Two-chip VCO linearly controls ramp's amplitude and frequency

by Forrest P. Clay Jr. and Mark S. Eaton
Department of Physics, Old Dominion University, Norfolk, Va.

This inexpensive ramp generator provides a proportional voltage control of both the period and amplitude of a waveform over a wide range and thus doubles as a linear voltage-to-frequency converter. Only a few active devices are needed: two operational amplifiers and a transistor.

Two dc differential control signals, V_{f_1} and V_{f_2}, are applied to op amps A_1 and A_2. The output from A_3 is $V_{C_1}+(V_{f_1}-V_{f_2}) = V_{C_1}+V_f$, where V_{C_1} is the voltage across ramp capacitor C_1 and after buffering becomes one of the inputs to A_3. The combination of all inputs to A_3 yields a dc bootstrap circuit with a controlled offset voltage. Thus, current $i_c = (V_{C_1}+V_f-V_{C_1})/R_1 = V_f/R_1$, and the voltage across the capacitor is:

$$V_{C_1} = \int_0^t (i_c/C_1)dt = (V_f/R_1C_1)t$$

C_1 is discharged through transistor Q_1 at time T when V_{C_1} equals the control voltage V_p, which is adjustable from 0 to 2.5 volts. thus $[V_{C_1}]_{max} = V_p = (V_f/R_1C_1)T$ and $f = 1/T = V_f/V_pR_1C_1$. The 311 comparator has a trigger output to synchronize external circuitry for easy operation.

The slope of the control voltage versus frequency in kilohertz is 1 for $1<V_f<10$ volts. This linear relationship holds even for slow ramps (increasing the value of C_1) with small values of V_f. Capacitor C_2 is selected to maintain the 311's output in a high state long enough so that C_1 may be completely discharged through Q_1 during the appropriate portion of the cycle. □

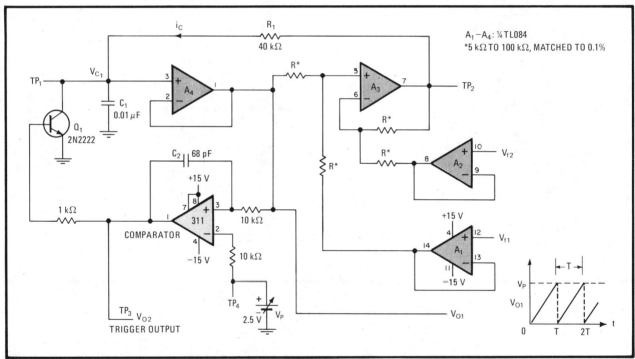

Potentially proportioned. A two-chip, one-transistor ramp generator is linearly adjusted by two dc control signals, with V_p setting the amplitude from 0 to 2.5 volts and V_f and V_p setting the frequency over the range of 0 to 10 kilohertz. Proportional control is achieved by placing ramp capacitor C_1 in the dc bootstrap circuit of A_3 and A_4, which ensures that constant current i_c is a function of only V_f and R_1.

Interleaving decoder simplifies serial-to-parallel conversion

by A. J. Bryant
Manelco Electronics Ltd., Winnipeg, Manitoba, Canada

Four low-cost chips combine to convert a 6-bit serial pulse train into its parallel equivalent in this decoder. Its circuitry is simplified because the bits are broken into two data streams so that it takes only two 4-bit shift registers and a latch to do the conversion.

A synchronous pulse (derived from a pulse detector that is not shown in the figure) is applied to flip-flops A and D (½74C73), initiating the conversion. The serial stream of negative-going pulses that will be decoded is also applied to flip-flop A, which serves as a divide-by-2 counter.

Both outputs of flip-flop A are then applied to two 4-bit shift registers (4015)—one through one-shot B, which serves as a clock for stepping the corresponding complementary signal of A to register L, and the other through one-shot C, which performs the same function of loading data into shift register H. As seen from the timing diagram, this asynchronous loading arrangement permits the 6-bit input stream to be split and then interlaced, with shift register L receiving low-order bits b_1 to b_4, and shift register H taking bits b_5 to b_6.

All 6 bits are then positioned onto the lines of the 74C174 6-bit latch and strobed onto the latch by one-shot D on the next recurring synchronous pulse. The pulse width of D must be thinner than the synchronous pulse so that its use as the data-valid output does not overlap a new frame. □

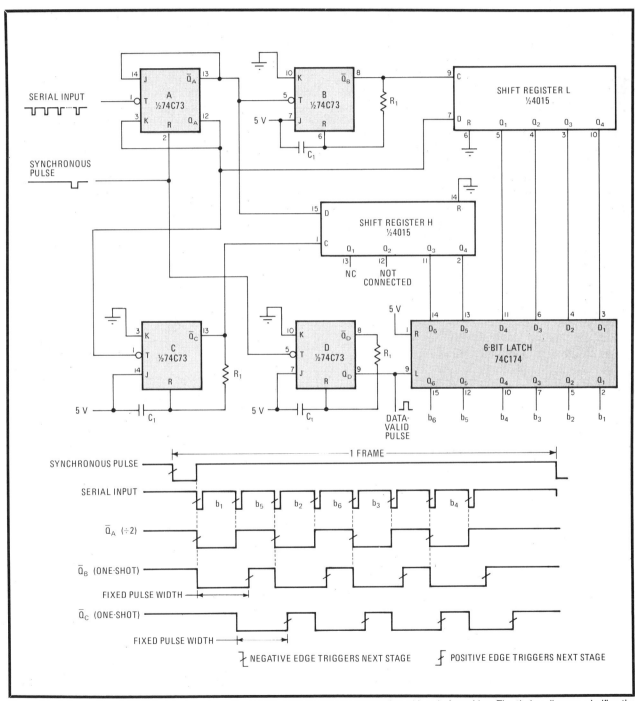

Scrambled. Separating 6-bit data into two streams enables serial-to-parallel conversion with only four chips. The timing diagram clarifies the bit-interleaving method used and details circuit operation. Timing components R_1 and C_1 for one-shots B and C should be selected to ensure that their on-time exceeds two data bits; R_t and C_t should be chosen to ensure there is no overlap with synchronous pulse into a new frame.

Decoder reworks binary input into hexadecimal or octal form

by Vaughn D. Martin
Cibolo, Texas

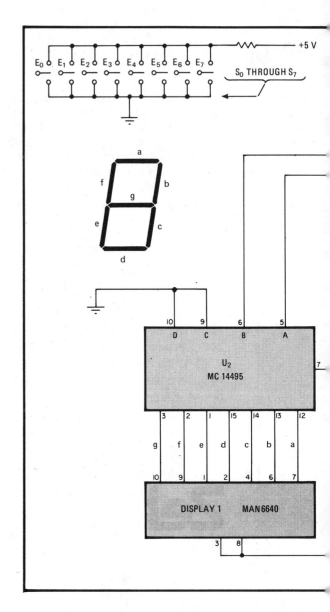

Many microprocessor- and digital-system applications require binary input data to be converted into hexadecimal or octal notation for faster program entry. This decoder, consisting of four integrated circuits, converts and displays a byte in either hexadecimal or octal form. Hexadecimal numbers B and D are displayed in lowercase to avoid confusion with an 8 and 0, respectively. The displays are a double- and single-digit common-cathode type with the driver's series-limiting resistors.

The data selector-multiplexer (U_1) and diodes D_1 and D_2 take a byte and convert it into groups of 4 bits and 3 bits for hexadecimal and octal notation, respectively. U_1 is a two-line-to-one-line data multiplexer and accepts 2 bytes. The control input to pin 1 selects data on the A or B inputs and passes it to the output.

When switch S_8 is in the octal position, a logic low appears on the D input of the binary-coded-decimal–to–seven-segment hexadecimal decoder (U_4) for a maximum 7-bit binary input. This low enables the A inputs of U_1 and directs the 2^3 bit to A_1 of U_1. In addition, A_4 is grounded because only 3 bits are permitted in octal. The last 2 bits at B_3 and B_4 are routed to U_2.

When S_8 is in the hexadecimal mode, the B inputs of U_1 are enabled. In this mode only two digits need be displayed. A logic high on pins 3 and 8 of display 1 disables it. U_3 and U_4 convert this byte into hexadecimal form and present it on display 2. □

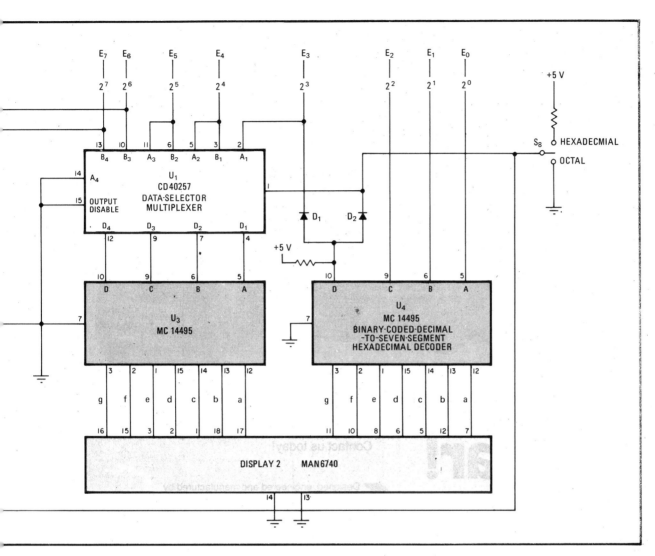

Decoder. This circuit converts a binary byte into hexadecimal or octal form and displays it. S_8 selects the mode. Data selector-multiplexer U_1 plus D_1 and D_2 arrange a byte into groups of 4 bits and 3 bits for hexadecimal and octal conversion, respectively.

Simplified multiplier improves standard shaft encoder

by Michael M. Butler
Minneapolis, Minn.

The pulse multiplier proposed by Amthor [*Electronics*, Sept. 11, p. 139] for increasing the resolution of a standard shaft encoder may be simplified, and improved as well, with this low-power circuit. Using only three complementary-MOS chips to derive a proportionally greater number of pulses from the encoder's output for a positioning up/down counter, it is relatively insensitive to encoder phase errors, uses no temperamental one-shots, and can detect the occurrence of illegal transition states generated by the encoder or circuit. It will serve well in electrically noisy industrial environments.

As in Amthor's circuit, the multiplier (a) used to drive the counter produces four pulses for each square-wave input of the two-phase encoder, whose outputs are displaced 90° with respect to one another. In this circuit, however, two 4-bit shift registers, three exclusive-OR gates, and an eight-channel multiplexer (b) derive the pulses. Previously four one-shots and 16 logic gates were required for the same task.

As seen, the clocked shift registers generate a 4-bit code to the multiplexer for each clock cycle. To ensure that no shaft-encoder transitions are missed, the clock frequency should be at least 8 NS, where N is the number of pulses produced by the encoder for each shaft revolution and S is the maximum speed, in revolutions per second, to be expected. A clock frequency of 1 megahertz or less is recommended for optimum circuit operation.

The code is symmetrical, and so only three exclusive-OR gates are required to fold the data into 3 bits. These bits are then applied to the control ports of the three-input multiplexer, which will generate the truth table shown. □

		\multicolumn{4}{c	}{MULTIPLIER LOGIC}	
A	B	\multicolumn{4}{c	}{Multiplexer code}	Count
		Old A	New A	Old B	New B	
0	0	0	0	0	0	no change
0	↑	0	0	0	1	down
0	↓	0	0	1	0	up
0	1	0	0	1	1	no change
↑	0	0	1	0	0	up
↑	↑	0	1	0	1	error
↑	↓	0	1	1	0	error
↑	1	0	1	1	1	down
↓	0	1	0	0	0	down
↓	↑	1	0	0	1	error
↓	↓	1	0	1	0	error
↓	1	1	0	1	1	up
1	0	1	1	0	0	no change
1	↑	1	1	0	1	up
1	↓	1	1	1	0	down
1	1	1	1	1	1	no change

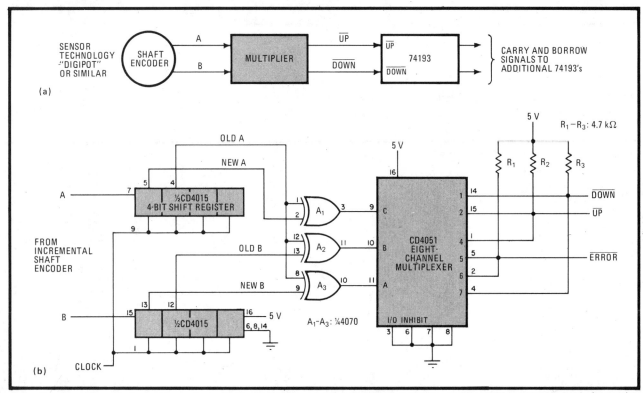

Accretion. Requiring only three chips, pulse multiplier gives better resolution to optical positioning systems that are driven by a shaft encoder. Low-power circuit is relatively insensitive to encoder phase errors. Stability is high because no one-shot multivibrators are used.

Decoder logs signals' order of arrival

by Claude Haridge
Ottawa, Ontario, Canada

This decoder indicates the sequence of arrival of up to four digital input signals and therefore serves as an excellent priority encoder. Alternatively, it can aid the technician in troubleshooting high-speed circuits. Using Schottky TTL devices to minimize propagation delays, the decoder can resolve two signals only 30 nanoseconds apart.

As shown, one flip-flop, three NAND gates, and four light-emitting diodes per input are needed to capture the corresponding signals and compare their arrival times. There are four such sections. In order to perform the time-difference checks accurately, the gates of each section are cross-coupled as shown in the figure, so that they provide an effective signal-lockout function. Polarity switches at each input enable the user to designate either the rising or falling edge of a signal as a valid gating stimulus.

A system reset brings the Q output of each flip-flop low. At this time, all the LED indicators are off. A valid trigger input sets its corresponding flip-flop, which then turns on its indicator—LED A, B, C, or D. Simultaneously, the gates leading to the remaining three LEDs of the activated section are enabled. These three LEDs are used to indicate the relative arrival of succeeding pulses.

Thus, the lighting of LED A, followed by the LED associated with the A>C output, indicates that a signal at input A arrived before a pulse at input C. In this case, note that the LEDs connected to the C>A, B>A, and D>A ports are inhibited from turning on until the next system reset. Succeeding pulses reaching the B, C, and D inputs in any order enable the corresponding outputs and lock out the appropriate LEDs until all four inputs have been detected. □

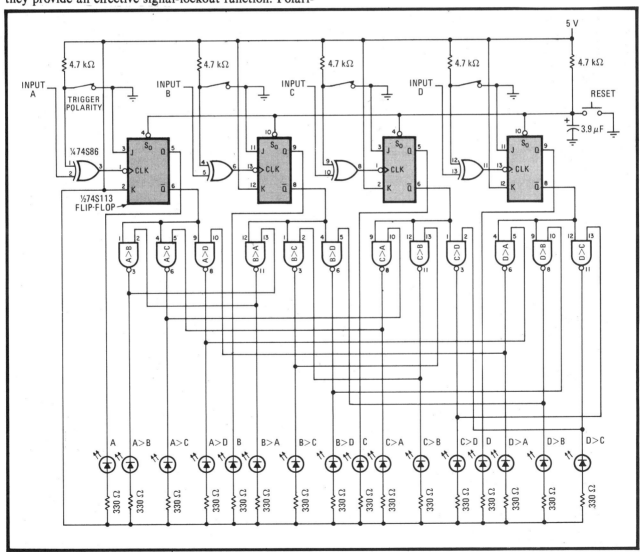

Signal sequence. Circuit indicates relative arrival times of four digital signal inputs. Using Schottky TTL, unit resolves any two signals separated by as little as 30 ns. Polarity switches at each input enable detection of signal's rising or falling edges.

RC oscillator linearizes thermistor output

by B. Sundqvist
Department of Physics, University of Umea, Sweden

Standard thermistors are seldom employed as temperature sensors because of the exponential relationship that exists between temperature and their resistance. However, a thermistor in conjunction with a simple RC oscillator will generate an output frequency that is proportional to absolute temperature. The linearity error is determined by the thermistor's resistance R_t, which may be derived from the exponential relationship $R_t = Ae^{(B/T)}$, where B is a constant of the circuit and T is the absolute temperature in Kelvin.

Initially, switch S_1 is closed and voltage $V_1 = V$. When S_1 is opened, the value of V_1 decreases according to the relation $V_1 = Ve^{-t/RC}$, where V is the supply voltage, and reaches V_0 when $t = t_0 = RC \ln(V/V_0) = RC(B/T)+k$, where k is a time constant. When V_0 is reached, the comparator output goes high and triggers monostable U_3 (a). This output has a duration of t_1 and resets the circuit by closing S_1. The resulting signal, f_{out}, is a pulse train with a frequency of $1/(t_0+t_1)$. However, by setting $t_1 = -k$, the relationship for f_{out} may be rewritten as $f_{out} = T/RCB$. The circuit can be calibrated in °F or °C by using a preset counter.

The circuit is best calibrated by measuring B for the hermistor. This value may then be plugged into the expression for f_{out} to determine the value of components R and C. Value t_1 is then set for a known temperature. The desired temperature range is selected by adjusting R_1 or V_{ref}, and the maximum error is ± 0.5 K over a range of 80 K. This error is mainly due to deviations from the assumed exponential relationship between R_t and T. □

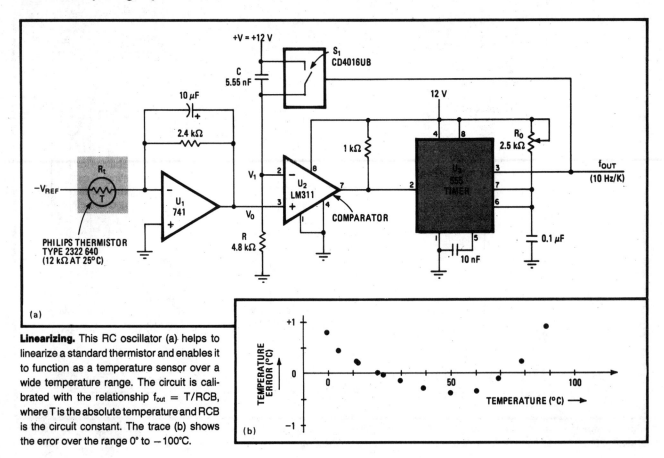

Linearizing. This RC oscillator (a) helps to linearize a standard thermistor and enables it to function as a temperature sensor over a wide temperature range. The circuit is calibrated with the relationship $f_{out} = T/RCB$, where T is the absolute temperature and RCB is the circuit constant. The trace (b) shows the error over the range 0° to −100°C.

Multiplier increases resolution of standard shaft encoders

by Frank Amthor
School of Optometry, University of Alabama, Birmingham

The resolution that can be attained by two-channel shaft encoders of the type used in speed controllers and optical-positioning devices may be increased by employing a digital frequency multiplier to derive a proportionally greater number of pulses from its TTL-compatible outputs. In this way, an up/down counter, which is normally driven by the encoder in these applications, can position the shaft more accurately and is more responsive to changes in speed and direction. Only two one-shot multivibrators and several logic gates are needed for the multiplier circuitry.

The circuit works well with a typical encoder such as the Digipot (manufactured by Sensor Technology Inc., Chatsworth, Calif.). In this case, the 128 square waves that are generated per channel for each shaft revolution (with output from the other channel in quadrature) are transformed to 512 bits per cycle.

When the shaft rotates in a clockwise direction, the output from port A of the encoder always leads the output from port B by 90°, and the logic will generate pulses only to the up input of the counter on both edges of both channel outputs. Thus, four pulses per square wave are generated. Rotation speed is limited by the duration time of the positive-edge–triggered one-shots, which should be kept to a few microseconds or less. Note that both the count-up and the count-down inputs of the counter are normally held high.

On the other hand, when the shaft's rotation is in a counterclockwise direction, the output of B leads that of A by 90°. In this case, four pulses per square wave are presented to the down input of the counter. □

Increments. Pulse multiplier yields 512 count-up or -down bits for 128 square-wave cycles (one revolution) from two-channel shaft encoder, for more resolution in speed controllers and optical positioning systems. Eliminating tinted area yields 256 bits per revolution.

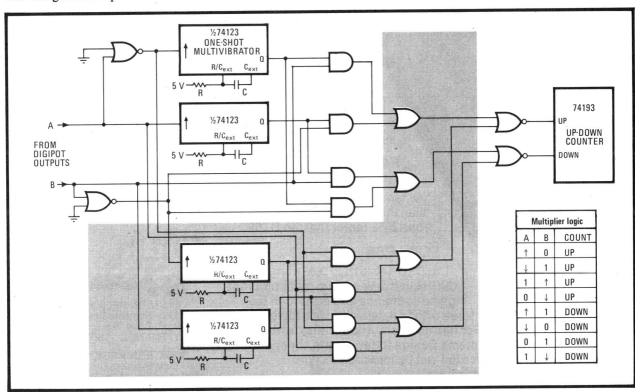

Multiplier logic		
A	B	COUNT
↑	0	UP
↓	1	UP
1	↑	UP
0	↓	UP
↑	1	DOWN
↓	0	DOWN
0	1	DOWN
1	↓	DOWN

Interfacing digital circuitry with plastic rotary encoders

by W. Berger
Kretztechnik, Zipf, Austria

Although potentiometers are handier tuning devices than keyboards and push buttons, they cannot be easily interfaced with digital and microprocessor-controlled equipment, as can the other devices. This circuit links plastic rotary encoders to digital circuitry with just a few integrated circuits. The advantages of this type of control include programmable resolution, interrupt operation, and low cost.

The 30-kilohm conductive plastic encoder (Fig. 1a) generates two phase-shifted signals $\overline{A}$ and $\overline{B}$ (Fig. 1b). The direction of rotation for the encoder may be determined by the phase angle between $\overline{A}$ and $\overline{B}$. Because both signals tend to drift, RC filters are placed at the input to minimize this noise. However, this chatter never occurs on both the signals simultaneously. In addition,

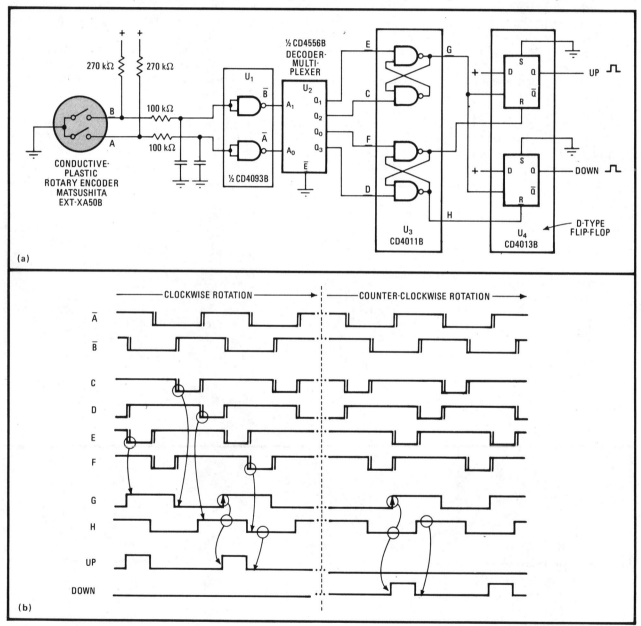

1. Rotary encoder. The circuit (a) interfaces a conductive-plastic rotary encoder with digital circuitry. The two phase-shifted signals that are generated by the encoder are also used to determine the direction of rotation. The up and down signals (b) generated by the digital circuit can be used for multiple purposes.

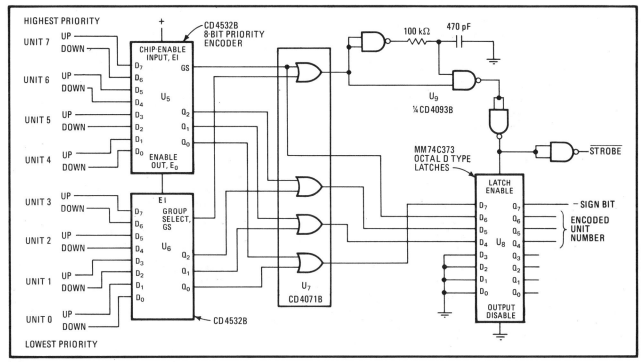

2. Interface. The circuit encodes eight rotary encoders into a 3-bit code that enables the rotary encoders to perform like a keyboard interface. Sixteen up and down pulses that are generated by the circuit are encoded into a 3-bit code using the 8-bit priority encoders U_5 and U_6 and OR gates U_7. Latch U_8 holds the output code when the strobe pulse is generated.

the encoder has a resolution of 50 periods per turn.

Demultiplexer U_2 decodes each of the four discrete states represented by the two signals generated with the encoder. The output of U_2 is used to set and reset the pair of flip-flops in U_3. The output at G clocks the D-type flip-flops (U_4), which generate the up and down pulses. These pulses are employed as interrupt signals for microcomputers.

An application for this technique is to encode the outputs of eight rotary encoders into 3-bit code. The encoders generate 16 up and down pulses (Fig. 2). These signals are further encoded into a 3-bit code by the 8-bit priority encoders U_5 and U_6 and OR gates U_7. Whenever one of the encoders is rotated, a strobe (low) pulse is generated that interrupts the microprocessor and sets latch U_8 so that it will hold the output code until the next up or down pulse arrives. □

8. FILTERS

Delay line eases comb-filter design

by Hanan Kupferman
Century Data Systems Inc., Anaheim, Calif.

Designing a multiple-frequency notch filter with standard band-stop transfer functions is complicated and tedious because the process needs repeating. In addition, filter matching is difficult for sections with different band-stop frequencies. Through exploiting the characteristics of a delay line, this comb filter is easily realized and provides a band-stop response for a fundamental frequency and its harmonics. The fundamental frequency corresponds to the delay time of the delay line.

The general scheme of the filter (a) shows that the delay line with a delay τ and characteristic impedance R is driven by and equals matched impedance sources R_1 and R_2. Use of input and output buffers in the circuit is optional. The fundamental frequency where the first notch occurs is $f_1 = 1/\tau$. Other band stops occur at $f_2 = 2/\tau$, $f_3 = 3/\tau$, and so on. Related to the rise time of the delay line, the filter's bandwidth determines how many

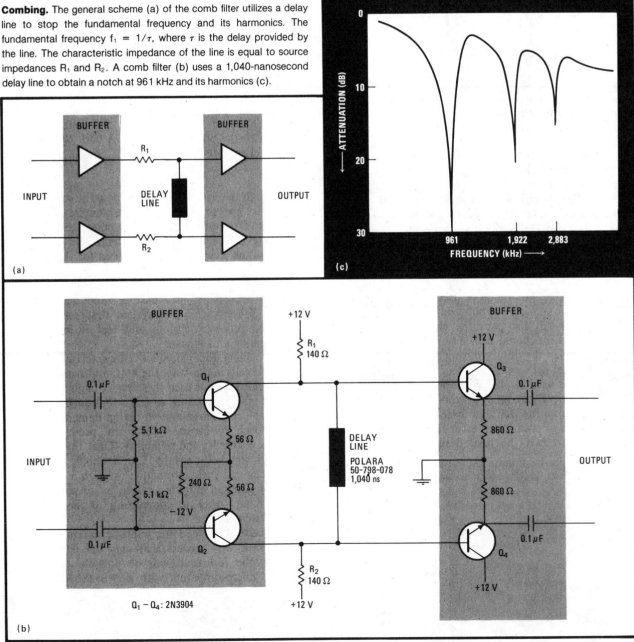

Combing. The general scheme (a) of the comb filter utilizes a delay line to stop the fundamental frequency and its harmonics. The fundamental frequency $f_1 = 1/\tau$, where τ is the delay provided by the line. The characteristic impedance of the line is equal to source impedances R_1 and R_2. A comb filter (b) uses a 1,040-nanosecond delay line to obtain a notch at 961 kHz and its harmonics (c).

harmonic frequencies will be attenuated. If the rise time of the line is faster, the bandwidth is wider, and therefore the stop-band range is higher.

This comb filter (b) is designed for a fundamental notch frequency of 961 kilohertz. The delay time of the delay line is 1,040 nanoseconds. Transistors Q_1 and Q_2 serve as input buffers, with Q_3 and Q_4 serving as output buffers. A degradation occurs in the attenuation of higher harmonics (c) and can be attributed to the rise time of the delay line. The magnitude of stop-band attenuation at higher frequencies can be improved by using a delay line with a faster rise time. □

Sample-and-hold devices ease tracking-filter design

by Ralph J. Amodeo and Gerald T. Volpe
University of Bridgeport, Bridgeport, Conn.

Tracking filters are often required for either passing or rejecting a carrier signal that varies over a certain frequency range. This design presents an active sample-and-hold resistor-capacitor filter that tracks an input carrier's frequency by using its own notch frequency. The circuit uses sample-and-hold devices, operational amplifiers, and RC elements. In addition, the filter is insensitive to component changes and uses equal-valued capacitors, which makes it attractive for monolithic integrated-circuit implementation.

This biquadratic filter (a) uses zero-order-hold ICs ZOH_2 and ZOH_3 to perform unit-delay function z^{-1}, where z is the Z-transform variable; ZOH_1 and ZOH_4 are used for data buffering. To obtain a second-order delay, ZOH_3 is strobed before ZOH_2 to ensure that the data at reference point N_2 is properly transferred to N_1 before N_2 is updated.

Also, ZOH_1 prevents the input voltage to ZOH_2 from changing until ZOH_1 is strobed. Without this feature, ZOH_2's output can modify its own input voltage, thus causing erroneous data transfer.

The timing generator (b) provides strobe pulses for the filter and operates at a sampling frequency that is 16 times the filter-notch frequency, f_o. The Harris 2820PLL and divider U_1 generates the clock frequency.

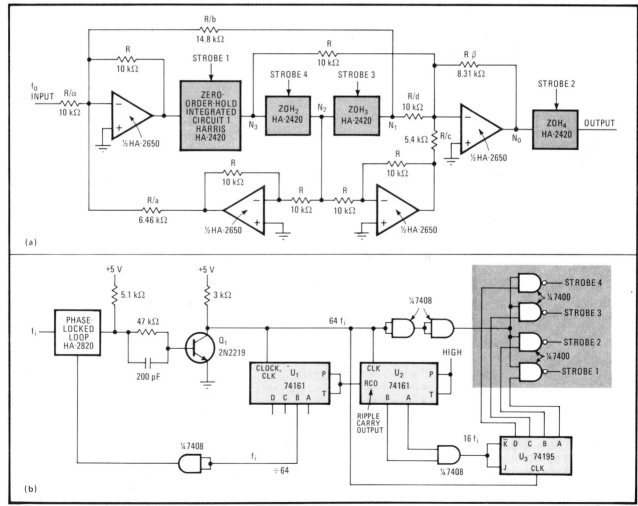

Tracking. The notch frequency of this tracking filter tracks the input carrier frequency and provides modulation information. The band-reject filter (a) is designed for a notch frequency of 13.5 kHz, Q = 1, and ξ = 0.5. The zero-order-hold devices provide second-order delays. The timing generator (b) provides strobe pulses for the filter.

Divider U_2 and shift register U_3 provide the four-phase strobe pulses.

Analysis of the filter yields a biquadratic Z-transformed transfer function:

$$H(z) = \frac{H_o(ez^2 - cz + d)}{z^2 - az + b}, \text{ where } H_o = \alpha\beta$$

Parameters a, b, c, and d control the filter's natural frequency, quality factor, notch frequency, and notch depth, respectively. In addition, parameter e is a constant, α is the filter's amplification factor, and β is the feedback factor.

The filter's gain (H_o) is controlled by input summing resistor R/α and feedback resistor $R\beta$. In order to obtain a notch-frequency response, a = $2e^{-\xi\omega_o\tau}\cos(\omega_o\tau[1-\xi^2]^{1/2})$, b = $e^{-2\xi\omega_o\tau}$, c = $2\cos\omega_o\tau$, d = 1, and e = 1, where ξ is a frequency constant.

With a sampling frequency of $16f_o$, $\omega_o\tau = \pi/8$. Therefore, as τ increases, ω_o decreases in order to maintain the value of $\omega_o\tau$. As an example, a second-order band-reject filter is designed with its notch frequency at 13.5 kilohertz, Q = 1, and ξ = 0.5. Substituting the data into the transfer equation yields:

$$H(z) = \frac{H_o(z^2 - 1.848z + 1)}{z^2 - 1.549z + 0.6752}$$

At dc (z = 1) the transfer-function factor is H(1) = $1.204H_o$. For this equation, a scale factor of H_o = 1/1.204 = 0.8306 is used. This factor reduces the equation by 0.8306. The reciprocals of a, b, c, and d are the fixed resistor values that are scaled by a convenient constant R. The practical resistor values obtained are as shown in the figure. □

High-pass Chebyshev filters use standard-value capacitors

by Ed Wetherhold
Honeywell Inc., Signal Analysis Center, Annapolis, Md.

Complementing the computer-generated design table for low-pass Chebyshev networks having standard-value capacitors [*Electronics*, June 19, 1980, p. 160], this listing is useful for building seven-element filters of the high-pass type that provide a roll-off attenuation of at least 42 decibels per octave. Both L and C values for 120 filters were found by an 85-line program written in Basic and were tabulated for an operating range of 1 to 10 megahertz. These components can be scaled for frequencies outside this range. Element values are specified for a source and load impedance of 50 ohms, but values for any input/output impedance are easily determined.

The high-pass filter (see table) uses a capacitive input and output configuration to minimize the number of inductors and to provide dc isolation with respect to ground. In this configuration and for the equally terminated condition, $C_1 = C_7$, $C_3 = C_5$, and $L_2 = L_6$. Given a standard capacitor value for C_3 and C_5, the program calculates all capacitor and inductor values for those unique values of the reflection coefficient that make C_1 and C_7 a standard value. The frequencies corresponding to the 3- and 50-dB attenuation levels are also listed.

To scale the tabulated frequency and component values into the 10-to-100- or 100-to-1,000-MHz frequency range decades, the user must multiply the frequency by 10 or 100, respectively, and divide all C and L values by the same number. Similarly, for the 1-to-10-, 10-to-100- and 100-to-1,000-kilohertz decades, the frequency is divided by 1,000, 100 or 10, respectively, and the component values are multiplied by the same number.

The following example demonstrates how to design for an I/O impedance other than 50 Ω. Consider a filter having a terminating impedance, Z_x, of 75 Ω and a 3-dB cutoff frequency, F_3^x, of 9.4 MHz. The user must:

- Calculate the impedance scaling factor, R = $Z_x/50$.
- Calculate the 3-dB cutoff frequency of a 50-Ω filter, F_3^{50}, from $F_3^{50} = R \cdot F_3^x$. If F_3^{50} is found to exceed 10, the table is scaled into the next frequency decade.
- Select the 50-Ω design from the table that most closely matches the calculated F_3^{50} value. The tabulated capacitor values are implemented directly in the new design.
- Find the new inductor values by multiplying the corresponding tabulated values by the square of the impedance-scaling factor, R.
- Find the corresponding 3- and 50-dB attenuation points by dividing the tabulated frequencies by the impedance-scaling factor.

Thus, for an F_3^x of 9.4 MHz and a Z_x of 75 Ω, R = 75/50 = 1.5, R^2 = 2.25, and F_3^{50} = 1.5(9.4) = 14.1 MHz. Because F_3^{50} is greater than 10 MHz, the 50-Ω tabulation is scaled to the 10-to-100-MHz decade by multiplying all frequencies by 10 and dividing all component values by 10. The user should select filter number 17 because its F_3^{50} value is closest to 14.1 MHz.

Therefore, $C_{1,7}$ = 360 picofarads, and $C_{3,5}$ = 120 pF. Inductor $L_{2,6}$ = $R^2(0.380)$ = 0.855 microhenry, and L_4 = $R^2(0.286)$ = 0.644 μH. Further, F_3^{75} = 14.1/1.5 = 9.4 MHz and F_{50}^{75} = 4.69 MHz. To construct the filter, polystyrene capacitors with a 2.5% tolerance, such as Mallory type SXM, are suitable. The inductors may be conveniently hand-wound on standard powdered-iron toroidal cores. □

DESIGN TABLE: HIGH-PASS CHEBYSHEV FILTERS

Filter No.	Frequency (MHz) 3 dB	50 dB	Reflection coefficient (%)	$C_{1,7}$ (pF)	$C_{3,5}$ (pF)	$L_{2,6}$ (µH)	L_4 (µH)
1	.849	.525	.254E 02	2400	1500	6.72	6.22
2	.958	.567	.141E 02	2700	1500	5.43	4.89
3	1.02	.583	.810E 01	3000	1500	4.92	4.31
4	1.06	.587	.471E 01	3300	1500	4.70	4.01
5	1.09	.584	.271E 01	3600	1500	4.63	3.83
6	1.11	.578	.151E 01	3900	1500	4.63	3.71
7	1.12	.568	.643E 00	4300	1500	4.70	3.62
8	1.14	.556	.233E 00	4700	1500	4.82	3.55
9	1.15	.544	.630E-01	5100	1500	4.97	3.50
10	1.16	.530	.455E-02	5600	1500	5.19	3.44
11	1.12	.679	.207E 02	2000	1200	4.90	4.50
12	1.21	.714	.128E 02	2200	1200	4.25	3.81
13	1.28	.729	.810E 01	2400	1200	3.94	3.45
14	1.34	.734	.411E 01	2700	1200	3.74	3.16
15	1.37	.727	.204E 01	3000	1200	3.70	3.01
16	1.40	.716	.949E 00	3300	1200	3.73	2.92
17	1.41	.703	.397E 00	3600	1200	3.80	2.86
18	1.43	.689	.138E 00	3900	1200	3.91	2.82
19	1.44	.670	.185E-01	4300	1200	4.07	2.77
20	1.46	.653	.206E-03	4700	1200	4.26	2.72
21	1.27	.787	.254E 02	1600	1000	4.48	4.15
22	1.44	.851	.141E 02	1800	1000	3.62	3.26
23	1.53	.875	.810E 01	2000	1000	3.28	2.87
24	1.59	.881	.471E 01	2200	1000	3.14	2.67
25	1.63	.877	.271E 01	2400	1000	3.08	2.55
26	1.67	.862	.111E 01	2700	1000	3.10	2.45
27	1.69	.843	.397E 00	3000	1000	3.17	2.39
28	1.72	.823	.108E 00	3300	1000	3.28	2.34
29	1.73	.803	.162E-01	3600	1000	3.40	2.31
30	1.76	.785	.314E-03	3900	1000	3.54	2.27
31	1.53	.951	.266E 02	1300	820	3.76	3.49
32	1.77	1.04	.129E 02	1500	820	2.91	2.61
33	1.85	1.06	.925E 01	1600	820	2.74	2.42
34	1.94	1.07	.478E 01	1800	820	2.57	2.19
35	2.00	1.07	.243E 01	2000	820	2.53	2.08
36	2.03	1.05	.117E 01	2200	820	2.54	2.01
37	2.06	1.03	.520E 00	2400	820	2.58	1.97
38	2.09	1.00	.152E 00	2700	820	2.68	1.92
39	2.12	.975	.978E-02	3000	820	2.81	1.89
40	2.15	.948	.554E-06	3300	820	2.95	1.85
41	1.90	1.17	.240E 02	1100	680	2.96	2.74
42	2.08	1.24	.155E 02	1200	680	2.53	2.29
43	2.20	1.28	.103E 02	1300	680	2.31	2.05
44	2.34	1.30	.464E 01	1500	680	2.13	1.81
45	2.39	1.29	.309E 01	1600	680	2.10	1.75
46	2.45	1.27	.131E 01	1800	680	2.10	1.67
47	2.49	1.25	.494E 00	2000	680	2.14	1.63
48	2.52	1.22	.148E 00	2200	680	2.21	1.60
49	2.54	1.19	.277E-01	2400	680	2.29	1.58
50	2.59	1.15	.326E-04	2700	680	2.43	1.54
51	2.32	1.43	.235E 02	910	560	2.42	2.23
52	2.55	1.51	.146E 02	1000	560	2.05	1.85
53	2.71	1.56	.893E 01	1100	560	1.86	1.64
54	2.82	1.57	.550E 01	1200	560	1.77	1.52
55	2.89	1.57	.338E 01	1300	560	1.73	1.45
56	2.98	1.54	.119E 01	1500	560	1.73	1.37
57	3.01	1.52	.664E 00	1600	560	1.75	1.35
58	3.05	1.48	.163E 00	1800	560	1.82	1.32
59	3.09	1.44	.203E-01	2000	560	1.90	1.29
60	3.14	1.40	.148E-03	2200	560	1.99	1.27
61	2.70	1.67	.257E 02	750	470	2.12	1.96
62	2.98	1.78	.165E 02	820	470	1.77	1.61
63	3.21	1.85	.964E 01	910	470	1.58	1.40
64	3.35	1.87	.573E 01	1000	470	1.49	1.28
65	3.45	1.87	.320E 01	1100	470	1.45	1.21
66	3.52	1.85	.174E 01	1200	470	1.45	1.17
67	3.57	1.83	.901E 00	1300	470	1.46	1.14
68	3.63	1.77	.181E 00	1500	470	1.52	1.11
69	3.66	1.74	.614E-01	1600	470	1.56	1.10
70	3.72	1.68	.124E-02	1800	470	1.65	1.07
71	3.24	2.01	.262E 02	620	390	1.78	1.65
72	3.59	2.15	.165E 02	680	390	1.47	1.34
73	3.85	2.23	.999E 01	750	390	1.32	1.17
74	4.01	2.26	.614E 01	820	390	1.24	1.07
75	4.15	2.25	.327E 01	910	390	1.21	1.01
76	4.24	2.23	.169E 01	1000	390	1.20	.970
77	4.31	2.19	.752E 00	1100	390	1.22	.944
78	4.36	2.15	.294E 00	1200	390	1.25	.926
79	4.40	2.10	.908E-01	1300	390	1.28	.913
80	4.49	2.02	.938E-03	1500	390	1.37	.889
81	4.14	2.50	.189E 02	560	330	1.30	1.19
82	4.49	2.62	.113E 02	620	330	1.14	1.01
83	4.71	2.66	.687E 01	680	330	1.06	.924
84	4.87	2.67	.386E 01	750	330	1.03	.865
85	4.98	2.65	.213E 01	820	330	1.02	.830
86	5.08	2.60	.926E 00	910	330	1.03	.803
87	5.14	2.55	.353E 00	1000	330	1.05	.786
88	5.20	2.49	.908E-01	1100	330	1.09	.772
89	5.26	2.43	.120E-01	1200	330	1.13	.761
90	5.33	2.37	.106E-03	1300	330	1.17	.748
91	4.69	2.90	.260E 02	430	270	1.22	1.13
92	5.18	3.10	.166E 02	470	270	1.02	.930
93	5.50	3.20	.110E 02	510	270	.927	.825
94	5.77	3.25	.663E 01	560	270	.867	.752
95	5.97	3.26	.362E 01	620	270	.838	.704
96	6.10	3.23	.193E 01	680	270	.832	.676
97	6.21	3.18	.867E 00	750	270	.840	.656
98	6.29	3.11	.344E 00	820	270	.859	.643
99	6.37	3.03	.744E-01	910	270	.892	.631
100	6.45	2.95	.628E-02	1000	270	.931	.620
101	5.96	3.65	.227E 02	360	220	.934	.861
102	6.45	3.84	.152E 02	390	220	.812	.734
103	6.88	3.96	.917E 01	430	220	.733	.647
104	7.16	4.00	.560E 01	470	220	.697	.599
105	7.35	4.00	.341E 01	510	220	.681	.571
106	7.51	3.96	.178E 01	560	220	.678	.548
107	7.64	3.89	.758E 00	620	220	.687	.532
108	7.73	3.80	.278E 00	680	220	.704	.522
109	7.83	3.71	.598E-01	750	220	.730	.513
110	7.92	3.61	.487E-02	820	220	.761	.505
111	7.44	4.53	.207E 02	300	180	.736	.675
112	8.09	4.76	.128E 02	330	180	.637	.572
113	8.51	4.86	.810E 01	360	180	.590	.517
114	8.80	4.89	.516E 01	390	180	.567	.486
115	9.05	4.87	.280E 01	430	180	.555	.460
116	9.22	4.82	.147E 01	470	180	.555	.445
117	9.34	4.74	.720E 00	510	180	.562	.435
118	9.46	4.64	.255E 00	560	180	.577	.427
119	9.58	4.52	.484E-01	620	180	.600	.419
120	9.70	4.40	.264E-02	680	180	.627	.412

State-variable filter has high Q and gain

by Kamil Kraus
Rokycany, Czechoslovakia

Although much work has been done in developing state-variable filters that use three or two operational amplifiers, earlier designs [*Electronics*, April 21, 1982, p. 126] suffer from a common disadvantage in that the filter's center frequency and the Q factor are interdependent. This new design provides a solution by introducing a parameter k in the transfer function of the filter.

The circuit uses three op amps, U_1 through U_3, to realize the filter. U_1 and U_2 function as negative and positive integrators, and U_3 completes the feedback loop. The second-order state-variable filter in (a) provides a bandpass output V_1 and a low-pass output V_2. Its transfer function is $V_1/V_{in} = -Qs/(1+s/Q+s^2)$ and $V_2/V_{in} = -1/k(1+s/Q+s^2)$ where $s = j\omega/\omega_0$, $1/\omega_0^2 = C_1C_2R_1R_2/k$, $Q = (C_2R_2/kC_1R_1)^{1/2}$, and $k = R_4/R_3$. The parameter ω_0 is the center frequency of the filter. Using $C_1 = C_2 = C$ in the above equations, $R_1 = 1/CQ\omega_0$ and $R_2 = kQ/C\omega_0$.

The filter in (b) gives a bandpass output V_1 and a high-pass output V_2. As a result, its transfer function can be written as $V_1/V_{in} = -Qs/(1+s/Q+s^2)$ and $V_2/V_{in} = -(s/k)/(1+s/Q+s^2)$ where $1/\omega_0^2 = kC_1C_2R_1R_2$ and $Q = (C_1R_1/kC_2R_2)^{1/2}$. Letting $C_1 = C_2 = C$ yields $R_1 = Q/\omega_0C$ and $R_2 = 1/\omega_0kQC$.

As an illustration, the filter in (a) is realized for a Q factor of 200 and a center frequency of 100 hertz. The design assumes a value of 1 nanofarad for C_1 and C_2. With the equations shown above, the values obtained are $R_1 = 8$ kilohms, $R_2 = 320$ kΩ, $R_4 = 100$ Ω, and $R_3 = 100$ kΩ. The gain provided by the filter is $A = 1/k = 1,000$. The circuit uses components with 1% tolerances to reduce frequency drift and enhance circuit reliability. □

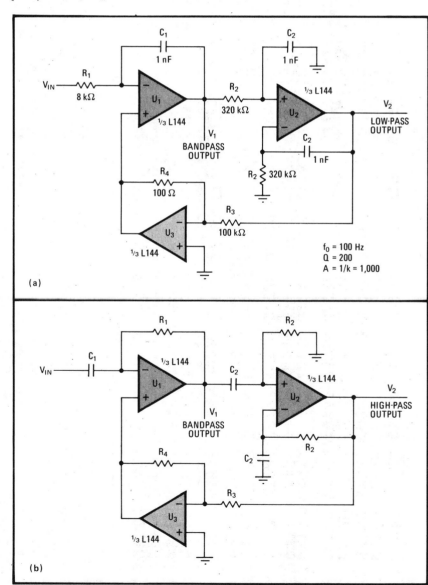

Independent. With inverter U_3 in the feedback loop of the state-variable active filter, the filter's center frequency and Q factor can be tuned separately. The filter as shown in (a) gives bandpass and low-pass outputs, and the version in (b) gives bandpass and high-pass outputs.

State-variable filter trims predecessor's component count

by James H. Hahn
Interface Technology Inc., St. Louis, Mo.

A state-variable active filter, even though it requires three operational amplifiers and eight passive components, is widely used because it is less sensitive than other filter designs to component changes, provides high Q and gain, and can operate at fairly high frequencies. A new design using fewer parts simplifies the standard filter[1] yet provides the same bandpass characteristics and low sensitivity.

A conventional state-variable circuit (a) is composed of summing amplifier A_1 and integrators A_2 and A_3. When high-pass output E_2 is eliminated and the first integrator is treated as the summing amplifier, the bandpass transfer function of the revised circuit (b) becomes:

$$H_{bp}(s) = \frac{E_3}{E_1} = \frac{-s/C_1R_1}{s^2 + sa + b}$$

where $a = \dfrac{R_2C_2(R_3+R_4) + R_4C_1R_5}{R_2R_5C_1C_2(R_3+R_4)}$

and $b = \dfrac{R_4(R_1+R_5)}{C_1C_2R_1R_2R_5(R_3+R_4)}$

In addition, the two-pole bandpass form of the circuit is preserved. If R_5 is eliminated and left an open circuit, the a and b terms simplify to:

$$a = \frac{R_4}{C_2R_2(R_3+R_4)} \text{ and } b = \frac{R_4}{C_1C_2R_1R_2(R_3+R_4)}$$

Comparing the transfer function to the general form:

$$H_{bp} = \frac{s\omega_o H_o/Q}{s^2 + s(\omega_o/Q) + \omega_o^2}$$

When H_o is the value of H_{bp} at $\omega = \omega_o$ and Q is the quality factor:

$$\omega_o = \left[\frac{R_4}{C_1C_2R_1R_2(R_3+R_4)}\right]^{1/2}$$

and $H_o = -Q^2 = \dfrac{-C_2R_2(R_3+R_4)}{C_1R_1R_4}$

Solving for R_1 and R_2:

$$R_1 = \frac{1}{\omega_o Q C_1} \text{ and } R_2 = \frac{R_4 Q}{(R_3+R_4)\omega_o C_2}$$

With the above equations, a two-pole active bandpass filter (c) is realized by choosing C_1, C_2, and $R_3:R_4$ for the given value of ω_o and Q and then computing R_1 and R_2. If the passband gain is high, an attenuator can be used at the input with R_1 serving as the attenuator's equivalent series resistance. The circuit can be further reduced (d) by letting R_4 be an open circuit and $R_3 = 0$ ohm. This modification gives:

$$\omega_o^2 = \frac{1}{C_1C_2R_1R_2} \text{ where}$$

$$R_1 = \frac{1}{\omega_o Q C_1} \text{ and } R_2 = \frac{Q}{\omega_o C_2}$$

The simplified configuration uses much less power and fewer components than the conventional circuit to achieve the same low sensitivities. If C_2 is greater than C_1, the circuit's near ideal performance will not be

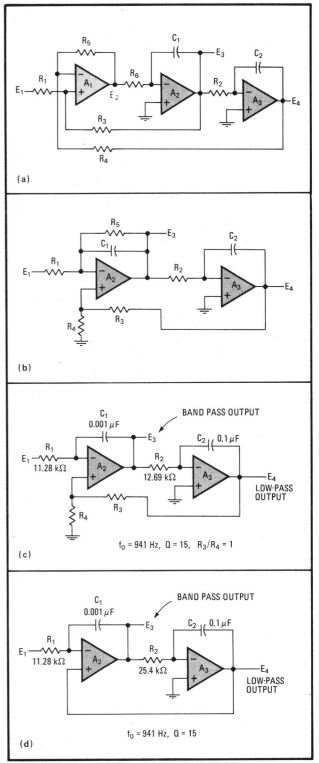

Reduced. The conventional state-variable filter (a) is simplified (b) when highpass output E_2 is eliminated and integrator A_2 is treated as a summing amplifier. This circuit is further reduced by creating an open circuit across R_5. A reduced two-pole bandpass state-variable filter for $f_o = 941$ Hz, Q = 15, and $R_3/R_4 = 1$ is shown in (c). A version with R_4 and R_5 open and $R_3 = 0$ is shown in (d).

degraded by second-order effects, such as finite amplifier gain. The filter is also useful for low-pass applications. □

References
1. J. Graeme, G. E. Tobey, and L. P. Huelsman, "Operational Amplifiers—Design and Applications," McGraw-Hill, 1971, p. 304.

Selecting the right filters for digital phase detectors

by Charles R. Jackson
E-Systems Inc., St. Petersburg, Fla.

Digital phase detectors are widely used in phase-locked loops because of their inherent high speed, large frequency coverage, and simplicity of operation. However, such detectors generate spurious alternating-current components that degrade the loop's performance. This design technique analyzes these spurious sidebands, enabling the designer to quickly project the performance of a PLL as a function of the frequency divider, the reference frequency, and the desired loop bandwidth. As a result, the proper filter for the circuit is precisely selected.

A single-loop frequency synthesizer (a) may be used to analyze the detector's spurious output. Spurious sidebands generated with the detector are minimized when a narrow bandwidth is set. As a result, phase tracking is degraded and lock-up slowed. A general expression for these spurious sidebands is S/C (sideband-to-carrier ratio) = 20 log ($\Delta f/2f$), where Δf = voltage-controlled–oscillator peak deviation and f = modulation rate. Simplifying the above expression in terms of the loop bandwidth (B), the numeric value of the frequency divider in the PLL feedback loop (N), and the reference frequency f_r, S/C = 20 log($4BN/2f_r$). This equation was developed using no low-pass filter in the generalized PLL and setting $f = f_r$.

On the plot of S/C in decibels versus the loop bandwidth for various values of N (b), B is expressed as a percentage of the loop reference frequency. In addition, the dotted lines are extensions of the equation for $\Delta f/f$ greater than 0.2.

For example, to find the loop bandwidth for S/C = −50 dB with the VCO operating at 3 megahertz and the reference frequency at 100 kilohertz, N must first be calculated. In this case it is equal to the

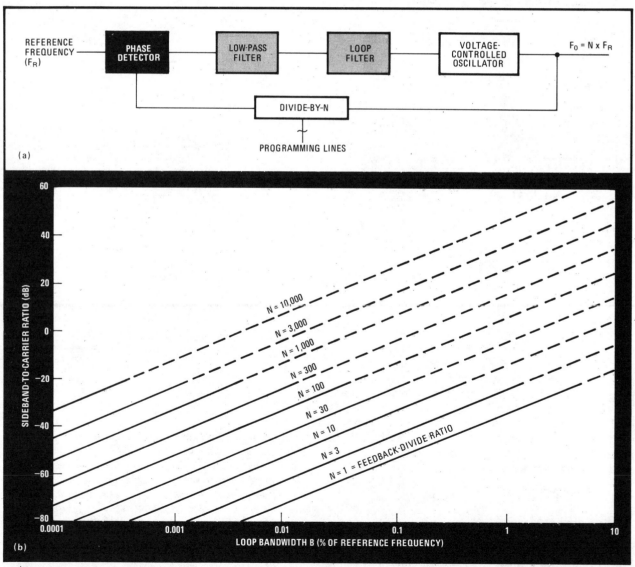

Analysis. The spurious output of a digital detector (a) is analyzed and an expression is developed for the sideband-to-carrier ratio. S/C in decibels is plotted versus loop bandwidth, which is expressed as a percentage of the loop reference frequency for various values of N.

$(3 \times 10^6)/(100 \times 10^3)$, or 30. The intersection of the -50-dB line and $N = 30$ (see b) gives a value of 0.0053% (f_r) for B. Hence the actual value of $B = 5.3$ hertz, or $5.3 \times 10^{-5} \times 100 \times 10^3$. However, if the loop bandwidth is increased to 500 Hz, a low-pass filter is needed to maintain a -50-dB value for S/C.

To determine the low-pass filter's characteristics, a new value of B must be calculated that is equal to 0.5%. The intersection of 0.5% and $N = 30$ gives $S/C = -10$ dB. Thus the low-pass filter needs to provide -40 dB. □

Logic-gate filter handles digital signals

by Andrzej M. Cisek
Electronics for Medicine, Honeywell Inc., Pleasantville, N.Y.

Performing the digital counterpart of electric-wave filtering in the analog domain, this unit can function as a low-pass, high-pass, bandpass, or band-reject filter of a square-wave pulse train. No RC integrating networks or comparators are needed, the all-digital filter being tuned simply by adjusting the reference frequency. Built originally for biomedical applications, it can find much broader use in the field of communications.

Consider the case of the band-reject filter shown in (a) in the figure. As seen with the aid of the timing diagram, the Q output of the edge-triggered set-reset flip-flop formed by the gates of the CD4011, A_1–A_4, and the 4013 D flip-flop (B_1) is brought high by the training edge of reference frequency f_0 and brought low by the falling edge of signal f_1. The combined output of the flip-flop and f_0 appears at gate C_1, moving low if $f_0 > f_1$. Similarly, C_2 moves low if $f_1 > f_0$. Therefore, signals from the output of the NAND gate formed by NOR gates D_1–D_4 appear whenever $f_0 \neq f_1$. Each pulse sets flip-flop B_2 high if it is not already so, permitting signal f_1 to pass through to the output.

Meanwhile, the 12-stage 4040 ripple counter advances on each pulse from f_0. The counter will reach the Q_n state if $f_0 = f_1$, because no reset pulse can emanate from gate D_4 under that condition. These events will disable gate C_3 and prevent f_1 from reaching the output.

The steepness of the filter's roll-off characteristic will be determined by which stage of the counter resets flip-flop B_2. The filter's reaction time to changes in the

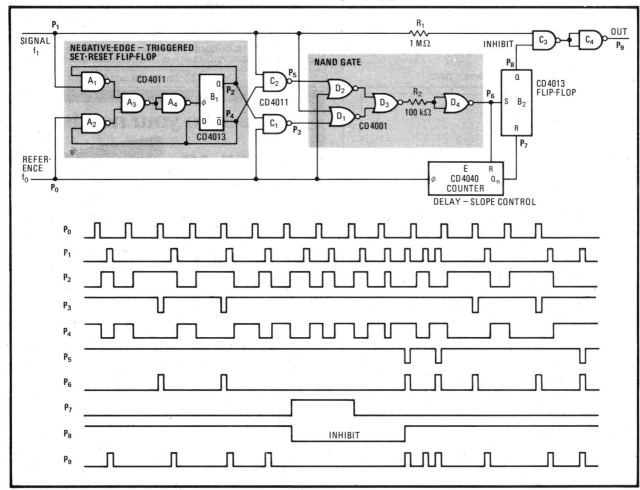

Digital damping. To perform band-reject function, this combinational logic circuit ascertains the frequency relationship of two square-wave signals. Tuning is done by adjusting the reference frequency. Selectivity is determined by tap position Q_n of the counter. Waveforms for given points in circuit show timing relationships. With minor changes, the filter is easily adapted for high-pass, low-pass, and bandpass duties.

input and reference frequencies will vary accordingly—that is, the steeper the slope, the longer the response time, this delay being the major drawback of the filter. The corner frequencies are $f_{1\,min} = (N-1)f_0/N$ and $f_{1\,max} = (N+1)f_0/N$, where N is the number of pulses of f_0 required for the counter to produce a reset pulse. The quality factor is $Q = f_0/\Delta f_1 = N/2$.

The stop-band filter can be easily modified to a band-pass type if the $\overline{Q}$ output of flip-flop B_2 is wired to serve as the inhibit line. To convert the filter for low-pass duties C_2 is removed and both inputs of D_3 are connected to D_1. In like fashion, C_1 is removed and both inputs of D_3 are connected to D_2 if a high-pass response is desired. Note that the NOR-gate circuitry is required to avoid any ambiguity of output state when pulses of input and reference signals overlap. Also, resistors R_1 and R_2 neutralize the effect of variable propagation-time differences of f_0 and f_1 through the gates ☐

Standard C, L-input filters stabilize hf transistor amplifiers

by Ed Wetherhold
Honeywell Inc., Signal Analysis Center, Annapolis, Md.

This display updates the design table for low-pass Chebyshev filters that use standard-value capacitors [*Electronics*, June 19, 1980, p. 160]. It is based on an inductive-input configuration designed to provide the response of the capacitor configuration while stabilizing transistor amplifiers that tend to oscillate when faced with a capacitive input[1]. The filters may be scaled for any frequency and input/output impedance that departs from their intended use in the 1-to-10-megahertz range with a 50-ohm source-load impedance; all designs yield more than 40 decibels of attenuation per octave.

Component values for these 30 seven-element designs (see table) were calculated using a simple Basic program. The passband response of all designs has equal ripple, and the reflection coefficient of the filters is less than 11% to minimize the voltage standing-wave ratio. For terminations that have an equal impedance, $L_1 = L_7$, $L_3 = L_5$, and $C_2 = C_6$.

To find the element values of a filter having an impedance level Z_x corresponding to a given frequency specification (and thus the frequencies corresponding to the equiripple point, A_p, and the 3-, 30-, and 50-dB attenuation points), the user must:

Filter No.	Frequency (MHz)				Reflection coefficient (%)	$C_{2,6}$ (pF)	C_4 (pF)	$L_{1,7}$ (µH)	$L_{3,5}$ (µH)
	A_p dB	3 dB	30 dB	50 dB					
1	0.921	1.08	1.50	1.98	3.57	4,700	5,600	6.33	14.58
2	1.014	1.18	1.63	2.15	3.89	4,300	5,100	5.89	13.37
3	1.087	1.29	1.81	2.40	2.88	3,900	4,700	5.06	12.04
4	1.197	1.41	1.96	2.58	3.41	3,600	4,300	4.81	11.15
5	1.065	1.45	2.14	2.88	0.61	3,300	4,300	3.59	10.35
6	1.328	1.54	2.12	2.80	4.16	3,300	3,900	4.58	10.29
7	1.179	1.60	2.35	3.16	0.64	3,000	3,900	3.28	9.40
8	1.425	1.68	2.35	3.11	3.12	3,000	3,600	3.95	9.27
9	1.528	1.86	2.63	3.49	2.21	2,700	3,300	3.36	8.32
10	1.634	2.06	2.96	3.95	1.43	2,400	3,000	2.83	7.41
11	1.906	2.07	2.75	3.57	10.65	2,400	2,700	4.31	8.19
12	1.859	2.27	3.22	4.28	2.04	2,200	2,700	2.71	6.78
13	2.137	2.53	3.53	4.66	3.12	2,000	2,400	2.63	6.18
14	2.291	2.78	3.94	5.23	2.21	1,800	2,200	2.24	5.54
15	2.452	3.09	4.44	5.92	1.43	1,600	2,000	1.89	4.94
16	2.859	3.11	4.13	5.36	10.65	1,600	1,800	2.88	5.46
17	2.849	3.37	4.71	6.22	3.12	1,500	1,800	1.97	4.64
18	3.126	3.84	5.46	7.26	1.93	1,300	1,600	1.59	4.00
19	3.475	3.90	5.29	6.91	6.53	1,300	1,500	2.00	4.17
20	3.269	4.12	5.92	7.89	1.43	1,200	1,500	1.41	3.70
21	3.985	4.61	6.37	8.39	4.16	1,100	1,300	1.53	3.43
22	3.538	4.80	7.06	9.49	0.64	1,000	1,300	1.09	3.13
23	4.274	5.05	7.06	9.33	3.12	1,000	1,200	1.32	3.09
24	4.633	5.53	7.78	10.29	2.72	910	1,100	1.17	2.81
25	5.053	6.12	8.64	11.47	2.30	820	1,000	1.03	2.53
26	5.581	6.70	9.44	12.51	2.54	750	910	.954	2.31
27	6.229	7.41	10.40	13.76	2.85	680	820	.881	2.10
28	6.791	8.12	11.41	15.11	2.68	620	750	.796	1.91
29	7.463	8.97	12.65	16.76	2.50	560	680	.711	1.73
30	8.176	9.85	13.89	18.41	2.44	510	620	.645	1.57

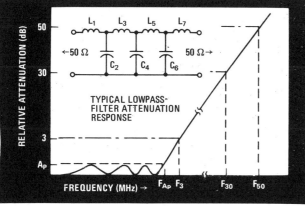

TYPICAL LOWPASS-FILTER ATTENUATION RESPONSE

- Calculate the scaled impedance factor, $R = Z_x/50$.
- Calculate the 3-dB cutoff frequency of a 50-Ω filter from $F_3^{50} = R \times F_3^x$, dividing Z_x by 10^n, where $n = 1, 2, \ldots$ if necessary to ensure that $F_3^{50} < 10$ MHz.
- Select from the table the design closest to that meeting the calculated F_3^{50} value. The tabulated values of C will be used in the new design, and the L values scaled.
- Calculate the exact value of $F_3^x = F'_3^{50}/R$, where F_3^{50} is the tabulated 3-dB frequency. In a similar manner, calculate the A_p, 30-, and 50-dB frequencies.
- Calculate the new L values for the desired termination impedance from $L = R^2 \times L_{50}$.

Consider a design example where $F_3^x = 6.0$ MHz and

$Z_x = 75\ \Omega$. Then $R = 75/50 = 1.5$, and $R^2 = 2.25$. Therefore, $F_3^{50} = 1.5(6) = 9.0$ MHz. Filter No. 29 is selected because its F_3^{50} value is closest to the desired value. Thus $C_2 = C_6 = 560$ picofarads, and $C_4 = 680$ pF. Inductors $L_1 = L_7 = R^2(0.711) = 1.60$ microhenrys, and $L_3 = L_5 = R^2(1.73) = 3.89\ \mu$H. The inductors will usually have nonstandard values, but this is no problem because any inductor may be conveniently hand-wound using iron-powdered cores that are commercially available. The exact f_A, and the 3-, 30-, and 50-dB frequencies are 4.98, 5.98, 8.43, and 11.17 MHz, respectively.

Although the design table addresses filters operating only in the 1-through-10-MHz decade, it is easy to scale the filter data for other frequencies. For example, the 10-to-100-MHz and 100-to-1,000-MHz decades may be derived simply by multiplying the frequency by 10 or 100, respectively, and dividing all C and L values by that same number. Similarly, for the 1-to-10-kilohertz, 10-to-100-kHz, and 100-to-1,000-kHz decades, the frequency should be divided by 1,000, 100, or 10, and the component values multiplied by the same number. □

References
1. R. Jack Frost, "Large-scale S parameters help analyze stability," Electronic Design, May 24, 1980, pp. 93–98.

9. FUNCTION GENERATORS

Canceling cusps on the peaks of shaped sine waves

by Stephen H. Burns
United States Naval Academy, Annapolis, Md.

Most triangle-to-sine-wave converters employed in a function generator are incapable of reducing cusps on the output peaks, so that discontinuities occur in the derivative. This design (Fig. 1a) subtracts a portion of the triangular wave from a shaped sine wave. As a result, the slope at the wave's peak can be reduced to zero with negligible harmonic distortion. In addition, the design is insensitive to the occurrence of small changes in triangular-wave amplitude.

Results for three different wave shapers are shown in the table. Because the differential amplifier has the lowest distortion and highest bandwidth, it is presented as an example (Fig. 1b).

Operational amplifier U_1 subtracts the triangular-wave input from the shaped sine wave, and resistors R_1 and R_2 control the amount of negative feedforward. The differential amplifier, functioning as the shaper, consists of transistors Q_1 and Q_2. Any differences in transistor characteristics, which develop even-order harmonic distortion at the output, is adjusted with resistors R_4, R_5, and R_6. Finally, R_7 and R_8 feed the shaped signal to the noninverting input of U_1.

The circuit's performance has been measured with a spectrum analyzer. It suppresses odd harmonics by more than 52 decibels. □

Shaper. This negative feedforward scheme (a) subtracts traingular components from sine waves and cancels cusps of their peaks. The scheme is implemented with a differential amplifier (b). Op amp U_1 subtracts the triangular wave from the sine-wave output, thereby suppressing odd harmonics by more than 52 dB.

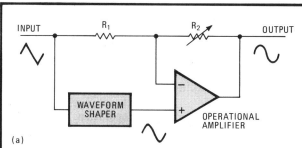

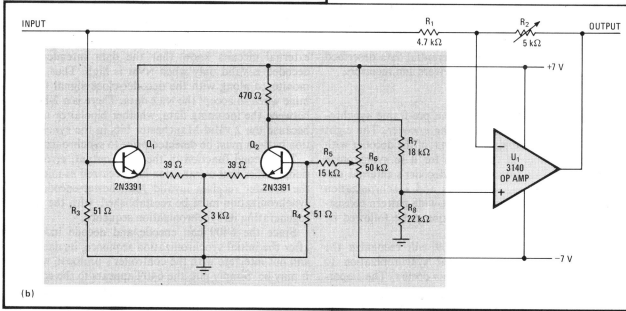

OPERATING CHARACTERISTICS OF THREE WAVEFORM SHAPERS				
Shaper	Triangular-wave input (V p-p)	Sine-wave output (V p-p)	Total harmonic distortion (%)	Frequency (kHz) at 2% THD
Four-diode type	3.45	2.400	0.7	150
Field-effect transistor	2.48	0.072	0.6	40
Differential amplifier	0.38	0.560	0.4	200

TTL can generate composite video signals

by Kenneth P. Evans
College of Medicine, University of Cincinnati, Cincinnati, Ohio

Obtaining a composite video signal from discrete components is a difficult task. However, exclusive-NOR gates having open collector outputs and TTL compatibility end the problem since they eliminate the need to bias high-speed transistors or to interface digital and analog devices in systems where only character or graphics information is used.

The circuit (a) produces a quality composite video signal. The current through resistor R_1 determines the output voltage. When the sync pulse appears at the input, gate U_{1-a} pulls the output to ground. During the display portion of the horizontal scan, the black output level of 0.25 volt results when the output of gate U_{1-b} is held low. This voltage shunts the current through R_3. When the output of U_{1-b} is high, the current from R_3 passes through diode D_1 to increase the voltage across R_1 to 1 v—the color white on the cathode-ray tube.

Additional features may be incorporated in the design as shown in (b). Reverse video is obtained by supplying a logic 0 to the input of U_{2-b} when characters are displayed. Also, tying the outputs of U_{2-b} and U_{2-c} together enables the circuit to blank the characters. In addition, U_{2-d} helps switch the character brightness between two levels. When the output of U_{2-d} is high, the current through R_4 increases the amplitude of the video-dot voltage across R_1. The low output impedance of this circuit is compatible with the 75-ohm input of standard video displays. □

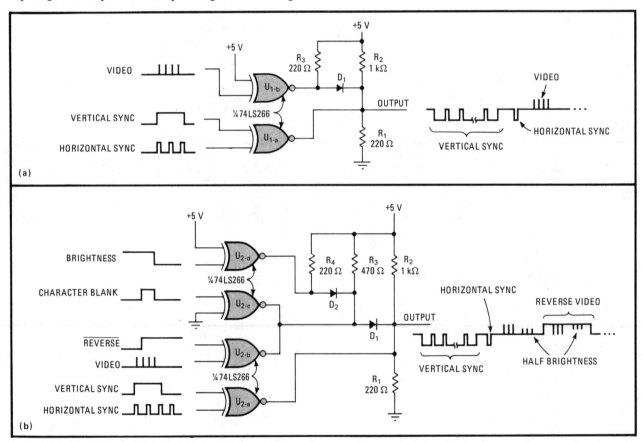

Composite video. Using exclusive-NOR gates with open collector outputs and TTL compatibility, the circuit (a) produces a quality composite video signal. The black output level is 0.25 volt and the white output level is 1 V. Adding more gates (b) provides such extra features as reverse video, character blanking, and two-level character brightness. In addition, the circuit's output impedance is low.

Generating triangular waves in accurate, adjustable shapes

by Virgil Tiponut and Adrian Stoian
Timisoara, Romania

Many precision instruments and generators require adjustable and well-defined triangular waveforms that have very accurate peak values. However, most of today's triangular-waveform generators lack features permitting simple parameter modification. This circuit allows these generators to overcome this shortcoming and provides the means for high resolution and speed

1. Precision. The circuit (a) built with comparators U_1 and U_2, constant current sources, and switch S_1 generates a triangular waveform (b) whose peak value, rise, and fall can be selected. The comparators determine the charge and discharge of capacitor C.

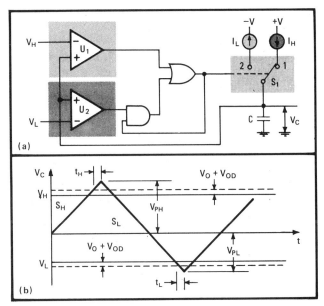

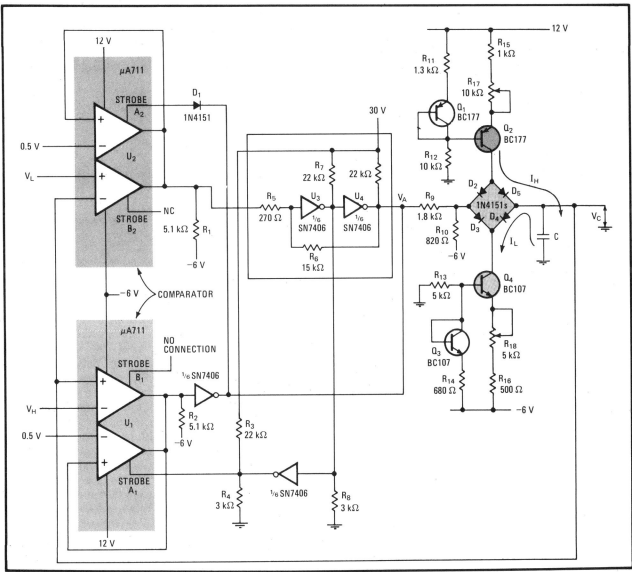

2. Adjustable generator. High speed and resolution, in addition to peak values that can be controlled, are obtained for this triangular-waveform generator by dual voltage comparators U_1 and U_2, which have positive feedback. Charging and discharging of capacitor C is controlled through strobe A_1 and strobe A_2. Inverters U_3 and U_4 are used to improve the switching time.

along with controllable peak, rise, and fall values.

Comparators U_1 and U_2 (Fig. 1a) having threshold voltages V_h and V_l, respectively, determine the charge and discharge rates of capacitor C. When switch S_1 is in position 1, the constant current I_h charges C. In position 2, the capacitor discharges current I_l. The resulting triangular waveform (Fig. 1b) shows that the peak, rise, and fall of V_c can be easily varied through the adjustment of threshold levels V_h and V_l and I_h and I_l.

The resultant error for V_h and V_l, respectively, is $e_h \leq V_o + V_{od} + t_h S_h$ and $e_l \leq V_o + V_{od} + t_l S_l$, where V_o is the input offset voltage of the comparators, t_h and t_l are the settling times for the high and low states corresponding to an overdrive voltage V_{od}, and S_h and S_l represent the slopes of V_c.

A dual voltage comparator μA711 having positive feedback provides high resolution and good propagation times (Fig. 2). When input voltage V_c goes above V_h, comparator U_1 is switched to a high state and remains in it, irrespective of the value of V_c, until a low level is applied to strobe input A_1. In this high state, current source Q_2 charges C with current I_h.

As a result, U_1 lowers the value of V_a, which, in turn, causes the capacitor to discharge I_l through transistor Q_4. While the capacitor is in the course of discharging, comparator U_2 switches to the low state because the voltage at strobe A_2 is low.

The discharge process continues until V_c drops below V_l. Once this drop occurs, comparator U_2 goes high. As a result, the output voltage of U_2 and consequently V_a will go high again. Thus the cycle repeats. Inverters U_3 and U_4 are connected as a Schmitt trigger to improve switching time, and D_2 through D_4 serve as a switch. Diode D_1 limits the strobe voltage when V_a is high. □

Doubling the resolution of a digital waveform synthesizer

by David M. Weigand
West Chester, Pa.

There seem to be almost as many ways of generating waveforms as there are applications. However, symmetrical waveforms such as sine or triangular waves generated by digital synthesizers can be improved by incorporating a simple operational-amplifier circuit. With this circuit, the resolution of the output waveform is doubled and its distortion substantially reduced.

The standard waveform generator (a) consists of a digital-to-analog converter driven by a read-only memory and a counter that provides sequential addresses for the ROM. It is modified (b) by replacing the M-bit counter with an (M+1)-bit counter and placing the op amp circuit between the converter and the low-pass filter, whose gain is also doubled. This modification halves the staircase step of the output waveform and thereby also doubles its resolution.

Only half of the sine wave is stored in ROM U_2. Input frequency f_{in} is doubled so that converter U_3 outputs two triangular waves of amplitude V_{dr} each quantized in N bits. Alternate halves of V_{dr} are flipped using transistor Q_1 and an (M+1) count of counter U_1.

The output of operational amplifier U_4 is alternately ground and $-(3/2)V_{dr}$. This output is continuously summed with V_{dr} using two 10-kilohm resistors, which results in $V_x = \pm V_{dr}/2$. The V_x signal is fed into the filter U_5 to obtain the desired output waveform V_{od} with greatly reduced distortion. □

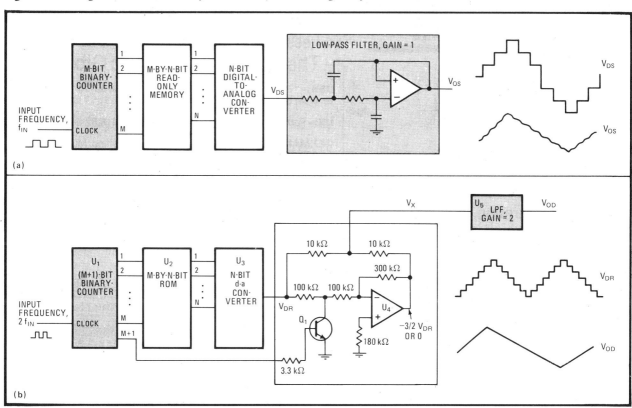

Higher resolution. The resolution of the standard ROM–plus–data-converter symmetric-waveform generator (a) can be doubled by incorporating in it a few simple changes—using an (M+1)-bit instead of an M-bit counter and employing an op-amp circuit between the converted and the filter (b). In addition, the input frequency is twice f_{in}. The output waveforms for the two circuits are as shown.

10. INSTRUMENT CIRCUITS

Measuring irregular waveforms by detecting amplitude changes

by Jeffrey Schenkel
Norwood, Mass.

When respiratory rates or other types of irregular waveforms are measured, it is often desirable to record just the change in the signal amplitude and leave out baseline variations and noise. This circuit does the job with only a baker's dozen of parts.

First, input waveform voltage V_{input} directs voltage V_x. In addition, the output voltages of operational amplifiers A_1 and A_2 are negative and positive power-supply rails. This arrangement forces diodes D_1 and D_2 into a reverse-biased condition and in turn couples the input signal to the op amps' inverting inputs.

As the input voltage increases, V_x eclipses the threshold voltage, V_{th}, which is set by an external source. Once this swing occurs, the output of A_2 begins to go negative until D_2 turns on, which prevents V_x from becoming more positive. V_{in} now begins to decline from its peak, D_2 shuts off, and V_x follows V_{in} in a negative direction until $V_x = 0$ volt. As a result, the output of A_1 begins to swing positive and turns on D_1, preventing V_x from going negative. Finally, when V_{in} reaches its negative peak, D_1 turns off and the cycle begins again.

Output voltages V_{A_1} and V_{A_2} of op amps A_1 and A_2 indicate negative and positive signal excursions (b). These signals are then combined and fed to the Schmitt trigger. The output of A_3 changes state only when V_x hits 0 V or the threshold voltage, and it completes a cycle when the input signal experiences both a negative and positive voltage swing that is equal to at least the threshold voltage.

V_{output} can be fed to edge-triggered circuitry for rate counting or other processing. As the system uses the highest peaks and lowest troughs to determine the excursion size, it is not fooled by small signal reversals. □

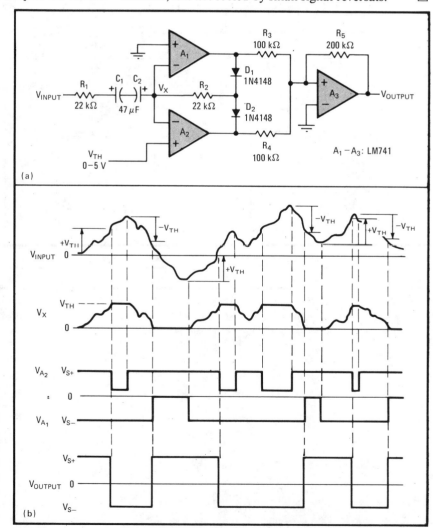

Detector. The output voltages of operational amplifiers A_1 and A_2 in the circuit in (a) are combined and fed to the Schmitt trigger A_3. The output of A_3, V_{output}, indicates the negative and positive changes in amplitude of the input waveform. Also, the threshold level is set by an external voltage. Various waveforms generated by the circuit that correspond to the input are shown in (b).

Four chips generate pseudorandom test data

by Wayne Sward
Sperry Univac Division, Salt Lake City, Utah

The inexpensive pseudorandom–bit-sequence generator shown here requires only four integrated circuits and will serve well in testing digital data links. Joined through a minimum of interconnections, it will furnish a recurring 511-bit string suitable for a variety of other applications as well.

The octal D-type latches in A_1, along with flip-flop A_2, form a 9-bit serial shift register that drives binary counter A_3, which detects the register's illegal all-zero state; exclusive-OR gate G_1 provides the required feedback connection. On power-up, the outputs of A_1 assume arbitrary logic values, and the 11-megahertz system clock steps the bits to the load input of A_3. When the first logic-1 bit cycles through to A_3, the counter presets to 6. As long as at least 1 bit in the shift register thereafter contains a logic 1, A_3 can never reach a count of 15. Consequently, its terminal-count output will always remain low.

If only zeros should appear in the shift register after power-up or during the course of operation, however, A_3 will eventually reach a count of 15, and its terminal-count output will go high. This action forces the three-state outputs of A_1 to become inactive, or an open circuit, and a logic 1 will be introduced (through resistor R_1) into the data stream, at the D input of A_2. This bit is detected by A_3 on the following clock cycle and normal operation is restored. Note that the circuit configuration eliminates the nine-input gate that is usually required at the input of the counter in order to detect the existence of an all-zero state.

The test code generated by this circuit is shown in the table, its format being compatible with the popular HP3780 bit-checking data unit. Of course, if the format of the test code must be altered, it may be changed simply by connecting the load input of A_3 to the appropriate stage of the shift register.

If the low-power Schottky chips used in this generator are all replaced by their standard Schottky equivalents, the bit rate can be extended to 26 MHz, though with a slight penalty in power dissipation. No additional wiring changes will be required. □

Self-correcting code. Simple generator provides a serial string of 511 pseudorandom bits suitable for testing digital data links at up to 11 MHz. The code, in a format compatible with the popular HP3780 test generator or data-bit checker, is easily changed by connecting the counter to the appropriate stage of shift register A_1–A_2. Bit rates to 26 MHz are achieved by replacing ICs with their Schottky counterparts.

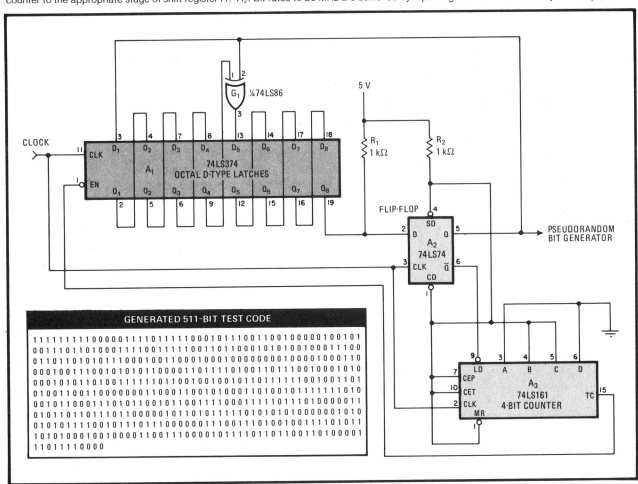

Wide-range capacitance meter employs universal counter

by Marvin Burke
Novato, Calif.

A wide-range capacitor meter (a) that eliminates the parasitic capacitance usually associated with capacitance measurements can be quickly built. This 1.0-picofarad-to-0.2-farad meter has a ±1% accuracy from 1.0 pF to 1 microfarad. It uses Intersil's universal counter, just a few integrated circuits and transistors, and a handful of passive components, yet can substitute for a more expensive hand-held device.

A square wave is generated by a relaxation oscillator formed from operational amplifier U_1. The period of this wave varies with the capacitance of unknown capacitor C_x. To avoid parasitic capacitance, the oscillator's output is fed to one-shot multivibrator U_2 through the driver consisting of transistors Q_1 and Q_2.

The pulse due to the parasitic capacitance at U_1's output, along with the one due to C_x, is shown in the graph (b). The value of C_x corresponds to $t_2 - t_1 = \tau$. The circuit thus compensates for the offset (t_1) produced with the oscillator and the driver by measuring only τ.

One-shot U_2 is adjusted with two potentiometers, one for each range of the two-range meter, so that its output is a pulse of interval t_1 that serves as universal counter U_4's A input. The oscillator's output, which is channeled through the driver, serves as the B input. The counter then takes these two inputs and measures the time interval τ between the falling edges of A and B.

Transistors Q_1 and Q_2 are used to increase the drive current, which enhances the circuit's range from 1 μF to 0.2 F. This range has an accuracy of ±3%. The ranges are selected by means of four-pole double-throw switch S_1, of which position 1 covers 1.0 pF to 1 μF and position 2 covers 1 μF to 0.2 F. The counter's light-emitting-diode display is calibrated to display the unknown capacitance with potentiometers R_1 and R_2. □

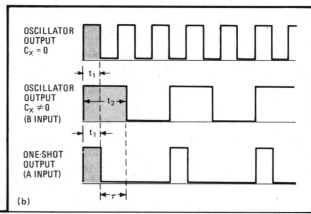

Capacitance meter. The circuit (a) uses universal counter U_4 to measure and display the unknown capacitance in digital form. It covers a range of 1.0 pF to 0.2 F in two steps with a four-pole double-throw switch. The time interval τ (b) between the falling edges of A and B inputs is calibrated in terms of unknown value. C_x.

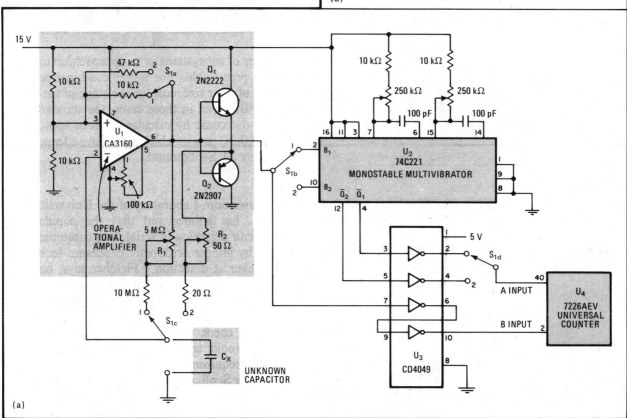

Building a low-cost optical time-domain reflectometer

by J. T. Harvey, G. D. Sizer, and N. S. Turnbull
AWA Research Laboratory, North Ryde, NSW, Australia

High-performance optical time-domain reflectometers require relatively complex optics and are fairly expensive—around $4,000 to $20,000. However, if a little accuracy can be sacrificed, this simple-to-use time-domain reflectometer, at a cost of only $500, can handily substitute for its more expensive brethren.

Designed primarily to locate faults in and measure losses of optical fibers, this device (a) uses a commercially available optical-fiber coupler, type TC4C. A light pulse is applied to the end of the fiber that is undergoing analysis, and imperfections in the cable are sensed by a photodetector. The detector [Electronics, Sept. 25, 1980, p. 161] measures the generated backscattered energy. Once detected with the TDR, this energy pulse is analyzed to determine the nature of the defect and its location on the fiber.

International Telephone & Telegraph's LA10-02 single heterostructure laser diode, which radiates at a wavelength of 905 nanometers, is used as the optical source. This diode is mounted into an Amp Inc. optimate connector and pulsed at a peak current of 25 amperes by six field-effect transistors that are connected in parallel. In addition, a fiber tail with a 100-micrometer-diameter core latched to port 4 of the optical coupler is plugged into the other end of the connector.

The receiving fiber tail is similarly coupled to a Siemens SFH202 p-i-n photodetector plugged into a photodiode amplifier. This 2.2-megohm transimpedance amplifier with a bandwidth of 5 megahertz uses in its input stage a MOS FET BF960, which is followed by a voltage amplifier with a gain of 30. In addition, proper layout reduces stray coupling and interference.

Port 2 of the coupler is index-matched with oil or a glycerin-water mixture that will absorb the reflection occurring at the end of the fiber tail. Port 1 is connected to the fiber under test. Because the fiber used in the coupler is wide, an Amp optimate connector bushing and a pair of AMP227285-4 connectors may be used as the joint between the two fibers. The connection is index-matched to minimize the front-face reflection.

An oscilloscope is used to read the location of splices or imperfections in the fiber under test. The magnitude of optical loss over a specified distance is obtained by estimating the backscatter (b) at each of the two points selected and calculating the loss—5 log V_1/V_2 decibels, where V_1 and V_2 represent the backscatter at the two selected measuring points. For example, when a laser with a peak power of 7 dBm is coupled into a test fiber having a core diameter of 50 μm, the fiber's backscatter at 1 kilometer was found to be 3 to 5 nanowatts and was observed for over 5 km of low-loss fiber. □

Inexpensive. This optical time-domain reflectometer (a) uses a commercially available Canstar 100-μm-diameter core step-index optical coupler to substantially lower the cost of the instrument yet maintains a satisfactory performance level. The scope photograph (b) shows measurable backscatter for 5 km of fiber under test.

Low-power thermometer has high resolution

by Mihai Antonescu
Federal Institute of Technology, Lausanne, Switzerland

An Intersil temperature transducer, an analog-to-digital converter, liquid-crystal-display drivers, and a few discrete components can yield a highly sensitive thermometer that consumes little power. The circuit draws 2.5 milliamperes from a 3-volt lithium battery and measures temperature with a resolution of 0.01°C.

A 5-kilohm potentiometer sets the analog-to-digital converter U_1's reference voltage to 1 v, while a pair of trimmer potentiometers in the input-voltage divider network are adjusted for temperatures between 0° and 100°C. So that the thermometer may have a resolution of 0.01°C and be very reliable, the resistors and potentiometers used are 1% metal-oxide and cermet types. The output of transducer AD590 is fed into a-d converter U_1, which is clocked by timer U_5. Drivers U_2 and U_3 can easily power conventional 4½-digit LCDs.

When measuring temperatures in a small chamber, the transducer radiates enough heat to introduce an error of 0.05°C. To eliminate this error, the battery is connected to an RC network and field-effect transistor Q_2 (see figure). After the start switch is pressed, this setup allows the circuit to draw current for 10 to 15 seconds, which is enough time for a stable measurement to be made and prevents the transducer from heating up. This feature also helps reduce power consumption.

Voltage converter U_4 provides a negative supply voltage for a-d converter U_1 and input amplifier U_5. The circuit also converts the 3-v supply to a higher voltage, required by temperature transducer AD590. □

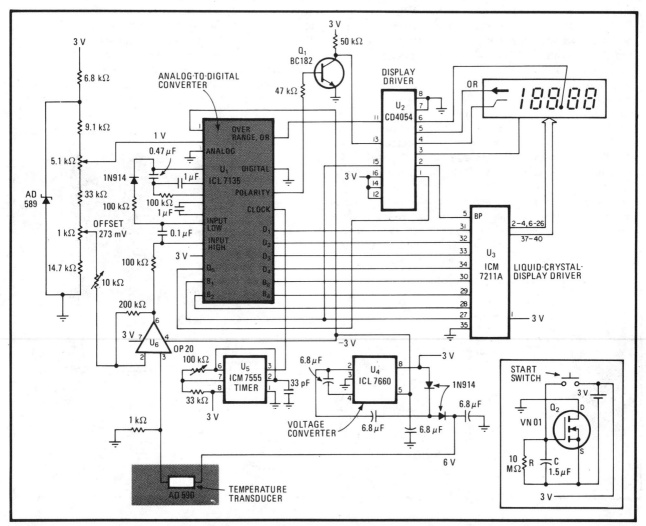

Precision thermometer. A direct-reading thermometer that measures temperature accurately is built from temperature transducer AD590, a-d converter ICL7135, a 4½-digit LCD display, and a handful of components. The circuit operates from a 3-V lithium cell and draws only 2.5 mA. The thermometer measures the temperature with a resolution of 0.01°C and displays the reading digitally.

Adapter equips HP analyzer for general ROM tracing

by Israel Gal
Liad Electronics, Moshav Yaad, Israel

A two-socket adapter turns the popular Hewlett-Packard family of HP1611A logic analyzers—which can normally be configured for debugging one specific microcomputer—into a general-purpose read-only-memory tracer. Thus, as this example shows, it can be used with the Z-80 Personality Module as a developmental tool for Intel machines, without the need for Intel's ICE series of in-circuit emulators.

Tracing is simply achieved by placing the microprocessor's address and data lines directly in parallel with those of the external program memory and disabling all other input lines to the Z-80. Placing a low-insertion-force socket, such as those produced by Textool Inc., at the Z-80 end enables fast connection to the conversion circuit. At the memory end, use of a hardwired 24-pin spring clip adds increased flexibility for in-field testing. Also recommended is a scope probe for latching onto the appropriate circuit point of the desired clock signal. The clock and its inverse signal are available, so that the user can synchronize the timing to each particular processor.

In operation, the logic analyzer will be synchronized as usual to accept address, data, and external information in every possible combination. Thus, most of the additional options of the analyzer, such as pretriggering, post-triggering, trace and count triggering, and trigger enable and disable can be utilized. The single-step and trace-then-halt options of the Intel 8031-8051 machines cannot be used here, although they will be functional on every other processor that has a wait line.

Using the HP1611A this way has several drawbacks, among them the fact that there is no disassembly—information is displayed in hexadecimal or octal format. And the information shifted onto available on-chip RAM or registers is not itself displayed, only the representation of the transfer as an operating code. Also, as a result of the clocking arrangement, there can be situations where the directive will be displayed twice (although the address is always correct). However, where low cost, convenience, and efficiency are important, this circuit is satisfactory. □

Translator. Parallel connection of data and address lines of external memory with HP1611A logic analyzer through appropriate interface eases debugging of any microprocessor-based system by a dedicated analyzer. Scheme has drawbacks—lack of disassembler, the fact that shifted information cannot be actually displayed, and occasional multiple display of op code directives—but is cheap, convenient, and efficient.

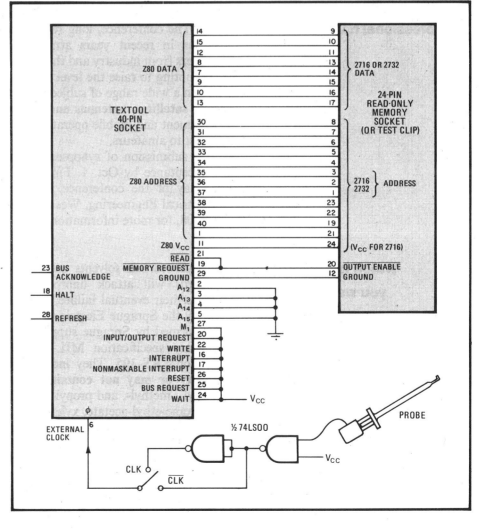

Instrumentation-meter driver operates on 2-V supply

by David M. Barnett and Everett L. Tindall
Martin Marietta Aerospace, Denver, Colo.

Complementary-MOS operational amplifiers and a bridge circuit, employed as a feedback loop, minimize the number of parts needed to build an accurate ac and dc meter that operates from a 2-volt supply. The meter has a wide bandwidth, consumes little power, and is used in testing cables in missile interfaces.

Three amplifiers, U_1 through U_3, are connected in a standard instrumentation configuration, while U_4 acts as the meter's driver. A full-wave rectifier bridge is used in the feedback loop of U_4. Since the operation of U_4 depends on the amount of current flowing through the bridge, diode voltage drops do not alter the circuit's performance. Under normal operation, current $I_{in} = I_{meter}$, and both are controlled with potentiometer R_8. Optional capacitor C_2 makes it possible to limit the meter's bandwidth.

The op amps are equipped with a programming pin that sets standby current I_q. Pulling these pins low sets I_q for maximum frequency response and current output. Because the circuit can handle only small currents, it may be necessary to use pnp transistors at the outputs of U_3 and U_4 if the full-scale meter current is high. Also, low-leakage diodes must be selected for the bridge.

The circuit requires a minimal adjustment. While R_6 allows adjustment of the input common-mode rejection ratio, R_{13} compensates for all op-amp input offsets, and R_8 provides a full-scale reading on the meter that corresponds to the full-scale input voltage. The circuit's 3-decibel bandwidth exceeds 50 kilohertz when C_2 is removed. The meter gives average readings for ac inputs. Its supply current is less than 8 milliamperes and can be reduced further if a wide bandwidth is not required by adjusting I_q. □

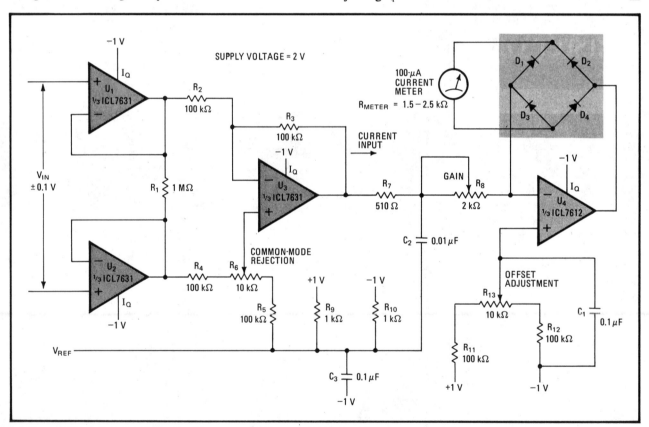

Low power. This instrumentation-meter driver uses complementary-MOS operational amplifiers U_1 through U_4 and a bridge rectifier to provide a design that can operate on a 2-volt supply. The meter scale is adjusted using resistor R_8, while capacitor C_2 permits bandwidth limiting. Under normal operation, current I_{in} equals I_{meter}.

Ac voltmeter measures FET dynamic transconductance

by M. J. Salvati
Flushing Communications, Flushing, N. Y.

Instruments used to test field-effect transistors and MOS FETs make either static transconductance measurements or no transconductance measurements at all. However, when this simple test circuit is combined with a standard ac voltmeter, the modified tester can directly read the small-signal dynamic transconductance of a FET. To accommodate low-transconductance FETs, the circuit requires that the voltmeter's most sensitive range be no higher than 3 millivolts full scale.

NAND gates U_{1-a} through U_{1-d} form the oscillator section of the test circuit that generates the test signal. This test signal is a square wave with a frequency of 2,500 hertz, whose amplitude is substantially reduced by the voltage divider consisting of resistors R_1 through R_3 and diode D_1. The diode also stabilizes this voltage by reducing any changes that are due to variations in the power supply.

The 5-kilohm potentiometer R_3 is adjusted until the test point shows an output of 10 mV. As a result, the FET under test produces an output voltage across the sampling resistor R_4, which is placed in the drain path of the FET. This output voltage varies at the rate of 1 mV per millimho of the FET transconductance. The same voltmeter as is used to set the test point at 10 mV must also be used to calibrate the transconductance of the FET. The n-channel FET shown in the circuit is only an example, for p-channel FETs can also be easily calibrated by changing the polarities of the gate and the drain supplies. A typical characteristic curve for the FET under test may be plotted that demonstrates the transconductance versus the gate bias. A similar curve for varying drain current can also be plotted. Gate and drain voltages are adjusted by separate power supplies. □

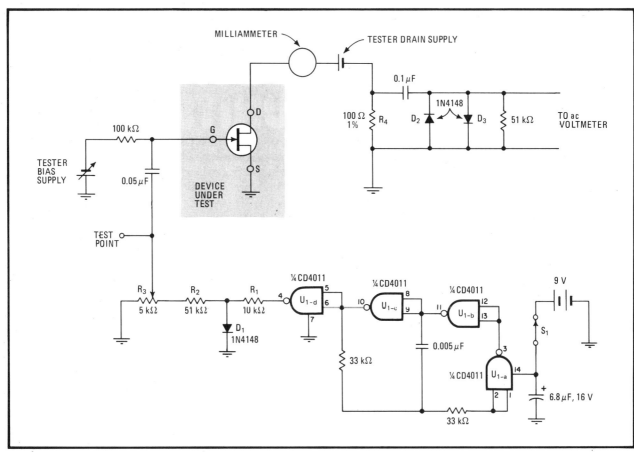

Dynamic. An ac voltmeter, when linked with this simple test circuit, can be calibrated to read directly the dynamic transconductance of an FET. The test signal, generated by NAND gates U_{1-a} through U_{1-d}, is fed into the gate of the FET under test whose output across R_4 is measured by the voltmeter. This output voltage varies at the rate of 1 mV per millimho of FET transconductance.

One-chip DVM displays two-input logarithmic ratio

by David Watson
Intersil Inc., Reading, Berks., England

The popular ICL7106 series of analog-to-digital converters that serve so widely nowadays as one-chip digital voltmeters can be easily converted to display the logarithm of the ratio between two input voltages, making them useful for chemical densitometry, colorimetry, and audio-level measurements. Only slight wiring modifications at the device's input and integrating ports are required.

Shown in (a) is the new configuration. The modifications from the standard a-d converter connection include the addition of a resistive divider, R_1–R_2, at the reference inputs, and the placing of resistor R_p in parallel with the device's integrating capacitor.

As shown with the aid of the timing diagram in (b), the time constant of the integrating network is given by $\tau = C_{int}R_p$, with the asymptotic endpoint voltage of the integration voltage being $V_{as} = R_p(V_1 - V_2)/R_{int}$, where V_1 and V_2 are the input voltages to be measured. The final integrator voltage therefore becomes $V_{int} = R_p(V_1 - V_2)(1 - e^{-T/\tau})/R_{int}$, where T is the fixed integration period.

During the deintegration portion of the cycle, the exponential decay moves toward the total voltage, V_{tot}, which equals $V_{int} + V_{ref}(R_p/R_{int})$. But $V_{ref} = kV_2$, where k is set by the resistive divider, so that $V_{tot} = R_p(V_1 - V_2)(1 - e^{-T/\tau})/R_{int} + R_p kV_2/R_{int}$. The integrator voltage actually crosses zero when the exponential waveform reaches $V_{final} = V_{ref}R_p/R_{int} = R_p kV_2/R_{int}$.

As seen, the time needed to reach the zero crossing is given by $T_{DEINT} = \tau \ln(V_{tot}/V_{final})$. Making $k = (1 - e^{-T/\tau})$ and $\tau = T/2.3$, it is realized that $T_{DEINT} = T \log_{10}(V_1/V_2)$. For this condition, $k = 0.9$, which is achieved by making $R_1 = 1$ MΩ and $R_2 = 9$ MΩ.

Theoretically, the system's full-scale output voltage is reached when $\log_{10}(V_1/V_2) = 2$, but noise will probably limit the range of the converter. Note also that the accuracy of the system is no longer independent of passive component variations. The simplest way to ensure that $k = 0.9$ is to use a pretrimmed divider. The system is calibrated by making $V_1 = 10V_2$ and by adjusting R_p until the display reads 1.000. □

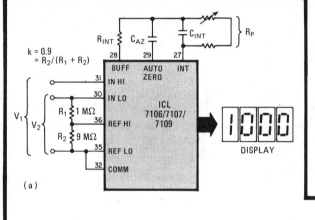

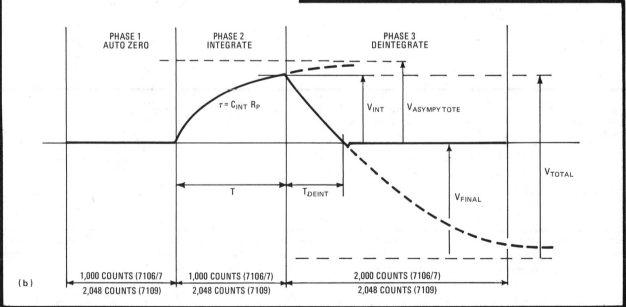

Log converter. ICL7106 analog-to-digital converter may be used to measure the logarithmic ratio of two input voltages. Modifying converter's input circuit (a) and integrating network and selecting suitable time constants ensure that its output is proportional to $\log_{10}(V_1/V_2)$. Timing diagram (b) clarifies circuit operation.

Meter measures processor's dynamic utilization capacity

by Henryk Napiatek
Lacznosci Institute, Gdansk, Poland

An ordinary milliammeter, calibrated in percentages, plays a key role in this simple one-chip, one-transistor indicator of the fractional utilization of a microcomputer's central processing unit in a real-time environment. As a result, the circuit (see figure) will be useful in optimizing system performance and debugging random process routines that typically occur in telephone- and vehicular-traffic applications.

The degree of utilization in processing data and handling interrupts versus the time the machine executes the scheduler's idle loop is simply measured by firing a monostable multivibrator with an output signal derived from the operating system's idle loop. The one-shot's pulse width is set equal to the execution time of the scheduler's idle loop, which generates one pulse for each loop's pass. The scheduler's idle loop is executed only if the processor does not process any data or handle any interrupts. The scheduler's idle-loop execution time is about 50 microseconds. Interrupts cause the processor to execute program routines concerning traffic changes.

The integrated ouput signal of the one-shot thus represents a fraction of the total time the CPU is not being used. This fraction will be indicated by a drop in the reading of the milliammeter that is connected to the inverting output circuit of an npn transistor.

A register-enable pulse or similar signal leaving the output bus of the appropriate system peripheral is applied to the 74123's input. This signal is essentially an idle CPU mark that is derived from the sample idle-loop routine of the scheduler and is written in macro-11 assembly language for the PDP-11/34 minicomputer (see program).

The pulses from the one-shot's output are amplified by transistor Q_1 and integrated by capacitor C_1 and the milliammeter's resistance and distributed inductance. The meter's reading thus reflects the difference between the circuit's 5-volt output limit, which represents 100%

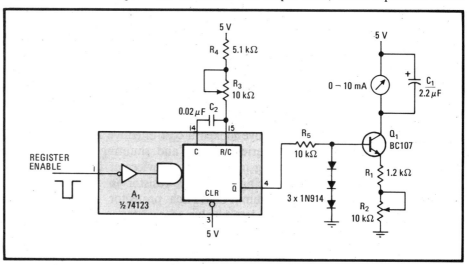

Indexing interrupts. This simple circuit determines the percentage of time the microcomputer's central processing unit is working on processing data and handling interrupts, thus serving as a low-cost optimization and debugging tool. A meter, calibrated directly from 0% to 100%, has typically a 2-µs integration time for rapidly following dynamic changes in machine capacity. The one-shot timer's pulse width is equal to the execution time of the scheduler's idle loop.

CPU utilization, and the total interrupt time, to yield an index of the CPU's actual use.

The circuit is calibrated by adjusting potentiometers R_2 and R_3. To calibrate the meter at full scale, the CPU's idle loop is halted (no input pulses) and R_2 is adjusted for a 100% meter reading. All external interrupts in the idle loop are then disabled (for example, the instruction CLR@#IDLESR should be replaced by the instruction CLR@#LIGHTS) and the routine run. The milliammeter is zeroed by adjusting R_3 for 0% processor utilization. In this case, the processor executes only the scheduler's idle loop. This design can be modified in hardware and software to accommodate indicators other than the milliammeter that can measure other parameters related to real-time operating systems. □

```
            IDLE LOOP OF PDP-11's SCHEDULER USED FOR CIRCUIT CALIBRATION

                    ; ETEXOS W.01/E/.04 OPERATING SYSTEM

        LIGHTS  =   177570      ; LIGHTS REGISTER
        IDLESR  =   160224      ; REGISTER IN SPECIAL INPUT/OUTPUT DEVICE
        PSW     =   177776      ; CENTRAL PROCESSING UNIT STATUS WORD
        PR7     =      340      ; PRIORITY 7
        R3      =       %3      ; CPU'S REGISTER 3
1$:     MOV #PR7+1, @ # PSW     ; EXTERNAL INTERRUPTS DISABLED, BIT C = 1
        MOV # IDLCNT, R3        ; ADDRESS OF IDLCNT 3-WORD VECTOR
        ADC /R3/+               ; BIT C+IDLCNT
        ADC /R3/+               ;
        ADC /R3/+               ;,COUNTING OF IDLE LOOPS IN IDLCNT VECTOR
        CLR @ # IDLESR          ; ONE-SHOT TIMER STIMULI
        CLR @ # PSW             ; EXTERNAL INTERRUPTS ENABLED
        BR 1$                   ; TO NEXT IDLE LOOP
IDLCNT: .WORD 0, 0, 0           ; IDLE LOOP VECTOR COUNTER
```

Pulse-width meter displays values digitally

by Paul Galuzzi
Beverly, Mass.

Built only with standard logic elements and solid-state displays, this meter provides digital readout of pulse width, a widely sought-after feature in instruments of this type. Measurement accuracy is one part in 10,000.

As the figure shows, this circuit is relatively straightforward, being made from several cascaded binary-coded decimal counters, flip-flops, BCD-to-seven-segment decoders, and the corresponding displays. In operation, the pulse (count) to be measured gates in the 100-kilohertz system clock, which steps the 74LS160 counter bank. After the count pulse goes to logic 0, the load pulse transfers the BCD data to the 74LS175 flip-flops that serve as a storage register.

At this time, the BCD data, whose value is directly proportional to the width of the count pulse, is decoded by the 74LS247 chips and displayed. A clear pulse then resets the counter bank, and the measurement cycle repeats.

The circuitry for generating the clear and load pulses is shown in the right-hand inset. As shown, sync flip-flop F_1 is enabled by the inverted count signal, thereby gating the 100-kHz system clock so that flip-flops F_2 and F_3 can shift through a two-stage cycle. Thus, the load and clear pulses are generated in sequence after the falling edge of the pulse whose width must be measured.

With the 100-kHz clock, the meter will display 100.00, its full-scale reading, for a pulse width of 100 milliseconds. Pulses as small as 10 μs in width can be measured accurately with the given clock rate, and more narrow pulses can be detected if the clock frequency is made proportionally higher. □

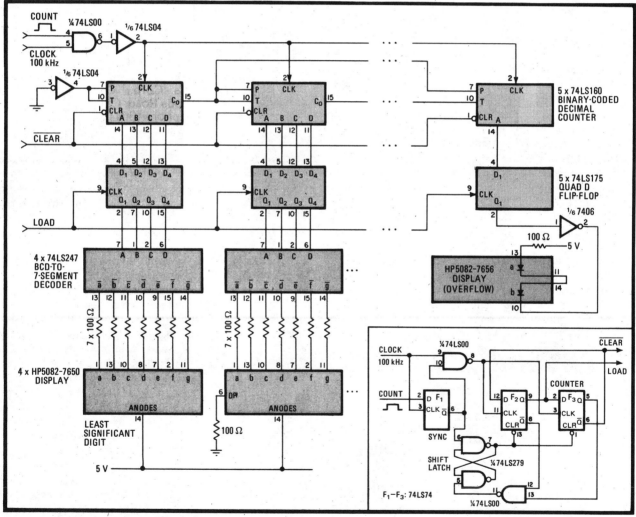

Count time. Digital meter measures pulse width by counting number of 100-kHz system clock cycles during time that pulse is present. Solid-state displays provide direct readout of time in milliseconds. Measurement accuracy is 1 part in 10^4 over range of 10 μs to 100 ms.

Reading a dual-trace scope with a ±0.025% accuracy

by Leonard Sherman
National Semiconductor Corp., Santa Clara, Calif.

Reading a voltage from an oscilloscope display by eye yields an accuracy to within about ±2% at best. But with the aid of a single, inexpensive integrated circuit and an ordinary 5½-digit voltmeter, a typical dual-trace scope with a delayed-sweep mode can be used to obtain readings accurate to within ±0.025% or better.

In the circuit shown, the LF398 sample-and-hold circuit is used to sample the voltage at the test point V_{in}, at a time determined by a gating signal from the oscilloscope. The retained voltage can then be read by taking a DVM input from pin 5. The circuit does not use input prescaling or offset trimming, so the LF398 input can vary between plus and minus 12 volts and the reading will typically be accurate to within ±3 millivolts. Prescaling or trimming can be added for a wider input range or greater measurement precision, respectively.

To gate the measurement, the scope is set up so that it provides a gating signal at a time determined by the delayed sweep signal. In the case of the Tektronix 453A, for example, the scope is set to operate in the A-intensified-during-B mode and the B gate output is used to trigger the sample-and-hold circuit. Any point along the A channel's displayed waveform can then be measured by simply adjusting the delay time multiplier.

Even at high sweep speeds or with short sample windows, the periodic nature of most measurements permits measurement accuracy to be maintained down to 200-nanosecond gate widths with the circuit. Typical applications would include precise rise-time determinations and peak readings. □

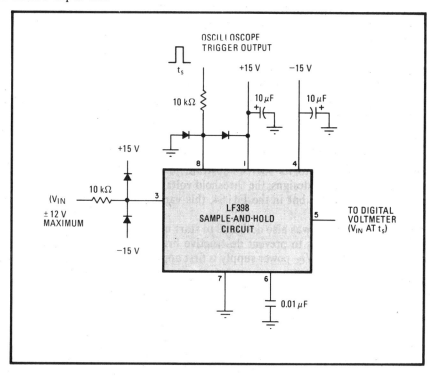

Sample, hold, and read. Based on an LF 398 sample-and-hold IC, the circuit above can be used to obtain precision measurements from a dual-channel scope with a delayed sweep that gates the measurement. V_{in} serves to drive the A channel input. All diodes shown are 1N4148s.

Tester checks functionality of static RAMs

by David J. Kramer
Sunnyvale, Calif.

Before a TTL-compatible random-access memory is inserted into boards, it is customarily checked out for functionality with a tester specifically designed for its particular bit size. However, this tester can examine all RAM sizes from 256-by-1-bit on up to 4-K-by-8-bit devices. It detects the bit sites that do not store a 1 or 0, faulty address inputs and buffers, and malfunctioning write or chip-enable inputs. However, it cannot deter-

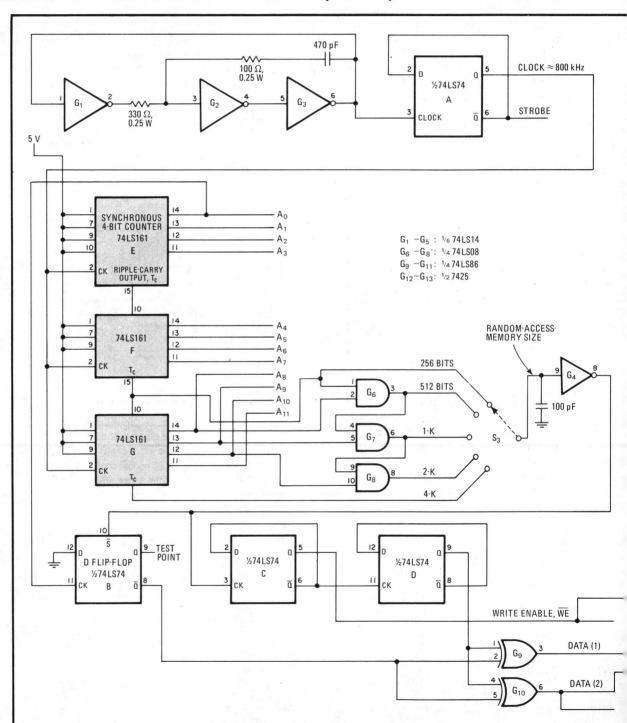

mine excessive supply current, slow timing, or other parametric faults.

The error-detecting circuitry (Fig. 1) basically consists of a clock, counters, a pattern generator, and exclusive-OR gates. Pinouts for bit-wide, nibble-wide, and byte-wide RAMs are shown in Fig. 2a, b, and c, respectively. The clock, composed of inverters G_1 through G_3 and flip-flop A, generates a 800-kilohertz square wave that is fed to counters E, F, and G. The counter outputs cycle through the addresses of the device under test. The pattern generator composed of flip-flops B, C, and D generates the $\overline{WE}$ signal and the data stream.

In each write cycle (Fig. 3) the data written into the bit site of the lowest and highest address is the complement of the data written into all other sites. Consequently, a faulty address buffer and decoder places this data in a site other than the lowest or highest and the error appears in the following read cycle.

The signals at the outputs of exclusive-OR gates J, K, and G are inverted and designated ERR_1, ERR_2, and ERR_3. This inverted ERR signal is fed to the one-shot L that is used as a pulse stretcher. The $\overline{WE}$ and STROBE inputs ensure that the one-shot is enabled only during a read cycle, when the RAM output data is valid. The RAM

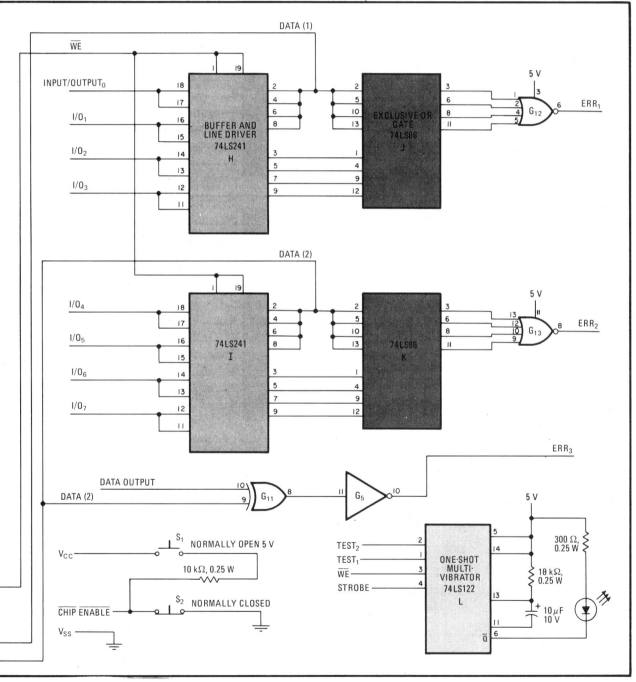

1. Tester. Error-detecting circuitry basically consists of a clock, counters, a pattern generator, and exclusive-OR gates. Inverters G_1 through G_3 and flip-flop A form the clock, which generates a 800-kHz square wave. Flip-flops B, C, and D constitute the pattern generator. Exclusive OR gates J, K, and G_{11} send their inverted ERR signals to one-shot L, used as a pulse stretcher.

2. Package. Pinouts for bit-wide, nibble-wide and byte-wide RAMs are as shown in (a), (b), and (c). The circuit also shows the connections of ERR signals to the appropriate pins of the one-shot L.

3. Waveforms. One test cycle consists of four test sequences in the following order: write DATA, read DATA, write $\overline{\text{DATA}}$, and read $\overline{\text{DATA}}$. XXF represents the highest address of the RAM under test. In each write cycle, the data written into the bit site of the lowest and highest address is the complement of the data written into all other sites. The Q output of flip-flop B is a test point.

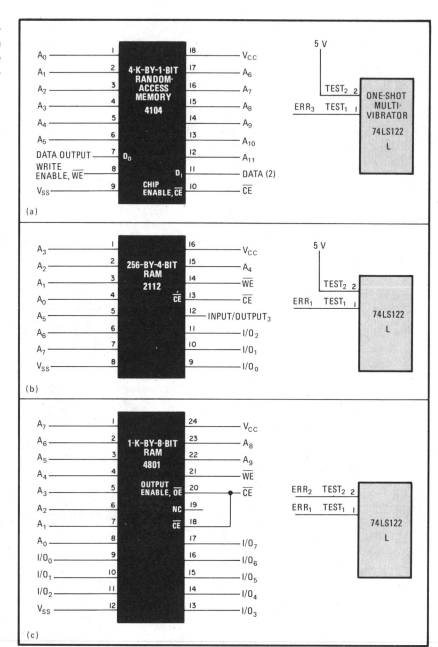

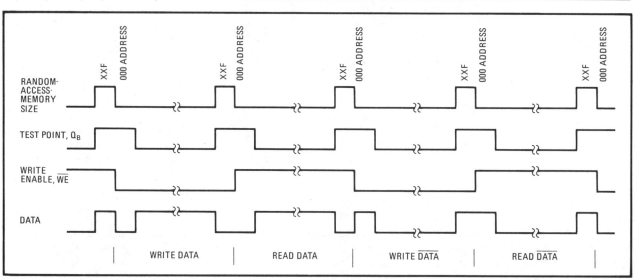

SIZE input of the pattern generator is connected to the appropriate size of the RAM by means of switch S_3.

For example, if a 2-K-by-4-bit RAM is to be tested, the 2-K line is connected to the RAM SIZE input. The ERR_1 line is then connected to the $TEST_1$ input of one-shot L, while the $TEST_2$ pin is held high. Finally V_{cc}, $\overline{WE}$ and $\overline{CE}$ are connected to the appropriate pins of the RAM under test and the tester is now ready for operation.

If a RAM is working properly, the light-emitting diode lights momentarily when S_1, which is normally open, is pressed. However, it stays on or flickers for a bad RAM. The $\overline{CE}$ line is tested by pressing normally closed S_2. If the LED does not light, there is faulty $\overline{CE}$ circuitry in the RAM that is keeping it enabled at all times. In addition, the LED will never light if a short exists between V_{cc} and V_{ss}, which pulls down the supply voltage. □

Continuity tester buzzes out breadboards's flaws

by Douglas Holberg
Texas Micro Engineering Inc., Austin, Texas

A continuity tester with a buzzer for performing breadboard checks is probably the least exotic but most needed tool in the lab. However, the performance of many buzz boxes falls short in a few areas, especially for checking boards with active components. This device, which uses two CA3096 transistor arrays, nine resistors, and one ceramic resonator, does not.

Ideally a continuity tester should draw no current when probes are open-circuited, draw low current through the probes when continuity is established, have no on-off switch, have a low-threshold sense resistance (100 ohms or less), disregard a silicon diode's pn junction as a valid response for continuity, and be compact or have a minimal amount of circuitry to increase reliability and reduce cost. This simple-to-build continuity tester has all of these desired characteristics.

Nine of the ten available transistors in the CA3096s are used and wired with the pins of the first device designated within color-coded circles. The wiring pins of the second device are indicated by gray circles.

Transistors Q_1 and Q_2 with resistors R_1 through R_5 form a bridge circuit that becomes active when current is allowed to pass through the probes. Q_1 provides bias for Q_2 and Q_3, with Q_3 providing the tail current for transistors Q_4 through Q_7 that comprise a differential amplifier. When the probes are open, Q_1 turns off, as does the remaining slaved circuitry. Thus, there is no current flow from the battery.

When the measured resistance across the probes becomes less than the value of R_5 (assuming a zero offset voltage in the differential amplifier), the output transistor, Q_8, turns on. Operating current to the oscillator circuit Q_9, R_6, R_7, and R_8 is then applied and the ceramic resonator will sound at a fundamental frequency of about 2 kilohertz.

The circuit draws about 18 milliamperes when sensing. About 4 mA flows through the probe when continuity is established. The probe current may be reduced by placing a low-value resistor between points A and B. □

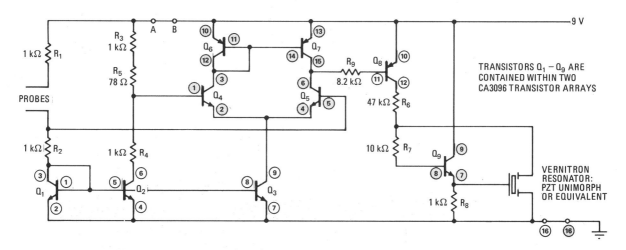

Check it out. A two-chip continuity checker for bench-testing breadboards has a low-threshold sense resistance and will not falsely indicate upon sensing pn diodes in the circuit under test. An oscillator in the tester resonates at 2 kHz. The circuit draws a maximum of 18 mA in operation, with only 4 mA flowing through the probe tips when continuity is established.

Hall-effect tachometer senses speed, direction of rotation

by Antonio L. Equizabal
Vancouver, British Columbia, Canada

Two Hall-effect sensors placed 90° apart and at an equal distance from the center of a rotating magnet-wheel assembly form the basis for this simple but reliable tachometer. As an added bonus, owing to the circuit's physical configuration, the direction of the rotating wheel can be detected as well.

As shown in the figure, small magnets are radially arranged on the wheel in an alternating north-to-south, south-to-north fashion. When the wheel rotates, each magnet periodically turns Hall sensors H_i and H_q either on or off magnetically, depending on the magnet's instantaneous orientation, thus generating output signals from the sensors that are in quadrature. The leading or lagging relationship of the H_i signal to the H_q signal, as detected by the 4013 flip-flop (working as a phase comparator), will indicate the wheel's direction of rotation. The frequency of the H_q signal (or for that matter, the H_i output) is directly related to the angular speed of the wheel and may be used to drive counters or frequency-to-voltage converters for linear servoamplifiers.

As for the constructional details, this tachometer's rotating magnet assembly is built using small cylindrical Alnico magnets, 3/16 inch in diameter by 1 in. long, glued into a machined polyvinyl chloride wheel with epoxy. The spacing (air gap) between the assembly and the Hall sensors will vary with the magnets' residual induction, and some tweaking is necessary. For GE519-1 magnets, a gap of 1/16 to 1/8 in. is satisfactory for the Texas Instrument's TL170C sensors used, which require a minimum of 350 gauss to change state.

The output frequency that is generated will be directly proportional to the number of magnets used. To ensure optimum tracking, however, their placement should meet the criterion that when one magnet is directly opposite sensor H_i, H_q should be midway between two other magnets of alternate polarity. Under that condition, the output frequency will be $f = NT/120$ hertz, where N is the number of magnets used, and T is the wheel speed in revolutions per minute. Thus, N will equal 6, 10, 14, . . . for magnets symmetrically arranged at angles of 60°, 36°, 25.7°, . . . respectively.

The Hall sensors are biased from a regulated 5-volt source for optimum performance. Input diode D_1 protects the circuit from any input-polarity reversal. The 1-kilohm load resistors bias the open-collector output of the sensors; with C_3 and C_4, they form an rf filter. Diodes D_2 through D_5 protect the input of the D flip-flop from inductive spikes. Noise reduction is ensured by the use of a shielded three-conductor cable. □

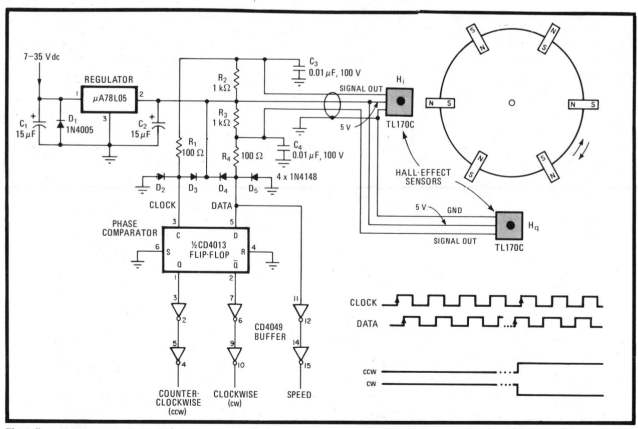

Fleet flux. Hall sensors, placed 90° apart, find wheel's angular speed by detecting instantaneous changes in field strength caused by rotating-magnet assembly. Output frequency produced by sensors is directly proportional to the number of magnets used. Leading and lagging relationship of sensors' quadrature output is also utilized to determine the direction of the rotating wheel.

Probing chip nodes with minimum disturbance

by Anthony M. Marques
The Charles Stark Draper Laboratory Inc., Cambridge, Mass.

Failure analysis of integrated circuits often requires the determination of voltages at internal nodes located on the silicon die. Making these measurements is difficult, however, because many circuit points will be sensitive to straightforward chip probing. The problems may be overcome, however, using two simple techniques.

Generally, internal node voltages can be measured with the aid of a probe consisting of a fine (0.15 mm in diameter), short, sharpened wire, a digital voltmeter, and a microscope to place the probe as desired. Typically, the capacitance of an isolated probe will be 10 pF, and the impedance of the combination probe and voltmeter facing any node will be 10 MΩ shunted by 100 pF.

When the probe is applied to certain points of especially sensitive circuits such as flip-flops and operational amplifiers, however, the node in question may be overloaded so that, in the case of the flip-flop, a state change may occur; or the output of an op amp may be set into oscillation. Just placing a probe near a sensitive point is often enough to upset the circuit.

One recognized but not widely used way to circumvent the difficulty is to place a precision voltage follower between the probe and the voltmeter, as shown in (a), in order to decrease the input capacitance presented to the node. In this circuit, it can be reduced to 30 pF or so. One drawback to this method, however, is that the open-circuit voltage on the probe will be a minimun of 4.6 volts, and this fixed potential may of itself upset the circuit under test.

The trial-and-error method shown in (b) avoids the problem. In this approach, a variable potential is externally applied to the probe in order to minimize the charge that flows into it when contact to the node is made. In simple terms, $\Delta V = Q/C$, where ΔV is the voltage difference between probe and node, and C is the capacitance of the probe/voltmeter system. Because C is fixed, charge, Q, can be minimized only by reducing ΔV, which is controlled by the external voltage source.

In use, the probe's applied voltage is varied until the probed circuit node does not change state (in the case of flip-flops) when the external potential is removed. As read by the digital voltmeter, little or no change in potential will be indicated. In the case of op amps, means must be found to check for oscillation during the measurement. Assuming no oscillation exists, the voltage at the node will be as indicated by the meter, once the external potential has been removed. Note that because the voltage difference between the source and node is in most cases less than the difference between the node and ground, less current will flow in establishing an equipotential condition.

With practice, node voltages can be successfully measured after two or three tries. It will be found that the external reference voltage will have to be brought to within about 0.5 volt of the potential at the node to be measured to prevent oscillation or a change of state at that node. This method is also useful in measuring ac node voltages, which requires node-synchronous ac signals. □

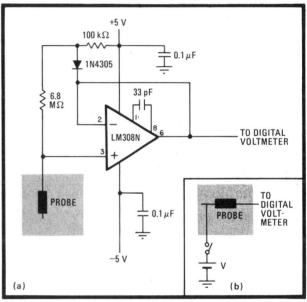

Balanced charges. High-impedance, precision voltage follower (a) reduces shunt capacitance across probed internal nodes of IC, despite inherent offset voltage (across probe) that may upset the circuit point under test. Applying external variable voltage to equalize probe and node potentials (b) solves problem by minimizing flow of charge into probe and permitting accurate voltage measurements.

Meter measures coatings magnetically

by Jules Schlesinger
UPA Technology Inc., Syosset, N. Y.

Utilizing the principle of reluctance to measure the thickness of nonmagnetic coatings (such as paint, rubber and metal plating) on a magnetic substrate, this meter will provide precise and stable readings over the range of 0 to 19.99 mils. The measurements are accurate to within ±1 microinch, which is adequate for most industrial requirements.

The heart of the instrument is its sensor, which is

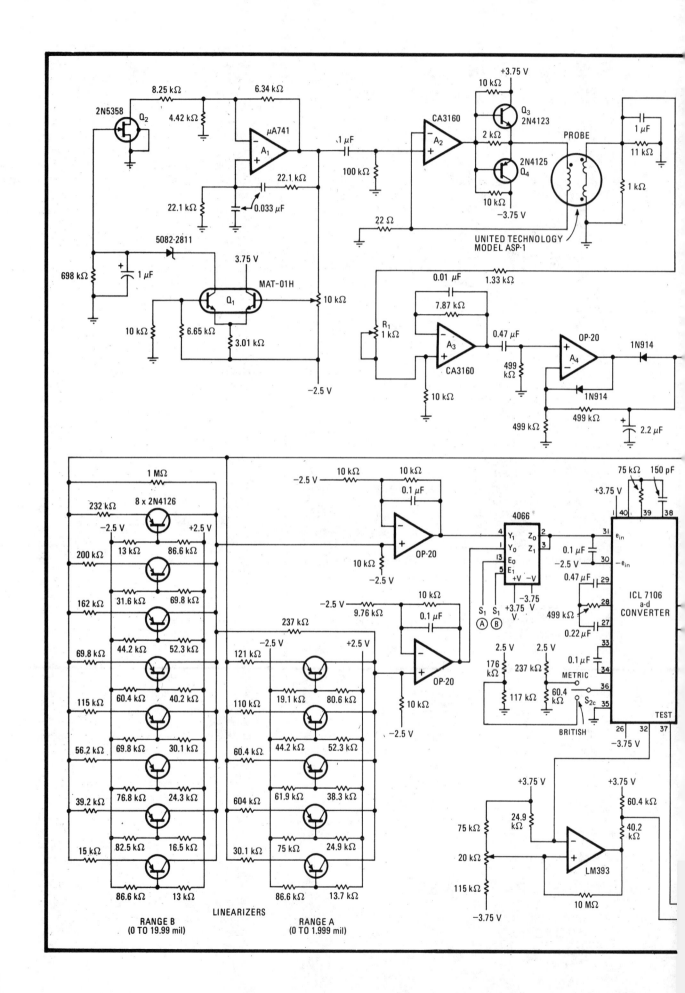

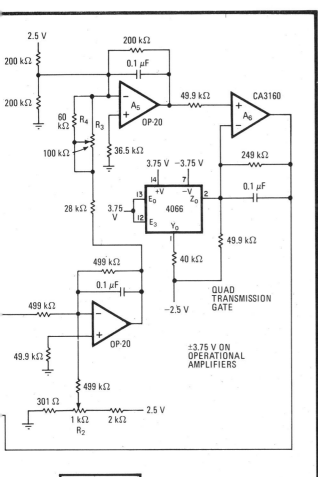

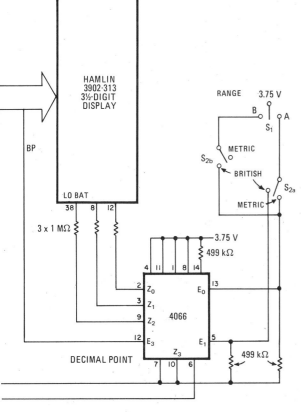

Field distance. Micrometer determines thickness of nonmagnetic coatings on magnetic material from difference in flux generated by series-opposing windings of output transformer, which constitutes the meter's probe. Unit's measurements are precise over the range of 0 to 19.99 mils and accurate to within ±1 microinch. At the flip of a switch, meter will display readings in micrometers.

basically a transformer having one primary and two secondary series-opposing coils and wound on a solid magnetic core. One end of the core protrudes from the sensor proper, becoming the probe that measures the contact surface. So, for the free-space condition, the output taken across the probe's secondary coils will cancel when it is excited by a low-frequency signal (below about 300 hertz to avoid generating eddy currents). Cancelation occurs because the output across one secondary coil is equal in amplitude and opposite in phase with the other. Otherwise, the output from the secondary coil closest to the contact surface will generate a voltage exceeding the potential produced by the other secondary winding. The magnitude of the root-mean-square difference will be a nonlinear function of the thickness of the magnetic substrate's coating.

In general operation, A_1 serves as the active element of a Wein-bridge oscillator for driving the probe. Q_1 and Q_2 in the feedback network provide automatic level control for stabilizing the oscillator's output. In addition, A_2, Q_3, and Q_4 provide constant-current drive for the meter's probe.

The output signal is fed to amplifier A_3 and ac-to-dc converter A_4 through potentiometer R_1, which is used to normalize, or zero, the probe for a steel contact surface. The signal is then passed through a chain of additional amplifiers, which provide a convenient means of setting system offset (R_2), the system gain for a known thickness (R_3 and range amplifiers A_5 and A_6), and system sensitivity (R_4).

A_6's output is then introduced into one of two transistor current switches, which provides a piecewise-linear fit for the probe's nonlinear output. After buffering and further amplification, the signal is presented to the 7106 a-d converter and its associated liquid-crystal display. Note that the reference voltage at pin 36 of the 7106 is selectable, so that the readings can be displayed either in British or metric units (that is, micrometers). The 4066 transmission gate and switches S_1 and S_2 are used to set the position of the decimal point. S_1 places the meter in the low-scale (0–1.999 mils, or 0–19.9 μm) or high-scale (0–19.99 mils, or 0–199 μm) positions. Switch S_2 selects the display mode—either British or metric. In terms of metric units, the meter's measurement resolution is ±0.1 μm.

Power for the equipment may be supplied by either eight nickel-cadmium batteries or an ac-line adapter/charger. In the latter case, it is suggested that integrated-circuit regulators be employed to provide stable ±3.75- and ±2.5-volt sources. Special attention should be paid to maximizing the suppression of line glitches and other transients. □

Charge integrator measures low-level currents

by Philip R. Gantt
Precision Monolithics Inc., Santa Clara, Calif.

Current cannot be measured accurately at picoampere levels by conventional measurement techniques. Yet in many cases, such small leakage currents can have a considerable effect on system performance. Measuring current at picoampere, and even femtoampere, levels is possible, however, using the method of charge integration. As shown here, a charge integrator can be easily used in conjunction with sample-and-hold amplifiers or analog-to-digital converters, or both, for such wide-ranging tasks as determining the leakage current at the gate of a field-effect transistor or measuring and recording small currents with the aid of an automated (microprocessor-based) test system.

The integrator, shown in Fig. 1, is one of the simplest circuits that can be configured with an operational amplifier. Its relative immunity to 60-hertz interference is due to the negative feedback provided through a capacitor (C). The accuracy of the basic circuit is limited mainly by two factors—the input bias current of the op amp and the user's ability to measure voltage with respect to time.

The output voltage of the integrator is given by:

$$V_o = -\frac{1}{C}\int_{t_1}^{t_2} I_{in}dt - \frac{1}{C}\int_{t_1}^{t_2} I_{b-}\,dt + V_{os}$$

where I_{in} is the driving current to the op amp, I_{b-} is the inverting input's bias current, and V_{os} is the amp's offset voltage. Thus, if $I_{in} = 0$, it is in principle possible for the circuit to measure the bias current I_{b-} or to accurately measure the leakage current, I_{in}, of an external device if $I_{in} >> I_{b-}$.

The integrator can be made practical for these low-level measurements by minimizing the leakage current of the entire circuit itself and by using an appropriate op amp. Installing Teflon standoffs at the op amp's inputs is recommended, and a low-leakage capacitor of the polystyrene type or, in high-temperature applications, a Teflon capacitor is necessary. Where printed-circuit boards are used, a high–surface-resistivity laminate such as Triazene is recommended to prevent leakage currents from flowing. The relay used to reset the integrator should be a high-impedance device of the glass reed type.

For measurements at several hundred picoamperes, a bipolar field-effect-transistor op amp should be used in conjunction with a 100-pF capacitor. Should measurements at lower levels be required, an electrometer amplifier with an input bias current of less than 0.1 pA is needed. Measurements down to 1 fA are possible using a discrete integrated-gate FET with the input biased at $V_{gs} = 0$ for low leakage.

Ideally, the input voltage as a function of time for a given input current will be a ramp. Since the V_{os} of the op amp is a constant and the terms relating to current are a change in voltage with respect to time, the offset voltage of the op amp is insignificant. Because the value of C will be known, $\Delta V/\Delta t$ can easily be observed on a storage scope or measured with the aid of a voltmeter and a stopwatch.

The anticipated current flow determines the capacitor value selected. A high-impedance reed relay placed across the integrating capacitor can be used to discharge it and reset the integrator for sequential measurements. For a 100-pA input current, a 100-pF capacitor would produce a voltage ramp of 1 volt/second, an easily measured value.

Because the noninverting input of the integrator is tied to ground, a potential at the input will normally cause the integrator to rail. To ensure against this, the input should be connected either to a point of zero potential

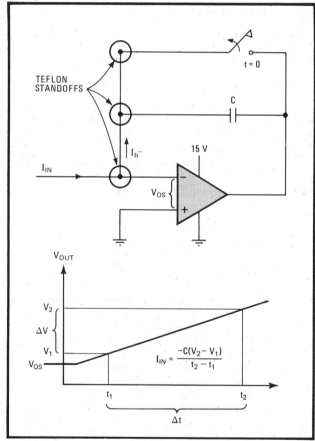

1. Current climb. Bi-FET integrator can measure femtoampere leakage currents. Particular attention is paid to reducing shunting errors by mounting feedback-to-input connections on Teflon standoffs and using a polystyrene or Teflon capacitor. With a 100-pF capacitor, 100-pA input current produces an easily measured ramp of 1 V/s.

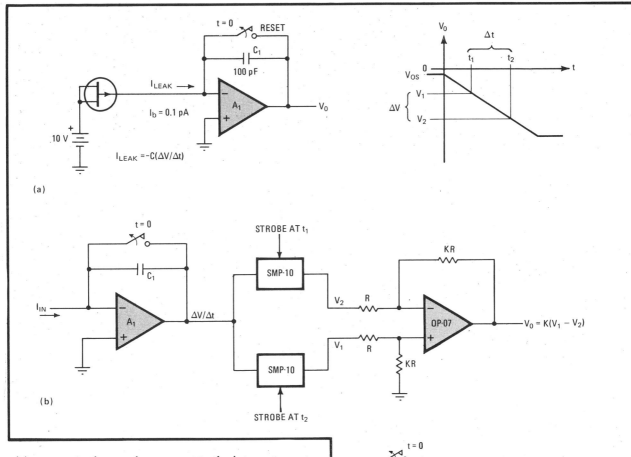

with respect to the supply common to the integrator or to a floating point that is electrically insulated from other potentials. The integrator input thus will always appear as a virtual ground with respect to the point to which it is connected.

In a typical application, such as measuring an FET's leakage current (Fig. 2a), the circuitry is virtually identical to that of the basic integrator. As noted, the slope of the linear ramp is directly proportional to the magnitude of the input current, I_{leak}.

An analog-to-digital converter will adapt the circuit of Fig. 2b to a microprocessor-based test system. As shown, sample-and-hold amplifiers convert I_{in} to a voltage at two specific time intervals. The OP–07 is biased as a differential amplifier and thus provides an output of $V_o = K(V_1 - V_2)$. With the aid of the a-d converter, therefore, the voltage corresponding to the input offset that is measured can be stored in memory. For greater accuracy, leakage error currents and droop errors can be calculated and then subtracted from measurements of device current. For measurements amid high noise, the integration period $(t_2 - t_1)$ should be a multiple of 60 Hz.

In Fig. 2c, three op-amp parameters are measured during one integrating time period. This circuit is especially suitable for measuring low I_b and low I_{os} values, as it avoids the resistive-type errors inherent in conventional measurements.

With the integrator in a voltage-follower configuration, the input offset voltage of the device under test can be measured by shorting the relays across the integrating capacitors. The V_{os} of the AD515 must be nulled to zero,

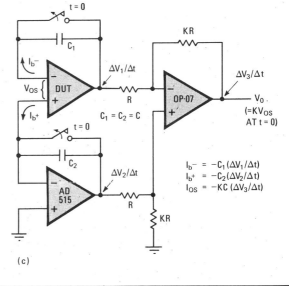

2. Applications. Charge integrator is ideal for measuring FET gate leakage (a). It is adapted to use with a microprocessor-based test system with sample-and-hold and differential amplifiers (b), which convert current into voltage so that measured values can be stored. More than one op amp offset parameter can be determined during a measurement (c). The circuit, which eliminates resistive-type errors, is well suited for measuring low currents.

or a relay having a low temperature-to-offset voltage dependence must be placed across the input of the device under test, in order to measure and subtract the errors on an automated basis. □

Thermistor gives thermocouple cold-junction compensation

by Harry L. Trietle,
Yellow Springs Instruments Co., Yellow Springs, Ohio

When measuring temperatures with thermocouples, either the reference or cold junction must be held at a constant temperature or electronic compensation for the temperature changes is needed at the cold junction. This cold-junction compensator (a) uses the latter approach and incorporates a precision thermistor having high sensitivity and accuracy.

The heart of the device is the thermistor bridge circuit

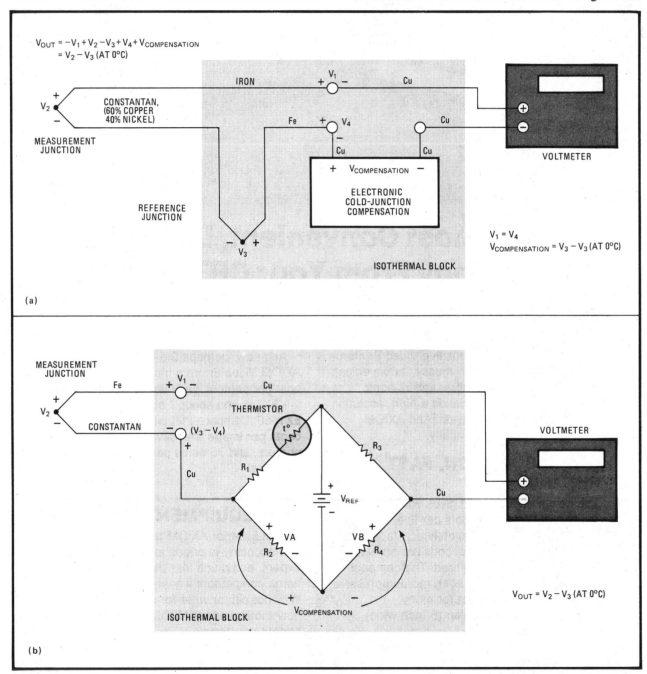

Compensation. The cold-junction compensator used in the general scheme (a) for thermocouple measurements is based on the thermistor bridge circuit (b). R_1 and R_2 must be selected so that the bridge output is near the reference junction's potential. The compensator's error is substantially reduced by using resistors and a bridge voltage accurate to 0.1%.

(b). The thermistor's nonlinearity characteristics are balanced by the resistance bridge, and with the proper choice of R_1 and R_2, the bridge's output closely matches that of the reference junction. The total series resistance of R_1 and R_2 necessary for a linearized bridge circuit is calculated by using the following equation:

$$R_1 + R_2 = \frac{R_{10}R_{25} + R_{25}R_{40} - 2R_{10}R_{40}}{R_{10} + R_{40} - 2R_{25}}$$

where R_{10}, R_{25}, and R_{40} are the thermistor's resistances at 10°, 25°, and 40° C, respectively.

With the resulting values of R_1 and R_2, the voltage drops across them between 10° and 25° C are the same as from 25° to 40° C. The average temperature sensitivity is then determined by calculating the voltage drop across $R_1 + R_2$ between 10° and 40° C. R_2 is selected so that the sensitivities of the thermocouple and V_A are equalized.

The table gives component values for the four most popular thermocouple types when a reference voltage of 1.000 volt is applied and a 10-kilohm precision thermistor is used in the bridge circuit. The thermistor's high sensitivity of about 4%/°C and ±0.2° C accuracy makes possible precise electrical compensation without calibration. The compensator's error may be substantially reduced by using resistors and a bridge voltage in the circuit that are accurate to 0.1%. The error in this case is less than ±0.55°C between 10°C and 40°C. □

RESISTOR VALUES FOR COMPENSATING THERMOCOUPLE COLD-JUNCTIONS

Thermocouple	R_2 (±0.1%)	R_4 (±0.1%)
Type E	46.03 Ω	23.66 Ω
Type J	39.02 Ω	19.88 Ω
Type K	30.51 Ω	15.53 Ω
Type T	30.66 Ω	15.86 Ω

Thermistor = YSI 44006 (18.79 kΩ at 10°C, 10.00 kΩ at 25°C, 5.592 kΩ at 40°)
V_{REF} = 1.000 V ±0.1%, R_1 = 7.15 kΩ ±0.1%, R_3 = 20.0 kΩ ±0.1%

Five-chip meter measures impedances ratiometrically

by N. E. Hadzidakis
Athens, Greece

Measuring both inductance and capacitance usually requires either a manual bridge, which is difficult to use, or a digital bridge, which is expensive. Recently, inexpensive hand-held capacitance meters have appeared on the market, but they employ a time-to-charge technique that cannot be applied to the measurement of inductance. The circuit shown here, however, utilizes a ratiometric method that is suitable for both types of measurement. Its only disadvantage is the requirement for one calibrated reference component per range. Still, it is inexpensive and easy to use.

Generally, the potentials across a reference and test inductor or capacitor, which are dependent on the frequency of the 8038 square-wave driving source, are applied to two ac-to-dc converters built around CA3130 operational amplifiers. The converters' output is then compared at the ICL7107 ratiometric converter. Because the value of the reference inductor or capacitor is a multiple of 10, the value of the test element can be read directly from the display.

In the case of measuring inductors, it can easily be shown that:

$$L_x = \frac{L_{ref}e}{2+1/e} + \frac{L^2_{ref}e^2(e+1)}{(2e+1) - R^2/4\pi^2 f^2}$$

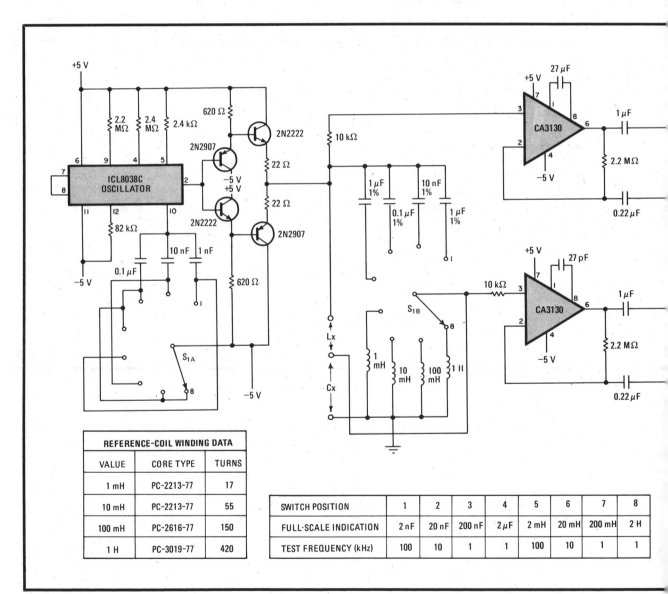

REFERENCE-COIL WINDING DATA		
VALUE	CORE TYPE	TURNS
1 mH	PC-2213-77	17
10 mH	PC-2213-77	55
100 mH	PC-2616-77	150
1 H	PC-3019-77	420

SWITCH POSITION	1	2	3	4	5	6	7	8
FULL-SCALE INDICATION	2 nF	20 nF	200 nF	2 µF	2 mH	20 mH	200 mH	2 H
TEST FREQUENCY (kHz)	100	10	1	1	100	10	1	1

where R is the dc resistance of L_x, f is the test frequency, and e is the display indication divided by 1,000 and with the decimal point disregarded. Because f is varied appropriately with range and the Q of L_{ref} is greater than 100 at 1 kilohertz—coils are hand-wound on Amidon pot cores using the largest-diameter wire possible (see table)—the equation given above reduces to $L_x = L_{ref}e$ for almost all practical measurements. For example, the error in measuring a 1-millihenry inductor having a Q of only 0.1 at 1 kHz will be less than 1%.

The concept of measuring capacitance is similar. Potentiometers R_1, R_2, and R_3 are used to cancel out the effect of parasitic capacitance at the converter's input terminals so that the display will read zero with no test capacitor connected. When that is done, the display will read $1,000C_x/C_{ref}$.

Construction is not critical. The only exception to that is the wiring to the input terminals, which should be kept reasonably short.

The calibration procedure for the circuit is straightforward. S_1 is set to 2 microfarads full scale and R_1 is adjusted for a zero display reading. Then S_1 is placed in the 20-nanofarad full-scale position and R_2 adjusted for a zero reading. Finally, the switch is set in the 2-nF full-scale position and R_3 adjusted for a zero display reading. In the prototype tested, no readjustment was necessary over a nine-month period of normal use, and accuracy was maintained to within 1%. □

Coils and capacitors. Ratiometric meter, easier to use than manual bridges and less expensive than digital types, measures inductances of reasonable Q over the 2-mH-to-2-H range and capacitances over the 2-nF-to-2-μF range to within 1%. Calibration, required for capacitance measurements, is easy and maintains long-term stability.

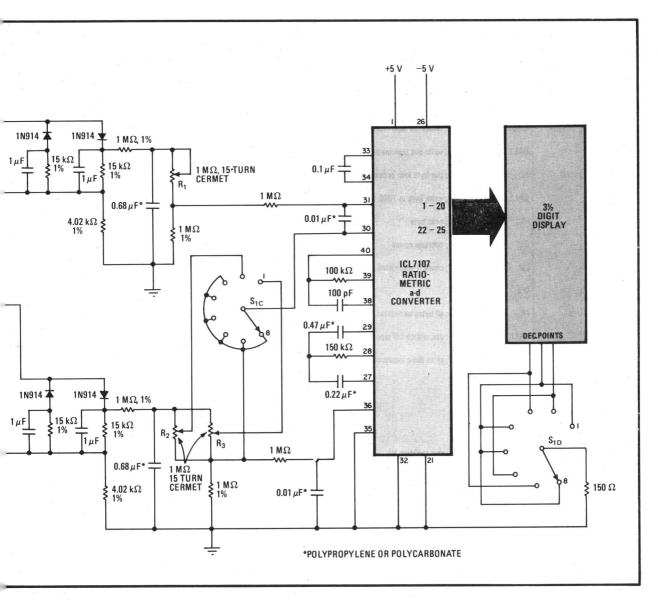

Hall compass points digitally to headings

by Gary Steinbaugh
Owens/Corning Fiberglas Corp., Technical Center, Granville, Ohio

Although the time-honored needle-type magnetic compass is very simple, inexpensive, and most efficient in terms of power consumption, two Hall-effect sensors and a few chips can be combined to make a compass with some important advantages of its own. Among them are direct digital readout of the magnetic heading, remote-sensing capability, the ability to interface with navigational computers, and the elimination of errors caused by acceleration or tuning of the body to which the compass is secured.

The general block diagram of such an instrument is shown in Fig. 1. Here, a 36-kilohertz system clock advances a 360-step counter, which in turn drives a sine and cosine digital-to-analog converter. The resulting 100-hertz outputs from the converter, $E\sin \omega t$ and $E\cos \omega t$, are then transformed into a current and introduced to the Hall-effect sensors, which are placed at right angles to each other. As a result, the output voltage of the west-east sensor is $kIB\sin \omega t \sin \theta$, and that of the north-south sensor $kIB\cos \omega t \cos \theta$, where I is the current corresponding to input voltage E, B is the strength of the incoming magnetic field, and θ is the angle of the field as measured with respect to the earth's magnetic north pole.

These two signals are then summed, with the result being $kIB\cos(\omega t - \theta)$. As seen, this voltage reaches its maximum value when $\theta = \omega t$. The peak is detected and used to latch the contents of the stepped counter at that instant, yielding an indication of the bearing θ (see excitation-response curves for several representative headings).

Figure 2 shows examples of how the individual building blocks of the system may be implemented. The sine and cosine converter (Fig. 2a) has a rather low resolution, but it is easily constructed and may be sufficient for simple applications. For 1° resolution, the designer is advised to use a microprocessor-based system, using lookup tables stored in read-only memory to drive two conventional d-a converters.

The voltage-to-current amplifier block may be combined with a special Hall sensor as shown in Fig. 2b to provide superior performance. Of particular interest is the F. W. Bell BH-850 flux-concentrating Hall-effect device, which will provide ±10 millivolts for a nominal ±½ gauss field (typical of the magnetic field encountered in North America), and an excitation current of 200 mA. This performance is far superior to that of the standard sensor, which will generate potentials only in the microvolt region for the same magnetic field strength. Although post-amplification with such a standard sensor is feasible, in practice the flux-concentrator type provides more linear performance and generates less noise.

The input and output leads of each Hall device must be fully isolated from each other for proper operation; this is best achieved by operating the device in the floating mode and using a differential amplifier (Fig. 2c). The input and output impedances of the Hall-effect sensors are low, and therefore a high-input impedance

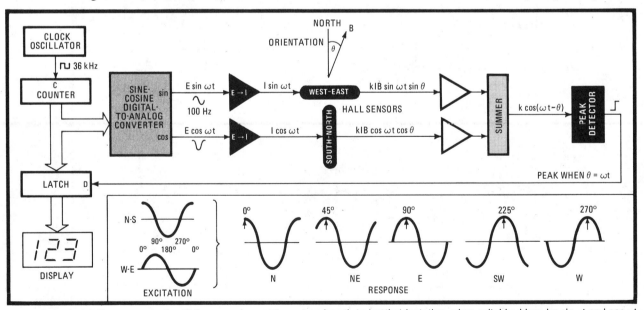

1. Heading north? Flux-concentrating Hall sensors in west-to-east and north-to-south orientation, when suitably driven by $\sin \omega t$ and $\cos \omega t$ sources, combine to indicate magnetic heading. When bearing $\theta = \omega t$, peak detector latches counter to provide digital readout of θ.

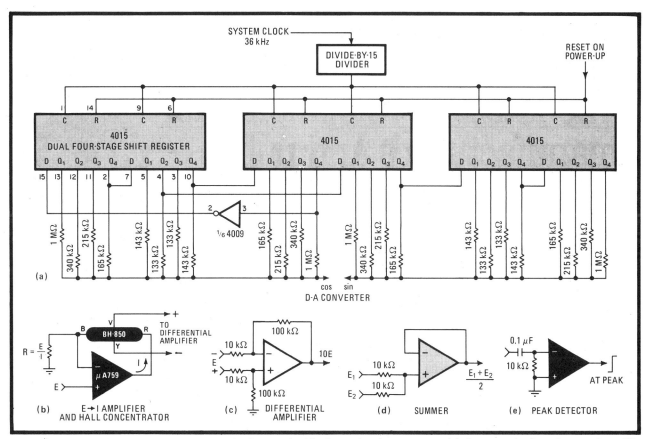

2. Implementation. Shift-register–type counters implement circuit's sine and cosine generators (a). Sensor and voltage-to-current amplifier may be united on one stage (b), with differential amplifiers (c) following. Summer (d) and peak detector (e) are conventional.

amplifier is not required, although the amplifier's common-mode rejection should be high. A factory-trimmed instrumentation amplifier is recommended.

The summer (Fig. 2d) and peak-detector (Fig. 2e) circuits are straightforward. Note that the peak detector is actually a differentiator and zero-crossing circuit and does not hold the peak voltage level as conventional detectors do.

In operation, the sensors should remain parallel to the earth's surface. In mobile applications, a gyroscopic platform may be required or another sensor used to permit measurement updates when the instrument passes through its level position. In fact, a third Hall sensor can be mounted perpendicular to the first two and its output used to correct the measured values for any tilt. □

Power-up relays prevent meter from pinning

by Michael Bozoian
Ann Arbor, Mich.

Sensitive microammeters with d'Arsonval movements are still manufactured and used widely today, but surprisingly, there has been little attempt to correct one defect in their design—they are still very prone to pointer damage from input-signal overload and turn-on/turn-off transients. Although ways of protecting the meter movement from input signals of excessive magnitude are well known and universally applied, no convenient means of preventing the pointer from slamming against the full-scale stop during power-up or power-off conditions has so far been introduced or suggested in the literature. However, the problem may be easily solved by the use of a 555 timer and two relays to place a protective shunt across the meter during these periods.

Basically, the 555 timer closes reed relay A's normally open contacts on power up and puts shunt resistor R across the meter for 5 or 6 seconds until the turn-on transients have subsided, as shown in the figure. The normally closed contacts of relay B are also opened at this time.

On power-down, relay B reintroduces the shunt to protect the meter from turn-off transients. Such a scheme is more effective than placing a diode across the meter, as is often done and is much more elegant and less bothersome than manually activating an auxiliary mechanical switch for placing R across the meter each time it is used.

R has been selected for a meter movement having a full-scale output of 200 microamperes and an internal resistance of 1,400 ohms. The complete circuit may be mounted on a 2-by-2¼-inch printed-circuit board. The only design precaution is to ensure that relay B is energized from a source that has a fast decay time during power-off conditions. Here, the voltage has been tapped from the meter's power-supply rectifier. □

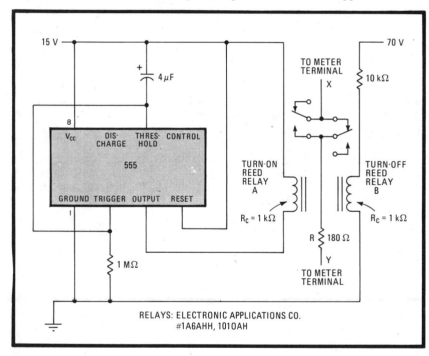

Shunted. Reed relays and 555 timer prevent d'Arsonval movement from slamming against microammeter's full-scale stop during power-up and power-down conditions by introducing shunt resistor across meter terminals until transients die out. Method does not degrade meter's accuracy or its transient response to input signals.

8048 system generates alphanumerics for CRT display

by J. C. Korta
Matra Harris Semiconductor, Paris, France

A simple controller built around the popular 8048/8748 series of 8-bit microcomputers can generate an alphanumeric display on an oscilloscope. It is an inexpensive way of adding a numerical readout to a frequency counter, a logic analyzer, a voltmeter, or a transmission link for handling plain-text messages between two receivers using cathode-ray-tube monitors at either end of a communications path. Owing to the flexibility of the

ASSEMBLY-LANGUAGE LISTING: 8048 REFRESH ROUTINE FOR GENERATION OF ALPHANUMERICS

```
                                    ;R1   CHARACTER COUNTER
                                    ;R2   NB OF CHARACTERS IN THE BUFFER
                                    ;R7   BUFFER START

30   99 FE   NXTLNE:  ANDL P1, #0FEH  ;SEND SYNC PULSE
32   89 01            ORL P1, #01H    ;
34   99 FE            ANDL P1, #0FEH  ;
36   FA               MOV A, R2       ;SET CHARACTER COUNTER
37   A9               MOV R1, A       ;TO THE NUMBER OF CHARACTERS
38   F9      NEXTCAR: MOV A, R1       ;IN THE BUFFER
39   6F               ADD A, R7       ;
3A   A8               MOV R0, A       ;SET DISPLAY POINTER TO BUFFER START
3B   14 80            CALL CHDISP     ;DISPLAY THE CHARACTER
3D   E9 38            DJNZ R1, NEXTCAR ;LOOP UNTIL END OF LINE
3E   04 30            JP NXTLNE       ;

                                    ;R0  DISPLAY POINTER
                                    ;R3  COLUMN POINTER
                                    ;R4  D/A COUNTER
                                    ;R5  DOT COUNTER
                                    ;R6  COLUMN COUNTER

80   BB 00   CHDISP:  MOV R3, #00H    ;CLEAR COLUMN POINTER
82   BE 08            MOV R6, #08H    ;SET COLUMN COUNTER
84   BC 00   NEXTCOL: MOV R4, #00H    ;CLEAR D/A COUNTER
86   BD 00            MOV R5, #08H    ;SET DOT COUNTER
88   54 F0            CALL COLUMN     ;LOAD COLUMN DOTS
8A   F7      NEXTDOT: RLC A           ;TEST IF DOT OR BLANK
8B   F6 98            JPC DOT         ;
8D   89 02   NODOT:   ORL P1, #02H    ;OUT "1" TO Z-AXIS ) DOT OFF
8F   1C               INC R4          ;INC D/A COUNTER
90   2C               XCH A, R4       ;
91   3A               OUT P2, A       ;OUT D/A COUNTER
92   2C               XCH A, R4       ;RESTORE COLUMN DOTS
93   ED 8A            DJNZ R5, NEXTDOT ;GET NEXT DOT
95   EE 84            DJNZ R6, NEXTCOL ;IF END, GET NEXT COLUMN
97   83               RET             ;RETURN IF CHARACTER OUT
98   1C      DOT:     INC R4          ;
99   2C               XCH A, R4       ;
9A   3A               OUT P2, A       ;
9B   2C               XCH A, R4       ;
9C   99 FC            ANDL P1, #0FCH  ;OUT "0" TO Z-AXIS ) DOT ON
9E   ED 8A            DJNZ R5, NEXTDOT ;
A0   EE 84            DJNZ R6, NEXTCOL ;
A2   83               RET             ;
```

ROM-ACCESS CODING

```
2F0  F0      COLUMN:  MOV A, &R0      ;GET CHARACTER ASCII
2F1  97               CLR CY          ;
2F2  F7               RLCA            ;MULTIPLY THE CHARACTER ASCII
2F3  F7               RLCA            ;BY 8 TO GET THE
2F4  F7               RLCA            ;START ADDRESS
2F5  4B               ORL A, R3       ;GET CURRENT COLUMN ADDRESS
2F6  1B               INC R3          ;INCREMENT COLUMN POINTER
2F7  F6 FB            JPC PG3         ;TEST IF CHARACTER IN PAGE 3
2F9  A3               MOVP A, @A      ;FETCH COLUMN FROM PAGE 2
2FA  83               RET             ;
2FB  E3      PG3:     MOVP3 A, @A     ;FETCH COLUMN FROM PAGE 3
2FC  83               RET             ;
```

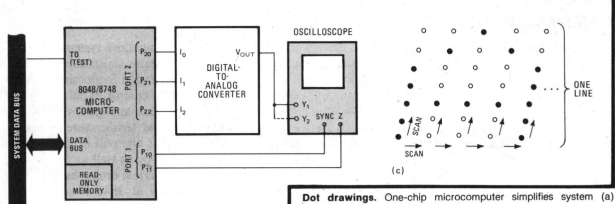

Dot drawings. One-chip microcomputer simplifies system (a) needed to generate alphanumeric symbols on oscilloscope screen. User-defined characters are stored (b) in internal read-only memory as is scanning routine (see program). Scanning of typical character, A, is done dot by dot, column by column (c).

ROM CHARACTER GENERATION			
Character	Hexadecimal address	Character	Hexadecimal address
@	200	(blank)	300
A	208	!	308
B	210	"	310
C	218	#	318
D	220	$	320
E	228	%	328
F	230	&	330
G	238	'	338
H	240	(	340
I	248	)	348
J	250	*	350
K	258	+	358
L	260	,	360
M	268	−	368
N	270	.	370
O	278	/	378
P	280	0	380
Q	288	1	388
R	290	2	390
S	298	3	398
T	2A0	4	3A0
U	2A8	5	3A8
V	2B0	6	3B0
W	2B8	7	3B8
X	2C0	8	3C0
Y	2C8	9	3C8
Z	2D0	:	3D0
[	2D8	;	3D8
\	2E0	<	3E0
]	2E8	=	3E8
(not used)	2F0	>	3F0
(not used)	2F8	?	3F8

(b)

8048/8748, only one external component is required by the system—a simple 3-bit digital-to-analog converter.

In general operation (a), the microcomputer and d-a converter present the dot matrix of any user-defined character to the scope. The matrix is stored in the microcomputer's read-only memory (b). Under software control (see program), line P_{10} synchronizes the scope sweep to each line to be displayed, with line P_{11} modulating the Z (intensity) axis of the scope in order to display the dot matrix of the character (c), in this case the letter A. Note that in typical applications, the erasable programmable ROM version of the 8048—that is, the 8748—would be used for character storage, the ROM of the 8048 being a mask-programmed device.

Scanning is done dot by dot, column to column, as seen in the program. The character table extends from address 200_{16} to $3FF_{16}$ for the 62-symbol array, with the exception of the space between addresses $2F0_{16}$ and $2FF_{16}$ that are needed to access the ROM. The main input/output routine and the refresh subroutine may occupy the space from 00_{16} to $1FF_{16}$, as shown.

The 8048's data bus will accept data or commands from any external device. Data may be stored in the device's internal random-access memory for later display. For maximum flexibility, a few internal registers should be used to store control words that specify certain modes of operation. For example, the register pointing to the beginning of the data buffer can be incremented to provide the scroll of the display.

As shown, the interface provides up to 32 characters in one line. A two-line display containing 16 characters per line may be implemented by connecting the Y_1 and Y_2 inputs of the scope to the output of the d-a converter and by switching to the alternate sweep mode of the scope. It will be noted that although a raster-scanning technique would permit use of a standard character-generator chip, the software would actually be more complicated, system speed would be reduced, and fewer characters per line would be displayed. Overall chip count would also be increased. □

Processed seismometer signals yield ground acceleration

by Thomas D. Roberts
Department of Electrical Engineering, University of Alaska, Fairbanks

The signals generated by a conventional electromagnetic seismometer can yield ground acceleration when transformed by this real-time signal processor (Fig. 1). Normally the output of the seismometer, which is proportional to the relative speed between its case and the suspended seismic mass, bears little resemblance to the input excitation. Thus the ideal but unlikely case of sinusoidal ground acceleration is the only condition in which the seismometer output is proportional to ground acceleration.

However, the output (Fig. 2a) of a typical short-period seismometer due to a step excitation has a natural frequency of 1 hertz with a damping factor of 0.4, and this can be obtained by removing a known weight from the suspended seismic mass of the seismometer. This signal is analyzed by the signal processor to enable the system to track the ground acceleration accurately.

Using the concept presented by Berckhemer and Schneider,[1] the processor's differential equation is:

$$K d^3y/dt^3 = d^2e/dt^2 + 2\zeta\omega_n \, de/dt + \omega_n^2 e$$

where
K = the gain constant
y = the vertical displacement of the seismic mass
ω_n = the natural frequency
e = the voltage output of the seismometer
ζ = its damping factor.

The processor provides the exact reciprocal of the seismometer transfer function. The true ground acceleration parallel to the seismometer axis is obtained by integration of the above equation, which yields:

$$\frac{d^2y}{dt^2} = \frac{1}{K}\left\{\frac{de}{dt} + 2\zeta\omega_n e + \omega^2{}_n \int_0^t e\,dt\right\}$$

This relationship is determined by the processor with

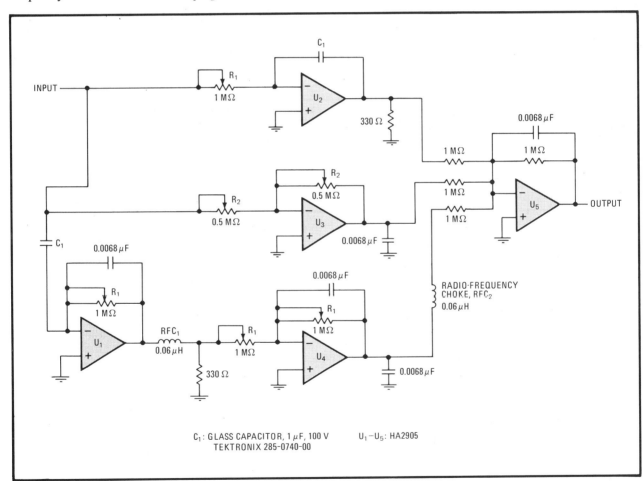

1. Seismic. This real-time signal processor tracks ground acceleration accurately. For accuracy and high performance, the circuit uses chopper-stabilized, high-input impedance operational amplifiers with excellent offset drift characteristics.

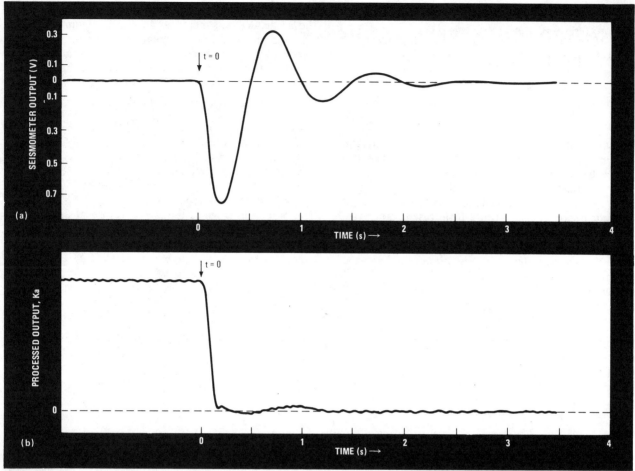

2. Acceleration. The waveform in (a) is the response of a short-period seismometer to a step-function input. The step response of the processed seismometer signal (b) resembles the step input excitation. The rise time is about 120 milliseconds, with 3% maximum overshoot.

satisfactory accuracy only if it uses modern, monolithic chopper-stabilized operational amplifiers with high input impedance and excellent offset characteristics. Op amps U_1–U_5 employ Harris Semiconductor's HA-2905.

The circuit malfunctions without the use of the recommended op amps. The right values for resistances R_1 and R_2 are obtained by trial and error for a given model of seismometer. The processed seismometer output is a step signal of high quality with a maximum overshoot of about 3% (Fig. 2b). This output response is due to a step input excitation of the seismometer. □

References
1. H. Berckhemer and G. Schneider, "Near Earthquakes Recorded with Long Period Seismometers," Bulletin of the Seismological Society of America, 1964, pp. 54 and 973.

Latch grabs glitches for waveform recorder

by David M. Smith
Storage Technology Corp., Louisville, Colo.

Especially at low sampling rates, recorders that sample asynchronously are very likely to miss glitches. Digital recorders in such circumstances use a latch to catch transients. This circuit is the analog equivalent of a latch. It works with most waveform recorders and is particularly effective when united with the Biomation series of machines. Suitably modified, it can work at frequencies as high as 50 megahertz.

An RC differentiator, a window detector, and a track-and-hold amplifier make up the latch, which should be installed between the recorder's input amplifier and its analog-to-digital converter, as shown in the right-hand figure. In operation, input signals differentiated by R_1C_1 that exceed a preset amplitude will reset the R-S flip-flop formed by two cross-coupled 7427 NOR gates, as can be seen in the left-hand figure. Latch sensitivity is determined by R_1 and is adjustable from 0 to ±0.3 volt.

In the reset state, the flip-flop turns on transistor Q_4, thus open-circuiting the diode bridge so that the instantaneous potential appearing at point A will be stored across capacitor C. When the system clock from the recorder's a-d converter arrives to initiate the next sampling interval, the flip-flop is cleared by means of the one-shot multivibrator (formed by the 7408 AND gate and R_2C_2), the diode bridge becomes functional, and the voltage across capacitor C will follow the variations of e_{in} until the next glitch. Provision is made to disable the

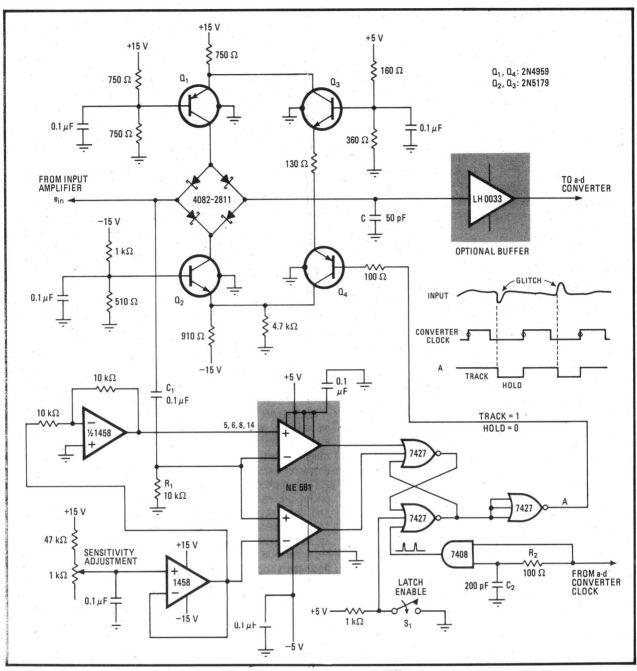

Capturing transients. Flip-flop–controlled sample-and-hold circuit adapts analog-waveform recorder for latch mode, permitting it to detect glitches despite a low sampling rate. With suitable logic-family substitutions and minimization of lead lengths, circuit functions of up to 50 MHz.

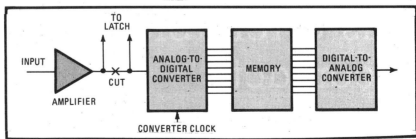

latch by means of switch S_1 to enable sampling of incoming waveforms in the traditional manner.

As for the circuit's operating limitations, the major factors determining its capture ability are the speed of the NE581 comparators and flip-flop and the turn-on times of Q_3 and Q_4. For the higher speeds, emitter-coupled-logic devices should be used and wiring lengths should be minimized. The 0.1-microfarad capacitor, C_1, should be reduced to 1,000 picofarads or so. Also, the addition of a high-speed, high-impedance buffer may be required, as shown, if the input impedance of the a-d converter is lower than 1 megohm at the operating frequency of interest.

Transistor probe simplifies solid-state gaussmeter

by Shanker Lal Agrawal and Rama Swami
Banaras Hindu University, Department of Physics, Varanasi, India

As a result of the subatomic energy exchanges that take place in many semiconductors because of particle-wave interaction, the electrical characteristics of the unijunction-transistor oscillator can be significantly changed by an external magnetic field. This property makes the low-cost unijunction transistor ideal for use as a probe in a gaussmeter or other flux-measuring instrument. Although it is not as linear, is not as easily calibrated, and does not provide the readout precision of some of the more elegant designs,[1] this circuit is simpler, just as sensitive, and virtually as accurate.

As shown, the gaussmeter is based on the comparator technique, wherein the frequency of the relaxation oscillator-probe is matched against a reference whose nominal frequency is about 400 hertz. Both generate positive-going spikes, which are lengthened by the pulse-stretching 74121 one-shot multivibrators. The NAND gate serves as a digital comparator, turning off the light-emitting diode when frequency $f_1 = f_2$.

In operation, the frequency of the reference oscillator is adjusted by resistor R until f_2 equals the free-running (free-field) frequency of the probe. Placing the oscillator-probe within the field to be measured will cause its frequency to change; potentiometer R in the reference must then be adjusted until the difference between f_1 and f_2 is minimized. The change in resistance of R from its nominal position may then be related to the strength of the magnetic field with the aid of the unit's individual calibration curve, which is shown at the right side of the figure.

As for calibration, at least three standard magnets will be required over the range of 0 to 1.5 weber per square meter, which is the range of the instrument. The circuit response is not likely to be linear above 1.0 wb/m², and depending on the particular UJT used, the standard-marker points will vary considerably as a function of R. Still, the circuit will hold calibration and will serve well in most general-purpose applications. □

References
1. Henno Normet, "Hall-probe adapter converts DMM into gaussmeter," *Electronics*, Jan. 3, 1980, p. 179.

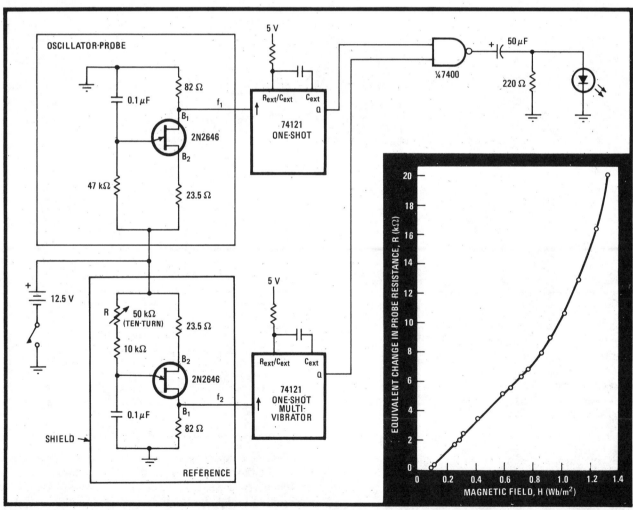

Flux finding. The low cost of the rudimentary unijunction transistor, which is sensitive to externally applied magnetic fields, makes it ideal for use in comparison-type gaussmeters. The circuit range is 0 to 1.5 Wb/m², with circuit response linear to 1.0 Wb/m².

Digital phase meter updates measurement each cycle

by R. E. S. Abdel-Aal
Department of Electronic Science, University of Strathclyde, Glasgow, Scotland

Because this meter measures the phase delay between two low-frequency square waves once every cycle, it is useful in applications where instantaneous readings of this delay are continuously required. The circuit resolution is within 1% for signal frequencies of up to 250 kilohertz.

Generally, the meter counts the number of pulses of a 25-megahertz clock for a time equal to the phase delay between the two incoming waveforms. Then it strobes the measured value into output latches once a cycle. The result is a continuously updated value expressed as a 15-bit binary number plus a sign bit.

To achieve this, the cycle is viewed as one that varies from plus to minus 180°. By using only one half of the cycle for measurement, the circuit is free during the other half to store the results in the output latches and to clear the phase counters for the next measurement.

The circuit automatically determines which of the signals is to be the reference, with the phase delay measured from the rising edge of the leading signal to the rising edge of the lagging waveform. The falling edge of the reference serves as the latching signal and to set up the counters for the next cycle.

In operation, the two incoming signals, A and B, are applied to two gates of A_1. Here, the complemented signals $\overline{A}$ and $\overline{B}$ are obtained with negligible differential delay. The other two gates in the chip generate gating signals corresponding to $A\overline{B}$ and $\overline{A}B$. Flip-flop A_2 determines which input signal is the reference.

If A leads B, then the Q output of A_2 goes low and gating signal $A\overline{B}$, together with input signal A, drives the 74LS157 selector chip, A_3. Otherwise, gating signal $\overline{A}B$ together with input signal B will be selected.

The selected phase-gating signal is used to enable a chain of synchronous counters, A_4–A_7, which are driven from a crystal-controlled 25-MHz clock built around three inverters in A_8. When the phase-gating signal drops, A_4–A_7 stop counting, holding their final result, which indicates the phase delay, at their parallel outputs. Following this, a short pulse from one-shot A_{9A} latches the results of the count in A_{10} and A_{11}. Then the pulse-counter chain is cleared by a second pulse from A_{9B}. To ensure a proper count and store cycle, the sum of the widths of the two short pulses should be less than half the period of the highest-frequency input signal. Also, the short pulse used to clear the counters should be greater than the clock period.

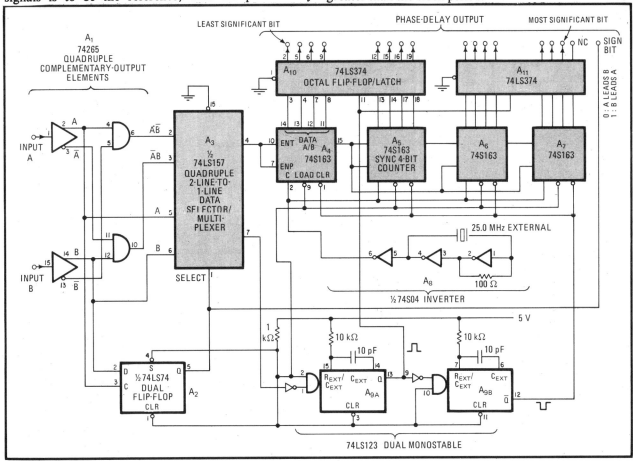

Instantaneous. Circuit continuously compares phases of two incoming square waves, providing a 15-bit and plus-sign output that has a resolution of $(f_{kHz}/250)\%$. With a 25-MHz clock, the practical upper frequency limits that can be handled for incoming signals is 250 kHz, with lowest-frequency boundaries being about 400 Hz. Lower limits can be reduced further by decreasing the clock frequency.

The upper limit on the frequency of the input signals is set by the resolution of the phase measurement that can be tolerated. With this circuit, the resolution is given by (f/250)%, where f is the frequency in kilohertz.

The lower limit of the signal frequency is set by the overflow of the phase counters before the end of half a cycle of the input signal (that is, the maximum phase delay measured). With a 25-MHz clock and a 15-bit binary number representing the magnitude of the phase (excluding the sign bit), the minimum input frequency will be $25(10^6)/(2(2^{15}-1)) = 381$ hertz. At low input frequencies, however, a lower-frequency clock can be used while maintaining good resolution, and thus the frequency limit can be brought down even further. □

Phase-locked loops replace precision component bridge

by Vilas Jagtap and Vidyut Bapat
Peico Electronics and Electricals Ltd., Pune, India

For accuracy and repeatability in measuring passive components, a resistor-capacitor bridge is difficult to surpass. But its one drawback is its prohibitively high cost. An inexpensive alternative is a circuit that uses two off-the-shelf phase-locked loops to perform this function accurately to within 0.1% and with a resolution of 0.01%.

As shown in the circuit, which is configured to measure capacitance, the 565 phase-locked loop, A_1, generates a frequency, f_{in}, corresponding to the component under test, C_x. This signal is then brought to the input of a second loop, A_2, which itself generates a reference frequency, f_{ref}, corresponding to component C_s. The output of A_2 then produces a signal proportional to the difference frequency. The difference frequency, $f_{in} - f_{ref}$, is amplified by the 530 operational amplifier, with the resulting signal applied to a zero-center meter that can be calibrated in terms of the percentage difference between C_x and C_s.

The frequency at which the 565s oscillate is determined by the capacitance between pins 1 and 9 (C_x, C_s) and the resistance between pins 8 and 10 (see the 565 data sheet). A wide range of values may be determined simply by adjusting the 4.7-kilohm potentiometer–2.2-kΩ resistor combination that is connected between these latter pins.

When resistances are compared, only four components need be changed. C_x becomes R_x, C_s becomes R_s, and variable capacitors replace the previously mentioned potentiometers.

Calibration is equally simple in either the capacitor-measuring or the resistor-measuring mode. Since the 565's frequency of oscillation can be within ±10% of a nominal value for a given set of frequency-determining components, both oscillators should initially be aligned by setting $C_x = C_s$ (or $R_x = R_s$). The potentiometers (or variable capacitors) should then be trimmed for a null on the meter. □

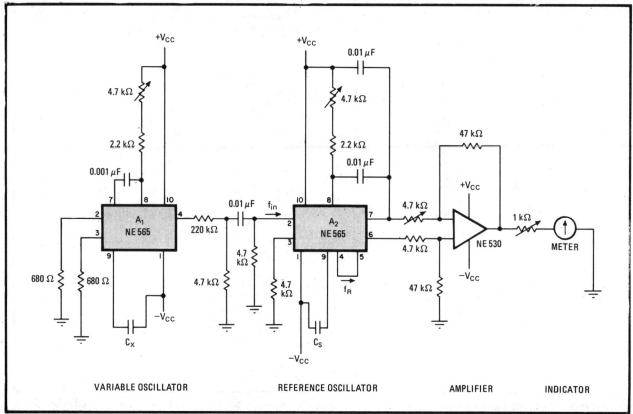

Matched? Phase-locked loops wired in series indicate percentage deviation between unknown and reference capacitors or resistors. A_2's output represents the difference between a standard and variable frequency, each of which is determined by C_s and C_x, respectively.

Four-chip meter measures capacitance to within 1%

by Peter Henry
Seattle, Wash.

Measuring capacitance over the range of 1,000 picofarads to 1,000 microfarads, this four-chip meter has an overall accuracy of 1%. Costing less than $20, the unit has a digital readout and is built from parts that are readily available.

Engaging switch S_1 momentarily fires timer A_1, whose pulse width is determined by capacitor C_1 and one of four timing resistors, R_1 through R_4. The timer enables an astable multivibrator, A_2, and resets a three-digit decimal counter, which then counts the number of pulses from A_2 until the timer runs out or a count of 999 (overflow) is reached. The number is then displayed.

A seven-decade range (see table) is achieved by switching in R_1–R_4, and R_5–R_6, which sets the frequency of A_2 to approximately 1 or 1,000 hertz. For best performance, R_1–R_4 should be hand-picked to achieve the desired 1:10, 1:100 or 1:1,000 ratios if an accuracy to within 1% is required.

Calibration is achieved with a known capacitance of

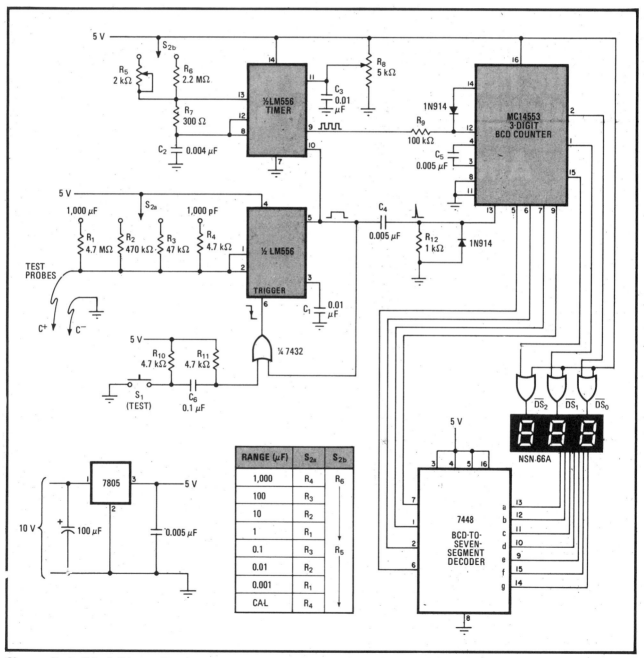

Charge check. This digital capacitance meter measures accurately over the range of 1,000 picofarads to 1,000 microfarads. The circuit displays the number of pulses from astable multivibrator A_2 in the measurement count period during which 555 timer A_1 is active. The measurements are repeatable to within a few counts on all but the lowest scale, giving an overall accuracy to within 1%.

approximately 0.3 μF. Potentiometer R_5 is adjusted for identical readings on the calibrating and 1-μF scale. Then potentiometer R_8 is adjusted for the same readings on both scales. This procedure is repeated as often as necessary for perfect agreement.

The circuit is very reliable and will yield repeatable measurements to within a few counts on all but the lowest scale. Differences in capacitance of 10 pF can be detected. Leaky capacitors may be difficult to measure because of their series resistance. □

Chip changes the colors of light-emitting diodes

by Marvin Burke
Novato, Calif.

A light-emitting diode that changes color as a function of the input voltage is useful for instrumentation and equipment displays, since both the presence of the light and the color convey information. A simplification of the circuit by Smithline,[1] this design needs no separate frequency generator and requires only a single supply voltage in the range of 5 to 16 volts.

As shown in Fig. 1a, the new design needs only one integrated circuit, six resistors, a capacitor, and a bicolor LED. Operational amplifier A_1, part of the quad LM324 package, interfaces with some external voltage source such as a voltage follower. Its output is the input, V_i, to A_2. For its part, A_2 is set up to act as a voltage-controlled pulse-width-modulation oscillator. The duty cycle of its output pulse varies exponentially with V_i as a percentage of B, the supply voltage. As seen in Fig. 1b, the ratio of A_2's pulse-width high level to its pulse-width low (PW_H/PW_L) follows the exponent of V_i linearly over two decades. That is more than enough to get a full

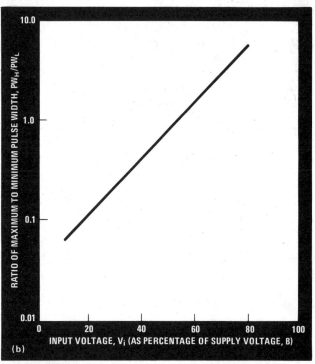

1. Multicolored outputs. Just one integrated circuit is required to construct a voltage-controlled multicolor light-emitting-diode display (a). Pulse-width modulation of A_2's output by means of input voltage V_i (b) is the key to obtaining any LED output color from green to red.

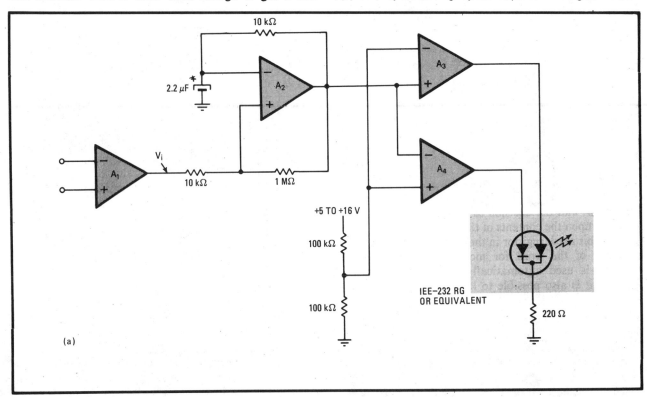

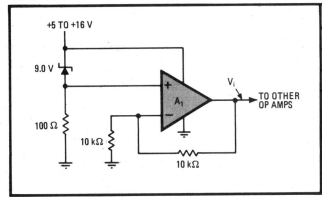

B	V_i	V_i/B	PW_H/PW_L	COLOR
13.0	8.0	0.62	2.00	GREEN
12.5	7.0	0.56	1.25	GREEN/YELLOW
12.0	6.0	0.50	1.00	YELLOW
11.5	5.0	0.43	0.60	ORANGE
11.0	4.0	0.36	0.40	RED

2. Watcher. The LED color modulator can monitor the supply voltage of a music synthesizer. Here, op amp A_1 replaces part of the same number in Fig. 1. The table shows yellow for the normal voltage (12 V), green for the high one, and red for the low one.

range of color output from the LED. The output of A_2 is fed to A_3 and A_4, which act as noninverting and inverting buffers, respectively, for A_2's pulse.

The buffers power the bicolor LED so that when the green LED is on, the red one is off, and vice versa. Since the overall color perceived is dependent on the relative power through each LED (provided they are flashing quickly enough for blending to occur), it depends strictly on the ratio of PW_H to PW_L. Thus, V_i, which determines PW_H/PW_L, is ultimately in control of the color. The color mix for a given range of V_i may be modified by placing separate resistors in the red and green leads of the LED and adjusting the ratio of the resistors until the desired effect is achieved.

A direct application of the technique is shown in Fig. 2. Here, A_1 is set up to monitor a synthesizer's supply voltage. With the parameters shown in the table, colors shift from green for high voltage to red for low voltage, keeping a close tab on the supply. □

References
1. Leonard M. Smithline, "Dual light-emitting diode synthesizes polychromatic light," *Electronics*, Aug. 16, 1979, p. 130.

'Dithering' display expands bar graph's resolution

by Robert A. Pease
National Semiconductor Corp., Santa Clara, Calif.

Commercially available bar-graph chips such as National's LM3914 offer an inexpensive and generally attractive way of discerning 10 levels of signal. If 20, 30 or more steps of resolution are required, however, bar-graph displays must be stacked, and with that, the circuit's power drain, cost and complexity all rise. But the techniques used here for creating a scanning-type "dithering" or modulated display will expand the resolution to 20 levels with only one 3914 or, alternatively, make it possible to implement fine-tuning control so that performance approaching infinite resolution can be achieved.

The light-emitting-diode display arrangement for simply distinguishing 20 levels is achieved with a rudimentary square-wave oscillator, as shown in Fig. 1. Here, the LM324 oscillator, running at 1 kilohertz, drives a 60-millivolt peak-to-peak signal into pin 8 of the 3914.

Now, the internal reference circuitry of the 3914 acts to force pin 7 to be 1.26 V above pin 8, so that pins 4 and 8 are at an instantaneous potential of 4.0 mV plus a 60-mV p-p square wave, while pins 6 and 7 will be at 1.264 V plus a 60-mV p-p square wave. Normally, the first LED at pin 1 would turn on when V_{in} exceeded 130 mV, but because of the dither caused by the ac component of the oscillator's output, the first LED now turns on at half intensity when V_{in} rises above the aforementioned value. Full intensity is achieved when V_{in} = 190 mV.

When V_{in} rises another 70 mV or so, the first LED will fall off to half brightness and the second one will begin

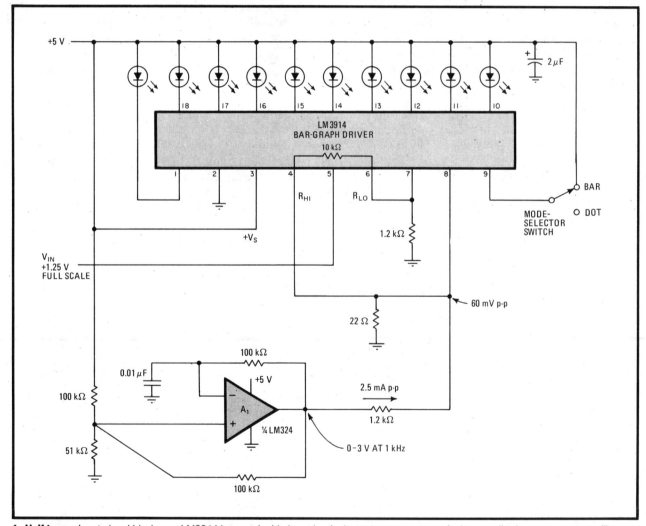

1. Half tones. Input-signal biasing on LM3914 bar-graph chip is set by the instantaneous output of a low-amplitude square-wave oscillator so that bar-graph resolution can be doubled. Each of 10 LEDs now has a fully-on and a partially-on mode, making 20 states discernible.

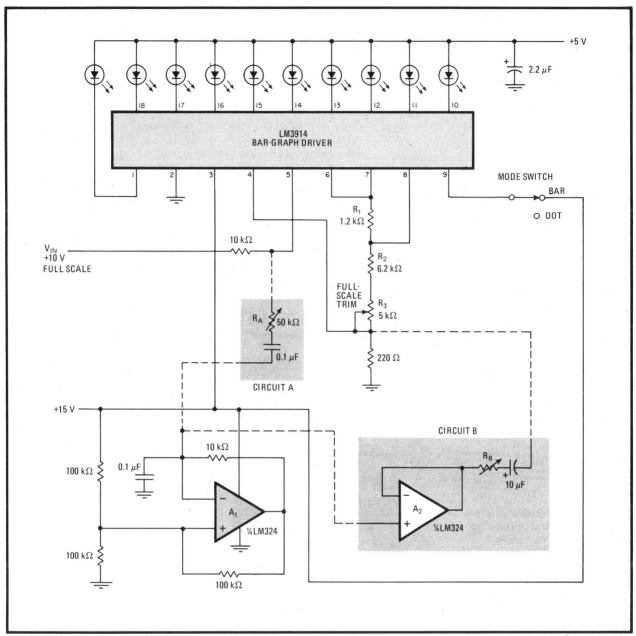

2. Spectrum. Greater resolution, limited only by the ability of the user to discern relative brightness, is achieved by employing a triangular-wave oscillator and more sensitive control circuitry to set the voltage levels and thus light levels of corresponding LEDs. Two RC networks, circuits A and B, provide required oscillator coupling and attenuation. B replaces A if oscillator cannot suffer heavy loading.

to glow. When V_{in} reaches 320 mV, the first LED will go off, and the second will turn on fully, and so on. Thus 20 levels of brightness are easily obtained.

Similarly, greater resolution can be achieved by employing a triangular-wave oscillator and two simple RC networks as seen in Fig. 2. Here, by means of circuit A, this voltage is capacitively coupled, attenuated, and superimposed on the input voltage at pin 5 of the LM3914. With appropriate setting of the 50-kilohm potentiometer, each incremental change in V_{in} can be detected because the glow from each LED can be made to spread gradually from one device to the next.

Of course, if the signal-source impedance is not low or linear, the ac signals coupled into the input circuit can cause false readings at the output. In this case, the circuit in block B should be used to buffer the output of the triangular-wave oscillator.

The display is most effective in the dot mode, where supply voltages can be brought up to 15 V. If the circuit's bar mode is used, the potentials applied to the LEDs should be made no greater than 5 V to avoid overheating.

To trim the circuit, set the LM3914's output to full scale with R_3. R_A or R_B should then be trimmed so that when one LED is lit, any small measured change of V_{in} will cause one of the adjacent LEDs in the chain to turn on. □

Light pen generates plotter signals

by E. Chandan and Agarwal Anant, *Department of Electrical Engineering, Indian Institute of Technology, Madras, India*

By tracing out the shape of any waveform and sending it to an X-Y plotter or other recording instrument, this photoresistive sensor, in conjunction with an oscilloscope, provides a convenient and inexpensive way to translate hand-drawn data and similar information to a remote location in real time. The sensor requires only three light-dependent resistors (LDRs) and two operational amplifiers. The scope supplies a pinpoint light source that follows the movement of the sensor and thereby generates the required X and Y signals to the plotter or recorder.

As shown in (a), the faces of three LDRs, each having a diameter of about 4 millimeters, are mounted on the tip of an ordinary pen so that their centers lie on the vertices of an equilateral triangle and in the same plane. The LDRs are then wired into the circuit whose schematic is shown in (b).

The pen is brought within a few millimeters of the scope and thus in proximity to the initially unswept scope trace, which is a spot of light on the screen when the scope is in the X-Y mode. The position of the LDRs with respect to the screen (R_3 at the top, R_1 and R_2 at the bottom) must remain fixed.

When the scope beam (shaded area) is in the center of the LDRs, their corresponding resistances are virtually equal, because approximately the same amount of light hits each device. Consequently, no error voltage (no X_{scope} or Y_{scope} signal) is generated, and the spot will not move with respect to the light pen.

If the pen is moved to the right or left or up or down, as shown, the resistances of the LDRs become unequal, an error voltage is generated, and the scope beam moves in a direction that minimizes the feedback voltage—toward the center of the LDRs. The spot thus follows the pen and in so doing creates the X_{scope} and Y_{scope} signals used to drive the remote recorder or plotter. Should the pen be removed from the proximity of the screen, the beam spot returns to its original setting.

As the schematic shows, when the spot hits the center of the sensor, $R_1 \approx R_2 \approx R_3$, and so $V_A \approx V_B \approx V_C$. Thus, neither 741 op amp generates any appreciable X_{scope} or Y_{scope} signal and the spot remains fixed. For the second condition, $V_C > V_B > V_A$, and a large positive voltage is generated at X_{scope} to push the beam to the right, while a small Y_{scope} voltage pushes the beam slightly upward. For the last condition, $V_C > V_A > V_B$, and the beam is pushed left and slightly upward.

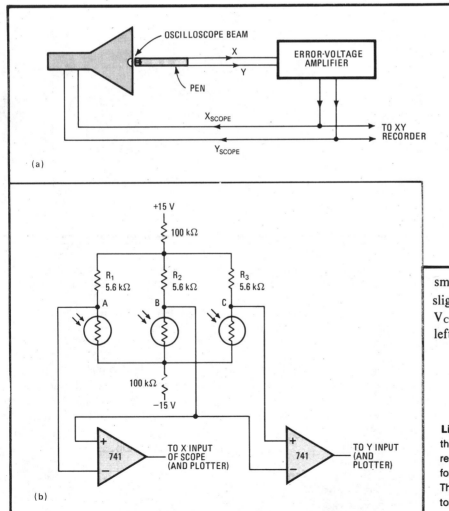

Light pen. An oscilloscope beam will track the movement of three pen-mounted photoresistors, generating a hand-drawn waveform to drive X-Y plotters and recorders (a). The feedback circuitry (b) uses photoresistors to adjust the scope beam to follow the pen's path of motion.

Versatile circuit measures pulse width accurately

by Kelvin Shih
General Motors Proving Grounds, Milford, Mich.

Designed to measure the width of a pulse, this circuit produces a dc voltage output that corresponds to the input pulse width—from 0.5 to 5.0 volts dc for a 1- to 10-millisecond input pulse width. Among its many potential applications, one of the most attractive is its use in measuring very low frequencies of under 1 hertz accurately and without waiting.

Schmitt trigger U_{1-a}'s output at point A is a pulse of the same width as the input and with a fixed amplitude of 7.5 v (Fig. 1a). The sample-and-hold pulse at point B is generated at the trailing edge of the input. After this signal is shifted in level from 0 to +7.5 v to ±7.5 v, the pulse is used to drive bilateral switch U_{3-b}. The signal at point C is the sum of the width of the input and the sample-and-hold pulse at B and, after level-shifting, resets integrator U_2 to zero when the input is low.

The integrator converts the input into a negative-going linear ramp that remains steady during the brief sample-and-hold period (Fig. 1b). During the rest of the cycle, the capacitor is discharged by U_{3-a} and the output of the integrator stays zero. Inverting amplifier U_4 changes the negative ramp into a positive one. The last stage, which is the sample-and-hold circuit, is used to store the peak

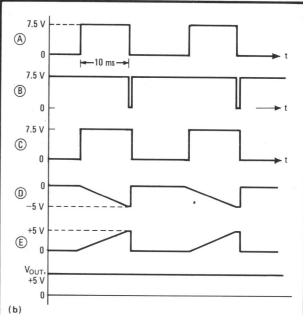

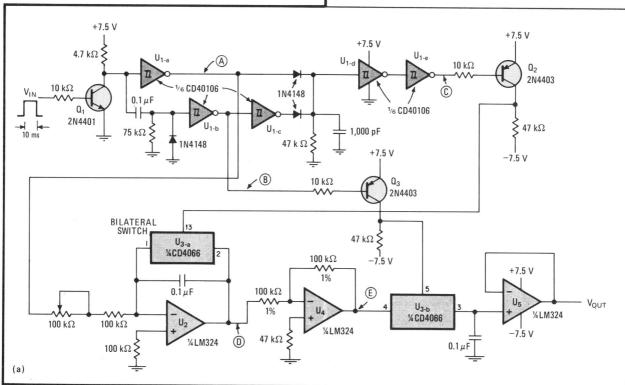

1. Measuring width. This pulse-width–measuring circuit (a) produces a dc voltage output that is proportional to the input pulse width. For the components shown, the output is 5 V dc for a 10-ms input pulse width. Integrator U_2 converts the input pulse width into a linear ramp (b) whose peak voltage is stored in the sample-and-hold circuit. The output voltage is constant if the pulse width is constant.

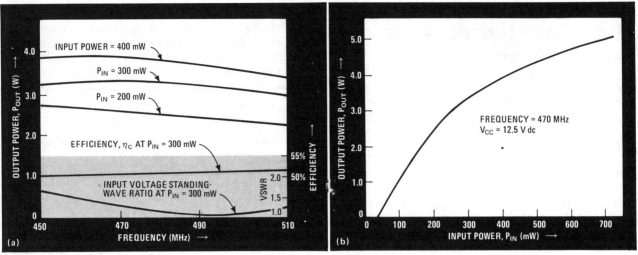

2. Broadband. Its high-frequency performance (a) shows that this amplifier can provide an output of more than 3.0 W above 490 MHz. The gain roll-off above 490 MHz is minimized by optimizing matching networks. The amplifier gain curve at 470 MHz is depicted in (b).

assembly, the thermal resistance of the transistor can be expected to be only 12° to 13°C/w. The gain curve (Fig. 2b) demonstrates typical performance of the transistor at ultrahigh frequencies. □

Digital weighing scale resolves quarter counts

by David Watson
Intersil Datel (UK) Ltd., Basingstoke, Hants., England

By using a display-driving counter and adding three flip-flops, it is possible to quadruple the resolution of a low-cost digital weighing scale (Fig. 1a). The rest of the scale's circuitry is standard, consisting of a strain-gage transducer, a complementary-MOS amplifier, and an analog-to-digital converter. Since the revised circuit is designed to resolve to a quarter of a displayed increment, the scale displays N/4 for an N-count conversion.

The a-d converter has three phases of operation. In the automatic zeroing phase, offset is measured and nullified and the busy output is low. During the integrating phase, the busy output goes high and the converter integrates the input signal for 10,000 clock cycles. Finally, the deintegrating cycle integrates the reference until the integrator returns to its starting, or zero-crossing, point. For a digital reading of N, the deintegration lasts for N+1 clock cycles, at the end of which the busy signal returns low. The busy output is therefore high for a total of 10,001 + N clock cycles.

The N/4 display is obtained by delaying for 10,001 clock cycles after the busy output of converter U_1 goes high and then enabling the counter U_2 to increment at ¼ the converter clock rate. The counter is halted as soon as the busy signal goes low and its contents are transferred to the display. The reset pulse now sets the converter ready for the next cycle.

During auto zero, the busy output is low (b) and the counter is disabled. The Q_3 output of flip-flop 3 enables gate G_3, which allows the U_1 clock pulses to go to counter U_2. The counter begins to increment at the start of the integrating phase. After 10,000 counts, the falling edge of carry output of U_2 clocks flip-flop 3 to its reset state. The other two flip-flops are now enabled and divide the input clock by 4 while G_3 blocks the direct clock input to the counter, which has now returned to zero. The strobe output of U_1 generates the pulse necessary to store the count in U_2 and then reset U_2. □

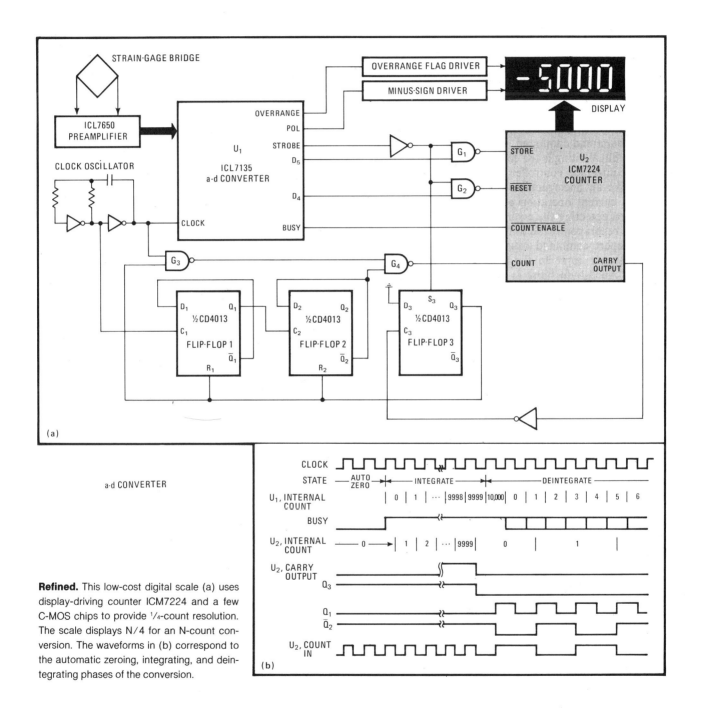

Refined. This low-cost digital scale (a) uses display-driving counter ICM7224 and a few C-MOS chips to provide ¼-count resolution. The scale displays N/4 for an N-count conversion. The waveforms in (b) correspond to the automatic zeroing, integrating, and deintegrating phases of the conversion.

11. LOGIC CIRCUITS

Treating the three-state bus as a transmission line

by William A. Palm
Magnetic Peripherals Inc., Minneapolis, Minn.

Though the input/output bus of today's typical small computer system would seem the ideal transmission link for passing data between the system's peripheral three-state TTL devices, difficulties may arise because the length of the bus is an appreciable fraction of the wavelength of the signal being propagated. The resulting loss of data caused by reflections along the bus can be virtually eliminated, however, by simply applying transmission-line matching techniques to the bus, by selecting the proper line drivers, and also by observing design practices that will spare the drivers from damage caused by line transients.

A first obvious option is to use an unterminated line. Pulse distortion from reflections, however, can be severe, depending on the line's length and the data rate. Lines which are a quarter wavelength long or odd multiples thereof at the frequency of interest will cause the worst reflections, and precautions should be made to avoid those lengths. In the typical case there will be ringing on the line and the line receiver will generate data errors.

If the data rate is relatively slow, the errors caused by ringing can be disregarded by sampling the data lines after the line transients have died out. The delay time required before looking at the lines is $t_d = 2Lt_p$ nanoseconds, where L is the line length in feet and t_p is the propagation time per foot.

Terminating a transmission line in its characteristic impedance is fundamental to the theory of transferring power efficiently and without line reflections, however. The figure illustrates a suggested technique of joining several daisy-chained TTL peripherals to a 100- to 300-ohm twisted-pair line. In practice, the line might be flat cable, which enjoys widespread use because of the physical ease with which it is connected to the circuitry.

Single-ended termination has its advantages, too, as in cases where enable or strobe lines are utilized. This method affords elimination of reflections with minimal power consumption. Termination is made as near to the output of the transmitter as possible, since the line receivers are often scattered at various positions along the line. Given proper termination, the system's limiting range factor then becomes the transmitter itself.

As for the hardware, several good bus transmitters and transceivers are available. Among them are the AMD8304, which offers an attractive solution to three-state buses of 20 ft or less in length. The 20-pin transceiver can sink 48 milliamperes and can handle flat cables having an impedance of more than 100 ohms satisfactorily. Used with a terminated 100-ohm cable it will have a saturation voltage of only about 0.4 volts dc at 55 milliamperes. Using higher terminating-resistance values will not appreciably change the saturation voltage, making it an excellent choice for a driver.

Another chip that meets the same line-driving requirements as the AMD8304 is the MC3482/6882. This useful device has latches and is available in inverting or noninverting versions but, unlike the more versatile

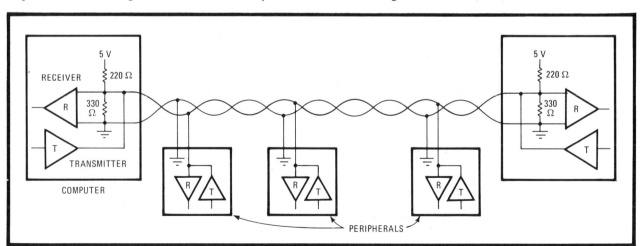

Termination. Ringing at receiver end of typical input/output bus will often leave inadequate 0–1 switching margin. Either singly or doubly terminated bus will reduce reflections. Line transceivers must be selected with care to handle driving requirements satisfactorily. Power sources associated with computer and peripherals should be connected to same line to reduce common-mode voltages on I/O bus.

AMD8304, is an eight-line transmitter.

The 74LS373 octal transmitter, complete with three-state latches, is another good choice. This device sinks only 24 mA so that terminating a line at both ends is not possible (though this is seldom a problem). The 75450 family of dual-peripheral drivers is useful for those applications that require a terminator at each end of the line. The 74LS245 transceiver is also excellent.

As for circuit details, there is one important point. Specifically, all 60-cycle power cords associated with the computer and its peripherals must be plugged into the same power-line bus. This arrangement will eliminate any possibility that common-mode voltages between the chassis of different units will appear on the line drivers or receivers. Often, these units have very limited common-mode range. In fact, negative transients that drop below -0.7 V can damage a chip permanently. Also, because all transceivers on a given chip are on a common substrate, a negative transient on any line will cause the other transceivers to change state. □

LED indicates timing error in emitter-coupled-logic one-shot

by M. U. Khan
Systronics, Naroda, Ahmedabad, India

Rather than scrutinize waveforms on an oscilloscope, it is possible to employ a simple circuit to monitor the output of a one-shot to determine if it is being triggered at the right time or if the output pulse width is correct.

If the clock pulse arrives while the normally low output $\bar{Q}_1$ or set line S_1 of the one-shot is at a logic 1 level, the $\bar{Q}_2$ output of the indicator flip-flop goes to a logic 0 level, turning on the light-emitting diode. Q_1 of the one-shot remains at the logic 1 level only in its quasi-stable state, whereas S_1 goes to the logic 1 level only in its recovery state. Thus whenever the one-shot is triggered too early—in other words, before recovering—the LED turns on. When the mistriggering is corrected, by reducing either the clock rate or the width of the one-shot, it automatically turns off.

Besides the LED, the indicator circuit consists of an MC 10103 OR gate and an MC 10231 D-type flip-flop, both emitter-coupled-logic devices. The circuit works satisfactorily up to 75 megahertz. For higher speeds—up to 100 MHz—propagation-delay compensation through an additional OR gate (dotted line) is needed. In the latter case, both the OR gates should be replaced by an MC 1660 dual four-input OR-NOR gate. □

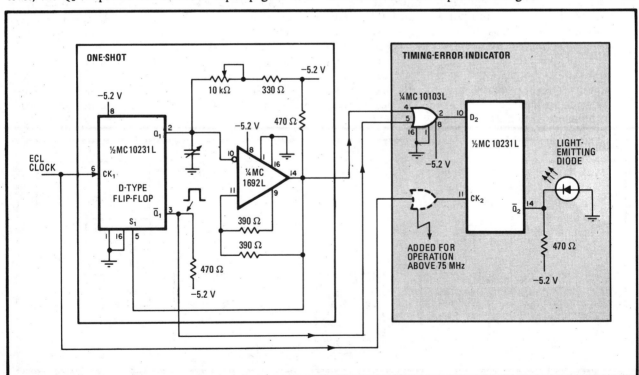

Hot shot. Spotting timing errors in a fast one-shot multivibrator is simplified with the addition of an error indicator circuit such as the one shown in the shaded area. If the clock rate is too high or the output pulse width is too small, the LED will indicate it.

Address checker troubleshoots memory drive, logic circuitry

by F. Chitayat
Canadian Marconi Co., Montreal, Quebec

Although error-detecting and -correcting schemes have tremendously enhanced the reliability of microprocessor-based systems, there is still a significant memory failure mode that these schemes cannot detect. If a memory address bit fails due to some problem in the driver or any memory input gate connected to it, the address will be inaccurately written or read. This address checker thoroughly checks for such failures and so complements existing error-correcting schemes. The circuit thus improves the system's troubleshooting capability.

A watchdog monitoring scheme, with the microprocessor programmed to make a certain periodic output that would then be monitored by specialized circuitry, would be only a partial answer, for it would not guard against the failure of more than a few address bits. Instead, the bits checked would generally be only those exercised by the watchdog program. Even if the program ran to 64 locations, at most 6 bits could be checked.

Using this simple circuit, however, the microprocessor clocks the address output onto a register and then reads it back through the data bus. Also, the microprocessor can be programmed to send several test patterns to the circuit by sliding a 1 or a 0 to ensure that every address bit is sent correctly.

The write operation causes the address being written to be stored in registers U_1 and U_2, with the most significant byte stored in U_1 and the least significant byte in U_2. This will not affect the actual writing of the data into the intended address. These registers are read by accessing predetermined memory locations preassigned as input/output locations. In the figure, the specific address locations are referred to as words A and B and are decoded in U_3. When each of these locations is read, the contents of the corresponding register are shifted onto the microprocessor data bus and subsequently read and checked by the microprocessor.

Since a write operation changes the contents of U_1 and U_2, these must be read immediately after a write operation has occurred at the memory-address test location—that is, before another write operation is begun. □

Foolproof. This address checker detects faults in memory addresses due to failures in the driver or receiver associated with each address bit. The memory address being written to is stored in registers U_1 and U_2. Since the registers are automatically clocked by a write operation, the error in the memory address is easily detected by the microprocessor when it reads those registers.

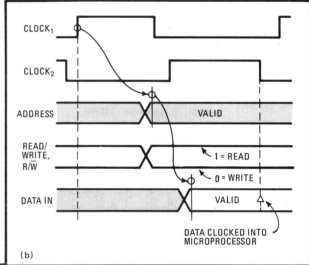

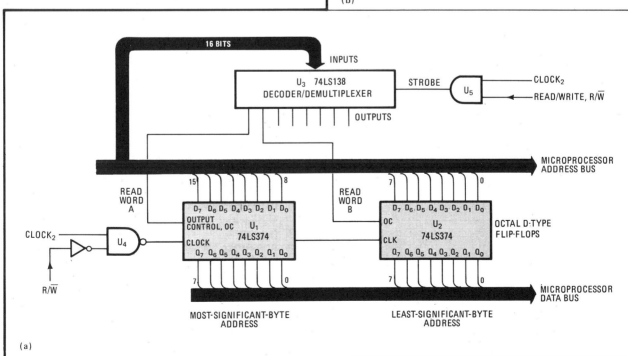

Protecting TTL gates from electrostatic-discharge pulses

by Peter Lefferts
Versatec, Santa Clara, Calif.

Though interference problems associated with electrostatic discharge have long plagued low-impedance logic, nothing much has been done to alleviate them. The movement of people, paper, and plastic packages can generate peak static impulses that can impair the functioning of logic gates 20 feet away. Also, if a gate output is spiked above 6 or 7 volts, the next gate output is distorted. The discussion and design presented allow the TTL gates to cope with the interfering pulses.

Unfortunately, these effects are not noticeable on an oscilloscope at normal logic rates. Only one high pulse out of 100,000 may have an electrostatic pulse riding on it, even when stimulated by a rapid sparker. These few distorted pulses are therefore passed off as random errors. However, in TTL circuits such low rates of errors can still cause annoyance, for just one error may result in a false form-feed that can cause an erroneous output and loss of information in a printer or plotter.

A driver stage for long logic lines such as clocks, resets, and initialization and busy commands is presented. This driver protects TTL gates from interference by making them less sensitive to any type of pulse. The combination of transistor Q_1 and light-emitting diode D_1 acts as a constant current source that loads the two parallel TTL gates (see figure). As a result, the impedance at the low logic level decreases and the speed and amplitude of the transition to logic high increases. Undershoots and overshoots due to line ringing decrease from 1 V or more to about ¼ V. There is a tremendous improvement in the output waveform, and interference of up to 10 V due to electrostatic discharge is easily eliminated by the circuit. The design ensures that the lines are well protected.

If the current is reduced by increasing the value of resistor R_1 to 160 or 180 ohms, the design also works well with the low-power Schottky TTL gates. Also, diode D_2 provides a 3.5-V clamp that minimizes the effect of electrostatic discharge and prevents Q_1 from saturating and introducing delays. This clamping technique yields impedances of about 3 Ω at high frequencies during the time when the signal is not in transition. □

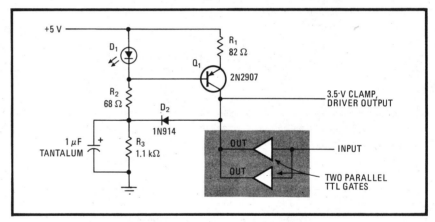

Static discharge. The circuit protects the TTL gates from interference due to electrostatic discharge. Transistor Q_1 and light-emitting diode D_1 are combined to function as a constant current source. This source loads the two parallel TTL gates and thereby reduces the impedance at logic low and enhances the speed and amplitude of the transition to logic high. Diode D_2 clamps Q_1 and prevents it from saturating and consequently from introducing delays.

Interfacing TTL with fast bipolar drivers

by J. A. R. Ball, P. J. Grehan, and P. Welton
Darling Downs Institute, School of Engineering, Queensland, Australia

Surprisingly, there are as yet no suitable integrated circuits for translating the 0-to-5-volt output swing of TTL into arbitrary bipolar levels. But even the discrete interfacing circuits that have appeared over the years will fall short in performance, especially if the requirement calls for a high-speed switch to drive the relatively high capacitance of a power device or load. The solution lies in modifying the typical textbook interface with a circuit that acts to decrease the input-circuit storage time of the output transistors but does not appreciably affect any other interface parameter or specification.

A slightly modified ±10-V TTL interface is shown in (a), which will be suitable for relatively high-speed switching at low to medium current (below 100 milliamperes). In this circuit, Q_1 turns on and remains in the active region when the TTL output exceeds 1.5 V. Providing the current drawn out of the base is sufficient, Q_3 will saturate and the voltage applied to the load will be almost 10 V. When the TTL output falls sufficiently, Q_1 and Q_3 turn off, and charge stored in the base of Q_3 escapes via resistor R_3. Transistors Q_2 and Q_4 comprising the other half of the circuit act in a complementary fashion, conducting when the TTL output falls below 1.5 v and applying −10 v to the load.

One disadvantage of this circuit is that it is possible for Q_3 and Q_4 to be conducting at the same instant during a change of state to cause a supply current spike whose magnitude may exceed the nominal load current by more than three times. Also, most of the power lost in the output transistor will be dissipated during a change of state when both are in the active region. Thus, the average dissipation will be proportional to the switching frequency. These problems may be minimized by increasing the zener voltage, V_Z, so as to increase the dead zone between the input threshold levels of the circuit. Switching speed may be increased by optimizing the value of the speed-up capacitors C_1 and C_2, operating Q_3 and Q_4 at very large base currents, and reducing R_3 and R_4 to minimize storage time.

This basic circuit can also be used to control far larger currents than 100 mA, providing appropriate output transistors are used. However, the storage time of these devices then becomes a major problem, and so speed is sacrificed. The circuit in (b) shows how to reduce the delay time by adding two transistors for supplying reverse base current to whichever output transistor is in the act of turning off.

Here, when the TTL output goes high, Q_1 conducts and Q_3 saturates as before, while Q_2 turns off and Q_4 begins to come out of saturation. In addition, the emitter current of Q_1 turns on Q_6, which provides a path for the escape of charge stored in the base of Q_4. This effectively shortens Q_4's turn-off delay.

When the TTL output goes low, then Q_1 and Q_6 turn off, Q_2 and Q_5 conduct, Q_4 saturates because of the base current supplied by Q_2, and Q_3 is rapidly turned off because of the action of Q_5. Discharge transistors Q_5 and Q_6 should be selected for high-speed saturated switching, so that they will not delay the turn-off of their associated output transistors.

Adding Q_5 and Q_6 will reduce the storage delay of the output transistors by a factor of from 2 to 4. The circuit in (b) provides a rise and fall time of about 80 nanoseconds for a load of 11 ohms (2-ampere load). The active pull-up output ensures the interface's low output impedance in either the logic 0 or logic 1 state. A further advantage is that the output voltage is specified within narrow limits in both states, unlike the case with totem-pole–type circuits. □

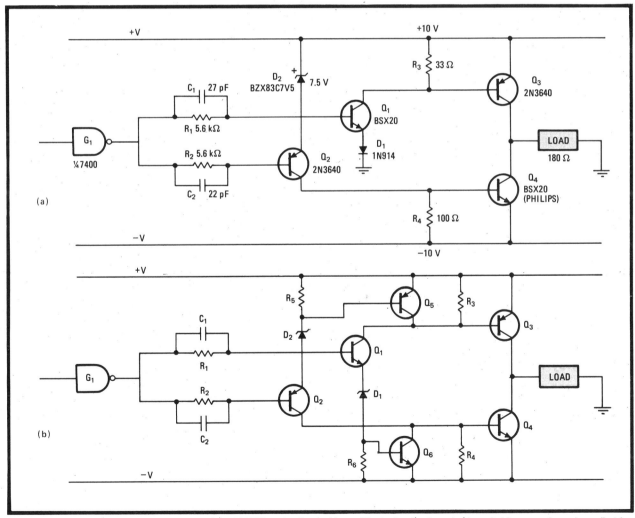

Conversion. Interface (a) for translating the 0-to-5-V TTL swing into arbitrary bipolar levels at moderate switching speeds works well at low load currents. For increased loads, circuit (b) offsets the large storage delay of the output transistors and reduces crossover switching.

Single chip solves MC6809 timing problems

by Kim A. Crane
Arizona State University, Mesa, Ariz.

Microprocessor-based systems often have several levels of peripheral decoding. Propagation delays due to these decoding schemes make the bus-timing parameters critical. As a result, they are difficult to keep within specifications from a worst-case–analysis point of view and introduce a timing fault TF.

TF is the amount of time by which the last opportunity to read or write valid data has been missed. However, this configuration uses only one chip and three resistors to eliminate this timing problem for the MC6809 processor when it is being interfaced with either M6800 or non-M6800 family parts.

The circuit (a) takes advantage of quadrature clock Q that is included in the MC6809 chip. The rising edge of this clock indicates that the address bus is valid, and its falling edge indicates that the data bus is valid. In addition, the read-cycle access time is extended by about 250 nanoseconds through enabling the decoders on the rising edge of this clock.

However, the main advantage lies in the write-cycle timing. On the falling edge of Q, during a write-cycle, the decoders that drive the appropriate chip-select high while enable E is only partially through its high cycle are disabled, causing the data to be written. Also, the data is valid until the falling edge of E occurs. This provides the chip-deselect signal with an additional 250 ns to propagate through the various decoding levels and still remain within worst-case timing margins. With this method, no timing faults will occur for up to 18 levels of LS type decoders connected in series (b).

Unfortunately, this write-cycle technique is not suitable for some M6800 family parts because of the fact that the data is written into these parts on the falling edge of E and not the rising edge of chip-select. Therefore, additional hardware is required for these parts. NAND gates U_{1c} and U_{1d} and resistors R_2 and R_3 solve this problem by extending the chip-select low during a write cycle to the falling edge of E without affecting the already correct read-cycle timing (c). □

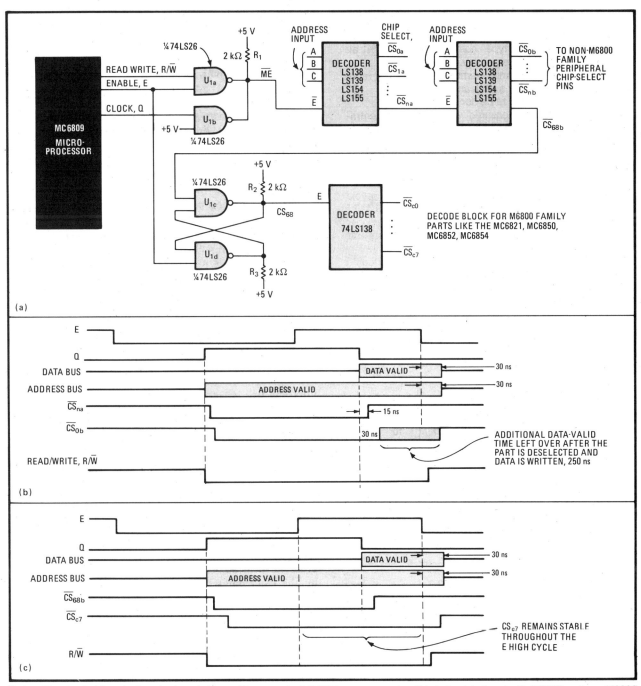

Delay. The circuit (a) ends decoder bus-timing problems on the MC6809 processor when it is being interfaced with M6800 and non-M6800 parts. Diagram (b) shows that a string of 18 74LS138 decoders can be used with this scheme without any timing-delay problems. For interfacing with some M6800 parts, logic extends the chip-select low during a write cycle (c) to the falling edge of E.

12. MEMORY CIRCUITS

Chip computer gains I/O lines when adding memory

by U. K. Kalyanaramudu and G. Aravanan
Bharat Electronics Ltd., Bangalore, India

With an extended 6800 microprocessor instruction set, 2-K bytes of read-only memory, 128 bytes of random-access memory, and 29 parallel input/output lines, Motorola's 8-bit single-chip microcomputer is a handy integrated circuit for designers to have in their kit. However, to supplement the MC68701C/6801C's on-chip memory is a complicated matter because I/O lines have to be sacrificed to the need for address and data lines for a memory interface. In this application, though, I/O lines do not have to be surrendered, for the microcomputer unit is interfaced with Intel's 8155-2 RAM.

Combining the microcomputer with the 256-byte RAM, which has I/O ports and a timer, actually gives the circuit an extra 22 parallel I/O lines and throws in a 14-bit programmable counter-timer into the bargain.

Microcomputer U_1's single read-write line is combined with the E clock to provide separate read and write lines as required by RAM U_2—NAND gates $U_{4\text{-}b}$ through $U_{4\text{-}d}$ help separate the two lines. The I/O-memory control signal for U_2 is obtained by connecting the line to the most significant address line of the microcomputer, while exclusive-OR gates $U_{5\text{-}a}$ and $U_{5\text{-}b}$ decode the chip-enable from address bits A_{14} and A_{15}. The microcomputer's address strobe is trimmed down to 200 nanoseconds by one-shot U_3 to meet the setup and hold time requirements of 8155-2.

The circuit can also be used for interfacing other peripheral devices, such as the 8755-2, which is an erasable electrically programmable ROM. □

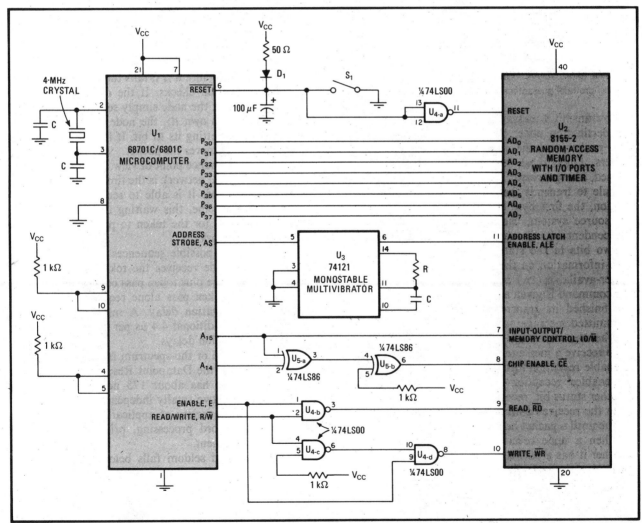

Expansion. Low-cost memory and I/O expansion for the MC68701C/6801C is achieved by interfacing the 8-bit single-chip microcomputer with Intel's 8155-2—a RAM with I/O ports and a timer. The address strobe is trimmed to 200 ns by one-shot U_3 to meet the address setup and hold time requirements of U_2. Read and write signals are generated by gating the E clock.

Stepper checks state of E-PROM's memory

by Steven Bennett
Harris Semiconductor, Melbourne, Fla.

Too often, ultraviolet-light–erasable programmable read-only memories have their contents blindly destroyed by users who cannot determine whether the memory contains valuable information or is totally blank. However, this circuit can scan each E-PROM location with a binary counter and so will distinguish memories that contain data from those that do not—all at a cost of around $8.

In use, the memory device is placed into the test socket and the momentary contact switch, S_1, is pressed. If as little as 1 bit of memory is stored in any of the E-PROM's locations (logic 0 for an E-PROM), the light-emitting diode will light.

A 2-kilohertz clock signal for the 12-bit binary counter, B_1, is generated by oscillator and buffer A_1 so that the addresses will cycle through the 2716 2-K-by-8-bit E-PROM in about 1 second. (Although this circuit was dsigned for the 2716, it may also be adapted for any type of memory, bipolar or MOS.)

If any bits in a given location are low, then a pulse will be generated at the E-PROM's output and will drive the 4068 NAND gate high. This pulse, which is generated at the NAND output, is stretched to 2 s by one-shot and buffer A_2 to drive the LED. □

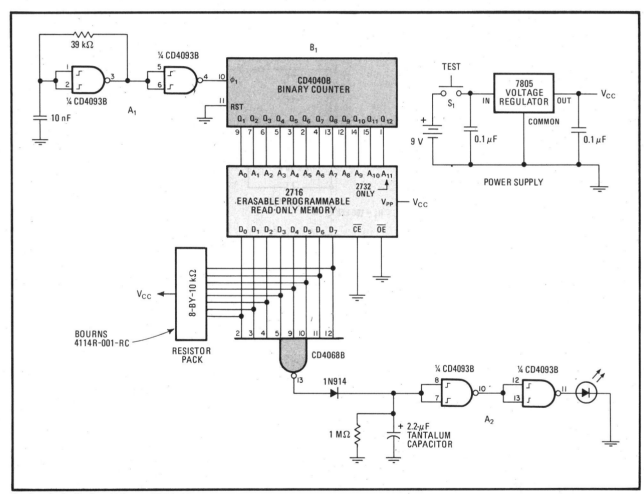

Seeing is believing Using a 12-bit binary counter, a tester of ultraviolet-light–erasable PROMs inspects each location of the device to determine if it contains data or is blank, thereby saving contents from accidental erasure. For a 2-K-by-8-bit E-PROM, the test takes about 1 second, with a light-emitting diode turning on if any memory location contains a data bit.

Transferring data reliably between core and CPU RAM

by Trung D. Nguyen
Litton Data Systems Division, Van Nuys, Calif.

Magnetic-core memories are still one of the most dependable ways to store large amounts of nonvolatile digital data (up to 1 megabyte). This circuit is a conduit for data transfer between the core memory and the random-access memory situated in the central processing unit. A word in the core is composed of 3 bits and is matched to the RAM's 18-bit words. Because of the low bits-per-word count in core, circuit reliability is high and costs are low.

The data that will be written into or read from core is temporarily stored in the data-register circuit. During the write cycle, input data D_{I/O_1} through $D_{I/O_{18}}$ is loaded into the core write-data register, which is composed of flip-flops U_1 through U_3 (Fig. 1). The two most significant bits—17 and 18—are used as parity bits for low-

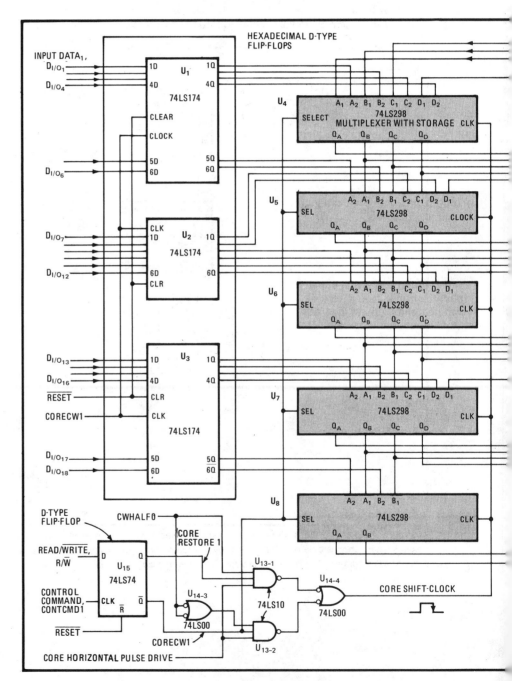

1. Shifting data. This circuit allows two-way transfer of digital data between magnetic-core memory (3 bits/word) and the random-access memory (18 bits/word) located in the central processing unit. Data shifting is controlled by the horizontal pulse-drive signal supplied to the core, and the direction of data transfer is controlled by the read/write pulse from the CPU.

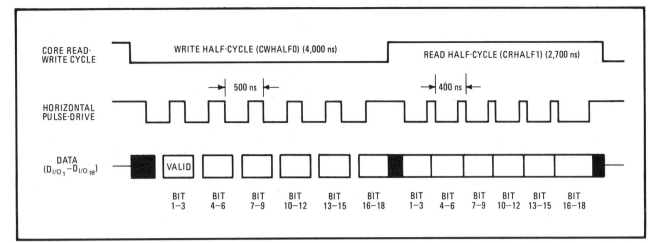

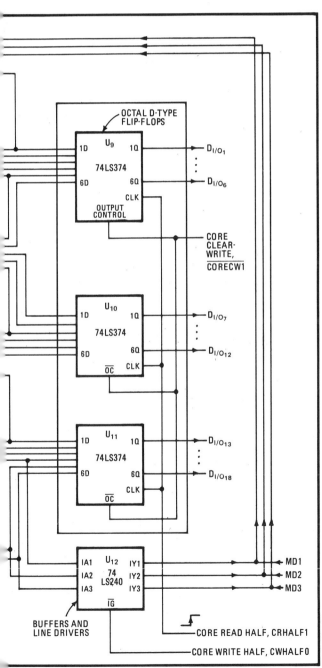

2. Timing. The horizontal drive signal supplied to the core is pulsed six times during each read or write half-cycle to shift data from the core to the RAM and *vice versa*. A typical R/W core-cycle lasts 6.7 microseconds, thereby permitting a data-transfer rate of 150,000 words per second with each word comprising 18 bits.

and high-order data bytes. In addition, on the leading edge of the core clear-write pulse (CORECW1), the data is transferred to the storage-multiplexer circuit, which comprises multiplexers U_4 through U_8. This pulse clears the location of the addressed word in the core memory.

Serial data is sent to the core read-data register (flip-flops U_9 through U_{11}) during the read cycle. This data travels through the storage-multiplexer circuit, which now functions as a data-shift register.

The word-select input is low during the read half-cycle. As a result, a 3-bit word from the core is shifted through the data-shift register by the core's horizontal pulse-drive signal and stored in the core's read-data registers. When the core clear-write pulse is low, this data reaches the RAM. The shift registers are clocked by the shift clock's (CSHIFTCH) negative-going edge, which occurs for every positive-going core horizontal pulse. The horizontal drive signal pulses six times for each memory half-cycle (Fig. 2).

In contrast, during the write half-cycle, the multiplexer word-select input is high and the RAM's 18-bit word is shifted through registers U_4 to U_8 by the horizontal drive pulse and stored in the core's write data register. When the core clear-write pulse is high, the data reaches the core memory.

For proper core-memory operation, the core read/write and horizontal pulse-drive signals are generated by the programmable read-only memory. Data shifting occurs on the leading edge of the core's horizontal drive pulse through its R/W cycle. In addition, the transfer of data during R/W operation is also controlled by NAND gates $U_{13\text{-}1}$ and $U_{13\text{-}2}$. The average data-transfer rate per second is 150,000 words, at 18 bits/word. However, if the number of bits per word differs from those mentioned, the horizontal drive pulse, which serves the core module as a timing signal, must be modified. □

RAM makes programmable digital delay circuit

by Darius Vakili
Bayly Engineering, Ajax, Ont., Canada

For applications lacking in long shift registers and for which bipolar components are too expensive, this simple random-access-memory circuit will program the delay of digital signals precisely and accurately—and without needing varying clock frequencies. The RAM's ready availability and low cost add to the circuit's attractions, especially when high signal speeds are involved.

A digital input code programs the output delay of the signal. The input signal is written into the RAM U_1 at an address generated by synchronous up-down counters U_6 and U_7 and by a fixed input digital code set by the dual-in-line-packaged switch. The stored signal is available at the output when the write-enable input to the RAM is high. This data is read out of the RAM from a location that corresponds to the address generated by the counters only. Therefore, the signal read from the RAM is delayed by a time period equal to the DIP switch's displacement value multiplied by the period of the clock used for generating the address.

The input clock is divided by 2 by latch U_9. The Q_2 output of U_9 clocks address counters U_6 and U_7 and also the select inputs of multiplexers U_2 and U_3. The NAND gates generate the RAM's read and write inputs and also ensure the data is written into the RAM when the control input of U_2 and U_3 is high and read out when it is low. □

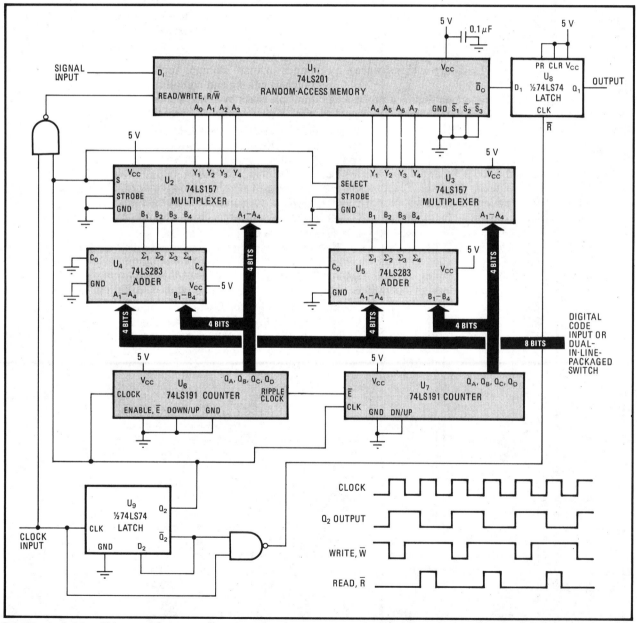

Programmed delay. Using random-access memory, this programmable-delay circuit accurately controls signal delay by means of a digital code generated by a DIP switch. Counters U_6 and U_7 generate the address for the RAM. U_9 provides the write and read inputs for the RAM.

Programmable source sets voltage of E-PROMs

by Ralph Tenny
George Goode & Associates, Dallas, Texas

Many erasable programmable read-only memories require varying voltages for different functions at pin 20. Intel's 2732 E-PROM, for example, multiplexes two functions on pin 20—the output enable and programming voltage input. It is desirable to have these inputs generated by a programmable source since the use of a relay to switch voltages is cumbersome, slowing down the circuit and adding a mechanical element to it. This programmable supply, controlled with two logic signals, provides an automatic selection of four voltages and has 0 volt as an off position.

This logic-controlled programmable voltage source (see figure) is composed of three integrated circuits, a voltage reference, and a few discrete components and has a slew rate of around 1 V per microsecond. However, this slew rate is limited by the operational amplifier (U_4). In addition, a constant slew rate for both the positive and negative swings of the supply is maintained by transistor Q_2, which pulls the output voltage down rapidly.

The binary-coded–decimal-to-decimal decoder in the circuit takes two logic input signals and converts them into four output signals. These outputs enable the four sections of switch U_3 that gate the operational amplifier's (U_4) input. The op amp along with transistors Q_1 and Q_2 forms a voltage regulator whose output is 11 times the input reference voltage. This 2.5-V reference is provided by U_1 and is tapped by potentiometers R_1, R_2, and R_3, which in turn produce the reference voltages needed to generate the desired programming voltages.

All the voltages needed for pin 20 are produced by means of this supply (as shown in the truth table). As a result, all 5-V E-PROMs, including the I2732A and I2764, may be programmed with this circuit. The supply is driven by the B port pins of the peripheral interface adapter MC6821.

The PIA lines that drive inputs A_1 and A_2 are terminated with pull-up resistors. This configuration produces a reset input of 11 (to U_2) that forces the programming voltage to 0 V. The performance of this circuit is adequate for all E-PROMs except for Motorola's MC68766, which requires a fast 12-V/µs slew rate. □

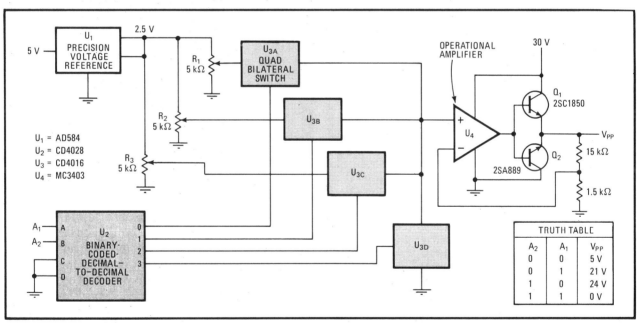

Programmable supply. Three ICs, a voltage reference and a few discrete components form this logic-controlled programmable voltage source with its approximately 1-V/µs slew rate. The supply is capable of providing four output voltages with 0 V as an off position.

Bipolar PROMs make versatile Camac instruction decoders

by István Hernyes and János Nagy
Central Research Institute for Physics, Budapest, Hungary

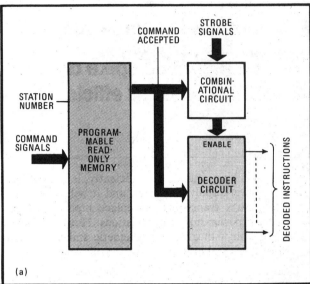

Whenever a computer is hooked up to peripheral devices using a data bus and there are more data lines needed than can be provided by an IEEE-488 interface system, the Camac instruction and interface standard may be inserted into the IEEE-488 to pick up the slack. Inexpensive and versatile, the computer-automated measurement-and-control instruction decoder may be easily built by exploiting the fast access time and low cost of bipolar programmable read-only memories. This principle may

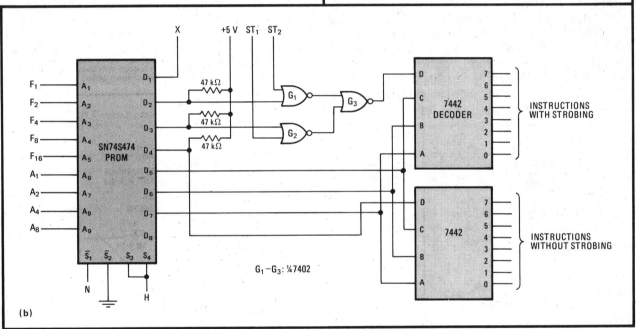

Decoder. A Camac instruction decoder (top) uses a bipolar PROM as a code converter. The PROM's output also controls the combinatorial circuit employed to strobe the decoded instructions. One such circuit (bottom) uses the SN74S474. Various pull-up resistors are connected to the pins D_2 through D_4 to guarantee high-level logic when the module is not selected by N.

also be applied to other bus-oriented systems.

Codes for every Camac instruction are programmed into the memory, and the decoder uses a bipolar PROM as a code-converter prior to the decoding circuit (see figure, top). The Camac module is specified by a station number N. Five function lines: F_1, F_2, F_4, F_8, and F_{16}, and four subaddress lines: A_1, A_2, A_4, and A_8 constitute a command for a specified module.

During each command operation, the Camac controller generates two sequential strobe signals ST_1 and ST_2 on separate bus lines. Whenever a module recognizes a command, it generates an active logic level on the X-line. The command signals are maintained for the full duration of dataway operation.

The station N and the function and subaddress lines are all tied to the address input of the PROM, which converts the 9-bit instruction into a 6-bit instruction for the decoder and combinational circuit. The N line may be connected to the enable input as well.

The output of the decoder circuit corresponds to different Camac instructions and one output of the memory represents an X-answer. The decoder circuit may be enabled for a full Camac cycle or the duration of a strobe signal.

The circuit in the bottom of the figure uses a 512-by-8-bit PROM that has three state outputs. The D_1 pin of the PROM gives the X answer when the Camac instruction is identified by the module. The combinational circuit has three NOR gates. The zero through seven pins of the upper 7442 decoder correspond to instructions that must be strobed by either ST_1 and D_2 (low) or ST_2 and D_3(low). Pins 0 through 7 of the lower 7442 generate instructions without strobing and are selected by pin D_4 of the PROM. □

Access control logic improves serial memory systems

by Robert G. Cantarella
Burroughs Corp., Paoli, Pa.

File storage systems used in large scientific processors or computers need serially organized dynamic random-access-memory systems to transfer blocks of data from one memory unit to another in a serial order. Dynamic RAM devices used in such secondary stores need to be refreshed in a fixed cyclic order to keep the cost and latency low. This refresh logic loop further reduces latency, while retaining the cost advantage of a serial organization, and refreshes the memory at twice the minimum rate. In addition, zero latency is guaranteed with block transfers of at least L milliseconds.

When an initial transfer request XFERRQST occurs, the requester's address RQSTADDR is loaded into the refresh counter, which resets the refresh loop. This setting initiates the transfer (XFERSTART) with zero latency. The refresh address is then held in the refresh counter. Because the refresh loop is cycling at twice the required frequency, all bits are properly refreshed within L ms. The reset is then enabled only after a full cycle has completed when the time-out down counter is loaded with L ms. However, this action only occurs when the time-out down counter is zero.

After the transfer request is initiated, access is granted either when the requester's address is identical to the refresh address or when the time-out down counter is zero. Thus latency is a function of the requester's address and the time since the last reset. As a result, a block transfer of L ms guarantees that the next transfer request will be granted immediate access.

The system performs like a random-access memory but retains the cost advantages of a serial-memory system. The constant refresh rate results in a steady current drain, thereby reducing the cost of the power-supply system and storage cards. □

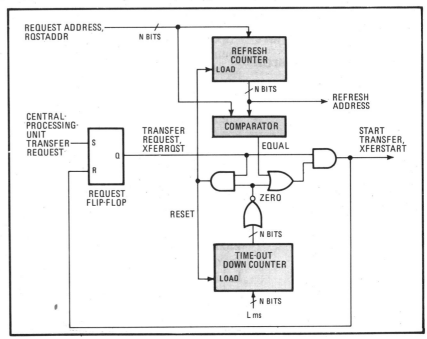

Access. The access control logic refreshes the memory at twice the requisite frequency to keep the latency low. Access is granted when the requester's address is identical to the refresh address or the time-out down counter is zero, whichever occurs first. Block transfer of L ms guarantees zero latency.

Deselected RAM refuses all data during power failure

by Dan Stern
Bayly Engineering, Toronto, Canada

A nonvolatile random-access memory is usually paired with a standby battery to prevent data loss during a power failure. However, when the power drops with such a circuit, the chip select ($\overline{CS}$) is placed in an undefined state that allows invalid data to be written into the memory. One cure for this condition is to deselect the battery-backed RAM when the main power supply drops below a specified threshold voltage.

During normal operation, memory in the battery-backup circuit is powered with the supply voltage minus the voltage drop across the germanium diode D_1, which is typically 0.2 volt. This diode along with resistor R_3 helps charge the rechargeable battery. The value of R_3 is selected so that a current 10% of the nominal charging rate will flow to the RAM.

The chip-deselect circuit (see figure), comprising transistors Q_1 and Q_2 and zener diode D_2, operates as two cascaded inverters while the main supply is present. If the main power supply drops below the threshold voltage, which is determined by D_2 and the voltage across Q_2's base and emitter, transistor Q_2 turns off. This action pulls $\overline{CS}$ to a logic high that is provided by the battery. As a result, the circuit deselects the memory chip. □

Subroutine tests RAM nondestructively

by Steve Strom
Motorola Inc., Semiconductor Group, Phoenix, Ariz.

☐ Nondestructive test patterns for random-access memories make it possible to check data on line even while executing programs. The RAM subroutine described here, which runs on the Motorola MC68000 16-bit microprocessor, tests a read/write memory of any length and requires only the starting and ending addresses of the storage block under scrutiny.

The program is said to be linear because the time required for the test goes up in a linear fashion as the number of locations increases. A linear pattern can never be as thorough as a higher-order test using, say, an N^2 pattern. But since the complexity of an N^2 pattern rises as the square of the input, testing time may become intolerably long. Quadrupling the size of the RAM increases the time for an N^2 test pattern by 16 but only by 4 for a linear pattern, possibly saving seconds to test a chip or minutes to test a larger system.

The test passes over memory three times. On the first pass, the subroutine calculates a checksum by exclusive-ORing each byte with data register D_0. The final result is kept in another data register, D_2.

In the second pass, another such checksum is taken, but the result is left in D_0. Also on this pass, after each byte is read, it is complemented, read again, and compared with the complement to insure that each location can store its own complement. When the top of memory is reached, the checksum is compared with the one taken on the first pass.

The last phase of the test is identical to the second, except that the program begins at the top of memory; as before, its checksum is compared with that obtained on the first pass. In the end, every location has been returned to its original state. If a failure was encountered during the second pass, the test routine sets a flag but continues testing to search for any additional faults downstream.

The program sets a flag whenever a failure is encountered. Before returning to the calling program, the subroutine tests the error flag and sets condition codes accordingly. Pinpointing the faulty component usually requires additional, off-line diagnostics.

With minor modification, the routine could provide some diagnostic information. For example, it could store the address of each faulty location, as well as the data written into and read from the RAM. In fact, the program listing provided stores the first bad address in an error register. ☐

NONDESTRUCTIVE RAM TEST SUBROUTINE

ENTRY REQUIREMENT:
A0 = BEGINNING ADDRESS OF RAM TO BE TESTED
A1 = ENDING ADDRESS + 1 OF RAM

RETURN CONDITIONS:
IF NO ERROR ENCOUNTERED — ZERO FLAG SET
IF ERROR ENCOUNTERED — ZERO FLAG NOT SET
& A2 = FAILED ADDRESS

TYPICAL CALLING ROUTINE:

```
      LEA   RAMBEG, A0
      LEA   RAMEND+1, A1
      BSR   RAMTST
      BNE   ERROR
```

```
RAMTST  MOVE.L  A0,A2       USE A2 AS A POINTER
        CLR.L   D7          USE D7 AS AN ERROR FLAG
        CLR.W   D0          USE D0 TO CALCULATE RAM CHECK SUM
EXOR    MOVE.W  (A2)+,D1    GET A BYTE FROM MEMORY
        EOR.W   D1,D0       EXCLUSIVE-OR IT TO D0
        CMP.L   A1,A2       SEE IF AT END OF MEMORY
        BNE.S   EXOR        CONTINUE UNTIL END OF MEMORY REACHED
        MOVE.W  D0,D2       SAVE EXOR RESULT IN D2

        MOVE.L  A0,A2       GO BACK TO BEGINNING OF MEMORY
        CLR.W   D0          GET READY TO RECALCULATE CHECK SUM
LOOP1   MOVE.W  (A2),D1     GET A BYTE FROM MEMORY
        EOR.W   D1,D0       EXCLUSIVE-OR IT TO D0
        NOT.W   D1          COMPLEMENT THE DATA
        MOVE.W  D1,(A2)     WRITE IT BACK TO MEMORY
        CMP.W   (A2)+,D1    SEE IF IT WAS STORED
        BEQ.S   PASS1       IF GOOD, THEN JUMP AROUND ERROR LOGGING
        TST.L   D7          SEE IF THERE WAS A PREVIOUS ERROR
        BNE.S   PASS1       IF SO, THEN JUMP AROUND ERROR LOGGING
        MOVE.L  A2,D7       RECORD FAILED ADDRESS
PASS1   CMP.L   A1,A2       SEE IF AT END OF MEMORY
        BNE.S   LOOP1       IF NOT, THEN STAY IN PASS1 LOOP
        CMP.W   D0,D2       SEE IF CHECK SUM CORRECT
        BEQ.S   END1        IF GOOD, THEN JUMP AROUND ERROR LOGGING
        TST.L   D7          SEE IF PREVIOUS ERROR ENCOUNTERED
        BNE.S   END1        IF SO, THEN JUMP AROUND ERROR LOGGING
        MOVE.L  A2,D7       SAVE FAILED ADDRESS

END1    CLR.W   D0          GET READY FOR NEXT CHECK SUM CALCULATION
LOOP2   MOVE.W  -(A2),D1    GET A BYTE FROM MEMORY
        NOT.W   D1          COMPLEMENT THE BYTE
        EOR.W   D1,D0       EXCLUSIVE-OR IT TO D0
        MOVE.W  D1,(A2)     WRITE IT BACK TO MEMORY
        CMP.W   (A2),D1     SEE IF IT GOT THERE
        BEQ.S   PASS2       IF GOOD, THEN JUMP AROUND ERROR LOGGING
        TST.L   D7          SEE IF THERE WAS A PREVIOUS ERROR
        BNE.S   PASS2       IF SO, THEN JUMP AROUND ERROR LOGGING
        MOVE.L  A2,D7       RECORD FAILED ADDRESS
PASS2   CMP.L   A0,A2       SEE IF AT BEGINNING OF MEMORY
        BNE.S   LOOP2       IF NOT, THEN STAY IN PASS2 LOOP
        CMP.W   D0,D2       SEE IF CORRECT CHECK SUM
        BEQ.S   END2        IF SO, THEN END OF TEST
        TST.L   D7          SEE IF THERE WAS A PREVIOUS FAILURE
        BNE.S   END2        IF SO, THEN STOP HERE
        MOVE.L  A2,D7       SAVE FAILED ADDRESS

END2    MOVE.L  D7,A2       POINT A2 TO FAULTY ADDRESS
        TST.L   D7          SET CONDITION CODES
        RTS                 RETURN TO CALLER
```

Random-access memories form E-PROM emulator

by David J. Kramer
Sunnyvale, Calif.

Many low-cost microprocessor kits that are used as educational tools possess a relatively small amount of on-board random-access memory and are incapable of transferring programs to erasable programmable read-only memories. This E-PROM emulator overcomes both of these limitations.

Four 2114 RAMs are connected in a 2-K-by-8-bit configuration (Fig.1). To emulate an E-PROM, this memory

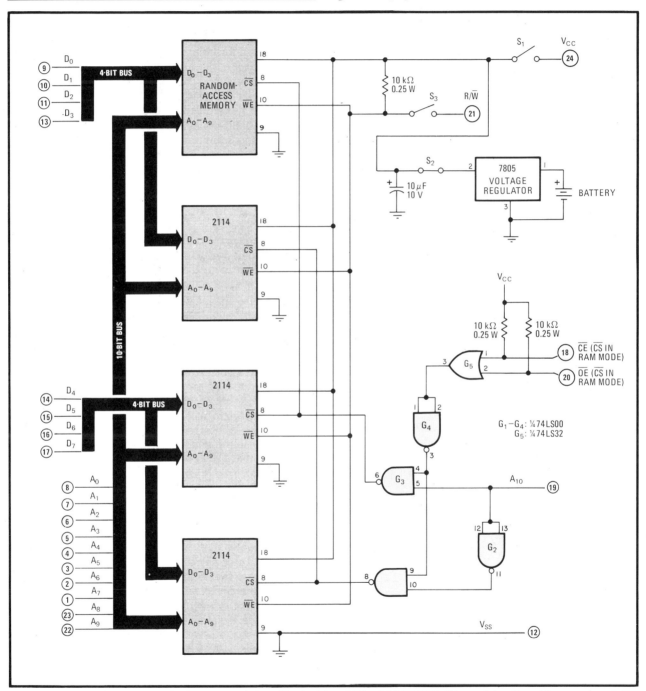

1. Emulator. Four 2114 RAMs are wired on a separate board to simulate a 2-K-by-8-bit E-PROM. Using switches S_1 and S_2, power supply V_{CC} for the emulator can be taken either from a trainer's kit or a battery. Numbers within circles denote pinouts for the 24-pin DIP.

2. Package. The pinouts of the emulator in both its modes—random-access memory and erasable programmable read-only memory—are as shown in (a) and (b). The only differences are in pins 18, 20, 21, and 24—in the E-PROM mode, pins 18 and 20 function as chip and output enable.

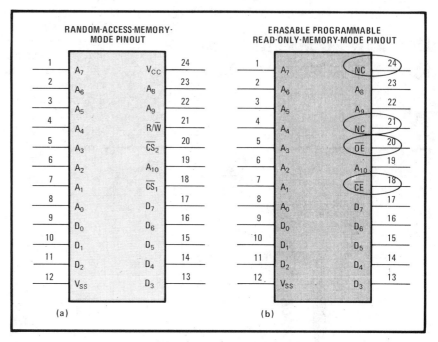

circuit is built on a separate board and has its leads terminating in a 24-pin dual in-line package. The power supply for the emulator is either from the microprocessor training kit (V_{cc}) or a battery and is coupled to the emulator by using switches S_1 or S_2. The $\overline{WE}$ line may be disconnected from the plug with S_3.

To function as an emulator, the 24-pin DIP must be connected to the appropriate address-decoding circuitry in the trainer. The memory is then plugged in and either S_1 or S_2 is closed (depending on the selected power supply). This puts the circuit in its RAM mode (Fig. 2a). In this mode, pins 18 and 20 function as chip-select controls. Pin 21 is the read-write input.

Once the data to be preserved is in the RAM, the battery should be (if not already) connected to the 7805 voltage regulator, and S_2 closed and then S_1 and S_3 opened. The circuit is now in its E-PROM mode (Fig.2b) and may be inserted in the master socket of any programmer that will handle a 2716 E-PROM.

The emulator will remain to be powered while the programming occurs. The type of keep-alive battery needed depends on the power requirements of the 2114 RAMs and the length of time the emulator is to remain in the E-PROM mode. □

Nonvolatile RAM provides on-board storage for computer

by Rex L. Berney
University of Dayton, Dayton, Ohio

The lack of convenient permanent memory storage presents a problem for those who use low-cost, single-board microcomputers. However, a nonvolatile random-access memory recently introduced by NCR can provide a microcomputer with permanent on-board storage. This new device has a shadow programmable read-only memory on the same chip. On command, the contents of the RAM can be stored into or recalled from this shadow PROM, a system that enables a nonvolatile RAM to work like a disk or cassette.

Nonvolatile RAM U_1 and 2716 erasable PROM U_3 are connected to the bus of Intel's SDK-8085 microcomputer along with several support devices for decoding and

PROGRAM LISTING FOR STORING AND RECALLING PROGRAMS TO CONTROL A 4485 NONVOLATILE RANDOM-ACCESS MEMORY FROM AN SDK-8085 MICROCOMPUTER

Location	Object code	Sequence	Source statement			Comments
		00220				; * * PRRECL * *
		00230				;
		00240				; Recall program — restores from the nonvolatile part of the
		00250				; 4485 nonvolatile RAM into the RAM part
		00260				; of this chip
		00270				; Note: RAM contents are overwritten
		00280				;
9000		00290		ORG	9000H	
9000	31C220	00300	PRRECL	LD	SP, 20C2H	; Set up stack pointer
9003	3E01	00310		LD	A, 01H	; Disable write enable to nonvolatile RAM
9005	D303	00320		OUT	(03H), A	; From computer
9007	D301	00330		OUT	(01H), A	
9009	3E03	00340		LD	A, 03H	; Pull write enable high
900B	D303	00350		OUT	(03H), A	
900D	D301	00360		OUT	(01H), A	
900F	3E07	00370		LD	A, 07H	; Pull write enable and nonvolatile high
9011	D303	00380		OUT	(03H), A	
9013	3E01	00390		LD	A, 01H	; Pull write enable and nonvolatile low
9015	D301	00400		OUT	(01H), A	
9017	3EFF	00410		LD	A, 0FFH	; Delay 256 loops
9019	3D	00420	LP1	DEC	A	; Actual recall at this time
901A	C21990	00430		JP	NZ, LP1	
901D	3E03	00440		LD	A, 03H	; Pull write enable high nonvolatile low
901F	D301	00450		OUT	(01H), A	
9021	3EFF	00460		LD	A, 0FFH	; Delay 256 loops
9023	3D	00470	LP2	DEC	A	
9024	C22390	00480		JP	NZ, LP2	
9027	3E07	00490		LD	A, 07H	; Pull write enable and nonvolatile high
9029	D301	00500		OUT	(01H), A	
902B	3E00	00510		LD	A, 00H	; Tristate port 1
902D	D303	00520		OUT	(03H), A	
902F	CF	00530		RST	08H	; Go to warm start RST1
		00540				;
		00550				;
		00560				; * * PRSTOR * *
		00570				;
		00580				; Storing program — stores the RAM contents of the
		00590				; 4485 nonvolatile RAM into the
		00600				; nonvolatile part of this chip
		00610				;
		00620				;
9040		00630		ORG	9040H	

```
9040   31C220   00640   PRSTOR  LD    SP, 20C2H     ; Set up stack pointer
9043   3E01     00650           LD    A, 01H        ; Disable write enable to nonvolatile RAM
9045   D303     00660           OUT   (03H), A      ; From computer
9047   D301     00670           OUT   (01H), A
9049   3E03     00680           LD    A, 03H        ; Pull write enable high
904B   D303     00690           OUT   (03H), A
904D   D301     00700           OUT   (01H), A
904F   3E57     00710           LD    A, 57H        ; Enable ± 22 volts
9051   D303     00720           OUT   (03H), A
9053   3E03     00730           LD    A, 03H        ; Pull write enable high nonvolatile low
9055   D301     00740           OUT   (01H), A
9057   3E13     00750           LD    A, 13H        ; Turn on −22 volts
9059   D301     00760           OUT   (01H), A
905B   3EFF     00770           LD    A, 0FFH       ; Delay 256 loops
905D   3D       00780   LP3     DEC   A
905E   C25D90   00790           JP    NZ, LP3
9061   3E03     00800           LD    A, 03H        ; Turn off −22 volts
9063   D301     00810           OUT   (01H), A
9065   060A     00820           LD    B, 0AH        ; Setup 10 loop counter
9067   3EFF     00830   AG1     LD    A, 0FFH       ; Delay 256 loops
9069   3D       00840   LP4     DEC   A
906A   C26990   00850           JP    NZ, LP4
906D   3E43     00860           LD    A, 43H        ; Turn on +22 volts
906F   D301     00870           OUT   (01H), A
9071   3ED0     00880           LD    A, 0D0H       ; Delay 1 ms
9073   3D       00890   LP5     DEC   A
9074   C27390   00900           JP    NZ, LP5
9077   3E03     00910           LD    A, 03H        ; Turn off +22 volts
9079   D301     00920           OUT   (01H), A
907B   3EFF     00930           LD    A, 0FFH       ; Delay 256 loops
907D   3D       00940   LP6     DEC   A             ; Between +22 volts
907E   C27D90   00950           JP    NZ, LP6       ; Pulses
9081   05       00960           DEC   B
9082   C26790   00970           JP    NZ, AG1       ; Do ten +22 pulses
9085   3E07     00980           LD    A, 07H        ; Disable ±22 volts
9087   D301     00990           OUT   (01H), A
9089   3E00     01000           LD    A, 00H        ; Turn off port 1
908B   D303     01010           OUT   (03H), A
908D   CF       01020           RST   08H           ; Warm start RST1
                01030
0000            01040           END
00000   Total errors

LP6      907D
LP5      9073
LP4      9069
AG1      9067
LP3      905D
PRSTOR   9040
LP2      9023
LP1      9019
PRRECL   9000
```

Note: Z80 Assembler, but machine code runs on 8080 and 8085 as well.

controlling the nonvolatile RAM and E-PROM (see figure). Several bits of port 1 of SDK-8085 are used to control the nonvolatile RAM.

When in the RAM mode of operation, the nonvolatile RAM has all the features of a static one. In this mode, the nonvolatile select line (pin 16 of U_1) is held high. In the erase, store, and recall modes of operation, the nonvolatile control line is low. The shadow PROM may be erased by applying a −22-volt pulse for 1 millisecond to pin 15 of U_1. The storage of RAM contents into the shadow PROM is achieved by applying ten 1-ms pulses of +22-v to pin 15. Lastly, the contents are recalled from the shadow PROM back into the RAM by holding pin 15 at ground and pulling the write-enable line low for 10

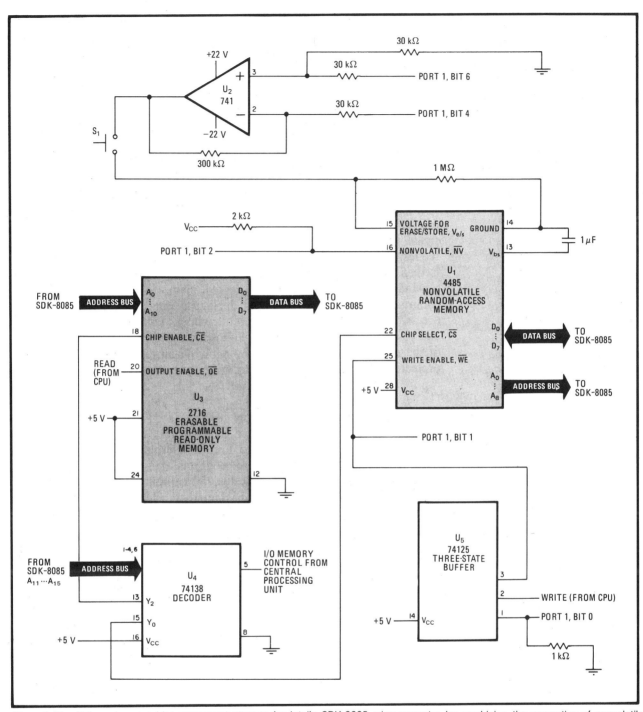

Nonvolatile. This system provides permanent storage for Intel's SDK-8085 microcomputer by combining the properties of nonvolatile random-access memory 4485 and E-PROM 2716 and linking them to the 8085 microcomputer unit. All the required nonvolatile-RAM control signals for erase, store, and recall are supplied through port 1 of the SDK-8085 microcomputer.

microseconds. It must be noted that for erase, store, and recall operations between the shadow PROM and the RAM, all the memory locations are moved simultaneously rather than 1 byte at a time.

Operational amplifier U_2 serves as a switch for the erase and store ($V_{e/s}$) voltages. Switch S_1 protects the programs stored in the nonvolatile part of U_1. The switch is held down when the contents of the nonvolatile section of U_1 are being changed. All the control signals for U_1 are supplied by port 1 of the microcomputer. In addition, correct sequencing and timing of these signals is accomplished through two programs contained in E-PROM U_3. Programs for controlling the store and recall functions of U_1 are listed in the table.

Decoder U_4 decodes U_1 for memory locations 8600H through 87FFH and U_3 for locations 9000H to 97FFH. Tristate buffer U_5 disables the write-enable line from the central-processing-unit bus so that it may be controlled directly from port 1. □

Simple interface links RAM with multiplexed processors

by Jeffrey M. Wilkinson
Harris Semiconductors, Melbourne, Fla.

Normally a sophisticated latch and logic circuitry are required to hook up extra random-access memory to a microprocessor whose memory has run short. However, users wanting to link multiplexed microprocessors with complementary-MOS RAMs will find the assembly much easier and cheaper with this interface, which has only one decoder circuit.

The design takes advantage of the address latches incorporated in the C-MOS RAM used, the Harris HM-6516. Circuit reliability is increased by the reduced parts count, while circuit size and cost along with the application-design time are all also reduced.

As an example, a 2-K-by-8-bit C-MOS static RAM is linked to an Intel 8085A microprocessor. At the start of the memory cycle, the input/output memory control (IO/$\overline{M}$) and the address-latch-enable (ALE) lines move low and the address decoder generates RAM-select. This low-going transition should occur while addresses on the bus are valid.

The decoder's output is fed to the RAM chip-enable pin ($\overline{E}$), which latches the chip's bus address. Address information is then latched in the on-chip address registers by the falling edge of $\overline{E}$. Also, the RAM's address and select inputs should occur simultaneously. After the cycle is complete and the addresses are removed from the bus, a memory read or write cycle occurs.

During the write cycle, the write-control ($\overline{WR}$) pin supplies a write pulse to the write-enable ($\overline{W}$) input on the RAM. The read-control ($\overline{RD}$) pin is held high so that the RAM's output is inactive during this cycle.

The read cycle is similar to the write cycle except that the $\overline{WR}$ line remains high during this interval. The read-control ($\overline{RD}$) line moves low, thus forcing the RAM's output-enable line also to go low. As a result, the RAM outputs are enabled and data can be transferred to the processor through the bus. □

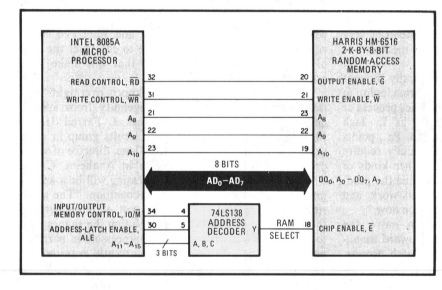

Hook-up. This one–decoder-circuit interface facilitates an easy exchange of data between C-MOS random-access memories having on-chip address latches and multiplexed microprocessors by carefully selecting memory elements that need to be tied to the processor. Circuit assembly time and costs are negligible.

13. MICROPROCESSORS

Redundancy increases microprocessor reliability

by Dan Stern
Stern Engineering, Willowdale, Ont., Canada

A classic way of increasing a microprocessor's reliability is to back it up with an additional identical microprocessor. This circuit generates a pulse to interrupt the processors (or, in this case, microcontrollers) whenever a faulty operation occurs. The pulse then initiates a software recovery routine for changing over to the properly functioning processor.

One-shots U_1 and U_2 continuously monitor the watchdog pulses generated by each microcontroller circuit (see figure). To convert these pulses into logic levels, time constants T_1 and T_2 must be greater than T_{wd1} and T_{wd2}, respectively.

Under normal operation, microcontroller No. 1 is assumed to be the active processor. Therefore when the power is turned on, the active microcontroller produces watchdog pulse T_{wd1} first. As a result, output Q_1 of U_1 is in a high state that prevents one-shot U_2 from being triggered. This enables three-state bus driver U_3 and forces U_4 into the three-state mode, thereby connecting bus No. 1 to the common bus.

When microcontroller No. 1 fails to produce a watchdog pulse, Q_1 of U_1 goes low and enables one-shot U_2, which allows watchdog pulse T_{wd2}, generated by the other microcontroller, to trigger U_2. This triggering changes the state of Q_2 to a high logic level, which in turn disables U_3 and switches the common bus to bus No. 2. Simultaneously, a high Q_2 disables U_1 and prevents the common bus from being switched back if unpredictable pulses occur at the first watchdog output.

Every time a changeover occurs, U_5 generates a short pulse to interrupt the microprocessors. As a result, the software recovery routine is activated and ensures a proper changeover to the functional microcontroller.

Each bus may contain as many bus drivers as required, depending on the number of lines per bus. Drivers U_3 and U_4 are switch-selected to operate as input, output, or bidirectionally. □

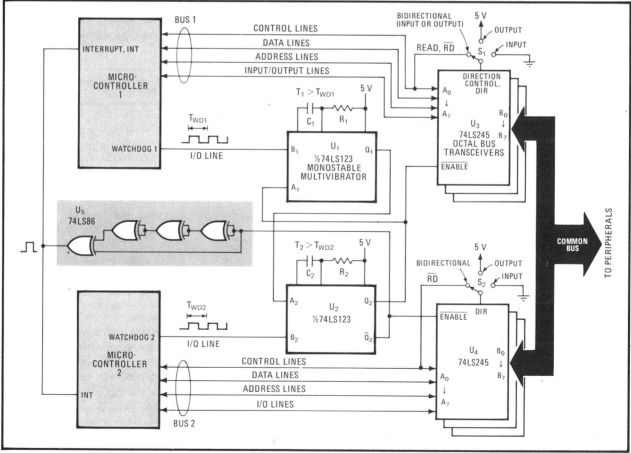

Reliable. Two identical microprocessor circuits are interfaced with a common bus, with microcontroller No. 2 functioning as a back-up circuit in the event of a faulty operation or failure. The watchdog pulses generated by the processors continuously monitor the circuit. When a changeover occurs, U_5 generates a pulse to interrupt the microprocessors and thereby activates a software recovery routine.

Macroinstruction for Z80 guarantees valid output data

by Daniel Ozick
Irex Medical Systems, Ramsey, N. J.

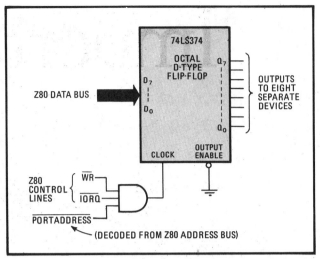

Output port. A standard 74LS-series octal latch is used as an output-only port in a Z80 microprocessor system. The Z80 instruction sequence guarantees that the output port is always in a valid state. The OUTI instruction latches data into the flip-flop.

Microprocessor-based systems often utilize 74LS-series octal latches as output-only ports (see figure). With this Z80 instruction sequence, a single output line can be made to change its state without affecting the states of the others. The program guarantees valid output data.

The standard practice has been to maintain an up-to-date image copy of the states of the output lines in the memory. When the state of an output line is to be changed, the image is fetched from the memory, updated with the desired bit change, transferred to the output, and stored back in the memory again. This procedure is adequate for systems without interrupts, but is unsafe when interrupt and background (noninterrupt) portions of a program share the same output port.

The conflict between interrupt and background can be avoided by using the safe macroinstruction shown in the table. In this sequence, the Z80 SET or RES instruction modifies the port image in memory in one uninterruptible instruction, ensuring that the port image is always valid. The output-and-increment (OUTI) instruction transfers the port image to the output port hardware in another indivisible instruction, ensuring that the image always matches the state of the port hardware. Thus, neither image nor port is left in an invalid state.

This sequence works well with Z80 systems, where interrupts cannot be disabled, and does not impose constraints on the assignment of devices to ports. □

SAFE PORT-OUTPUT MACROINSTRUCTION FOR THE Z80 MICROPROCESSOR	
Source statement	Comments
PortOut MACRO, PortAddress, BitNum, BitVal	; microsoft macro definition
	; input arguments are the address of the port (value ; from 0_{16} to $0FF_{16}$), the number of the bit to be ; modified, (values from 0 to 7), and the new bit ; value (0 or 1)
	; example of macro invocation using symbolic ; constants for port address and bit number: ; PortOut LightPort, ErrorLight, 1
PUSH BC PUSH HL	; save the registers to be used in the instruction ; sequence (optional)
LD HL, PortImage	; use HL as a pointer to the port image
IF BitVal EQ 0	; assemble the next line if the bit to be outputted ; has a value of 0
RES BitNum, (HL)	; in one indivisible instruction, reset the bit in the ; port image pointed to by HL
ELSE	; assemble the next line if the bit to be outputted ; has a value of 1
SET BitNum, (HL)	; in one indivisible instruction, set the bit in the ; port image pointed to by HL
ENDIF	; end of the conditional assembly
LD C, PortAddress OUTI	; (C) gets (HL), HL gets HL+1, B gets B−1 in one ; indivisible instruction, the port image held in ; memory is transferred to the port
POP HL POP BC	; restore the pushed registers ;
ENDM PortOut	; end of the PortOut macro definition

Interfacing C-MOS directly with 6800 and 6500 buses

by Ralph Tenny
George Goode & Associates Inc., Dallas, Texas

Most complementary-MOS devices used with the early members of the 6800 and 6500 microprocessor family are interfaced with the processor bus through programmable interface adapters because bus output data is available for only a short period of time. However, C-MOS devices can be added directly to 6800 and 6500 systems with this circuit. It is assumed some memory-address space is available.

Bus input timing is no problem. It is such that C-MOS latches with output-disable capability can be enabled with an AND/NAND gate of an address-decoding strobe and READ/$\overline{\text{WRITE}}$ signals.

When unassigned blocks of memory-address space exist, C-MOS latches and registers can serve as output registers residing on the processor bus. This technique is demonstrated through a 256-byte block of unused memory that interfaces bus data with a two-character hexadecimal display that is resident on the microprocessor bus (Fig. 1a).

With this design, the high-order address lines are decoded for port selection and the low-order address lines are used to transmit data to that port. Thus the 4-bit binary word is displayed as the equivalent hexadecimal word on the processor bus. The timing information (Fig. 1b) uses worst-case delays associated with C-MOS parts.

Address lines A_4 to A_7 can be decoded through the use of a dual four-input OR gate for U_2 so that a 16-bit output port occupies 256 bytes of memory-address space. The four C-MOS D-type registers U_5 through U_8 use the basic strobe to clock data into the latch (Fig. 2). The four address lines are now used as device enables, so that 4-bit data is entered into one of the four C-MOS parts, resulting in a 16-bit output port. □

2. Expansion. A 16-bit C-MOS parallel output port uses low-order address lines for both data and address. The four C-MOS devices U_5 through U_8 use the strobe to clock the data into the latch. A 16-bit output occupies 256 bytes of address space.

1. C-MOS interface. The logic in (a) allows a C-MOS latch to reside directly on the 6800 and 6500 microprocessor buses. It uses unassigned memory-address space to mix address and data information for a port residing on the processor bus. The bus timing (b) shows that microprocessor data is available for a very short time.

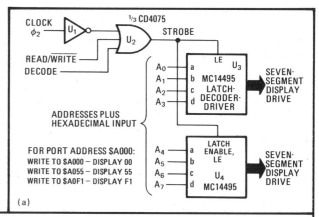

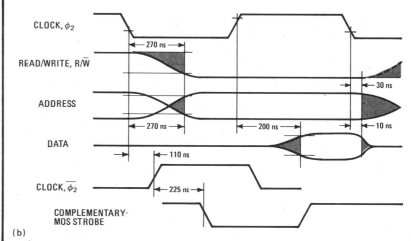

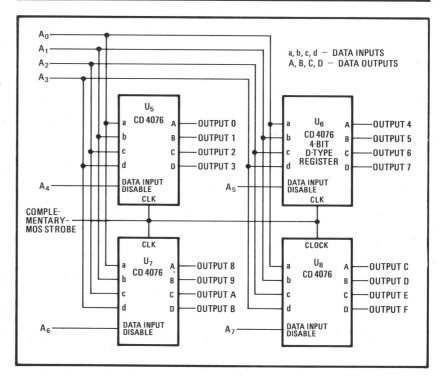

Static RAM relocates addresses in PROM

by Irving Gold
Basic Systems Corp., Santa Clara, Calif.

With programmable read-only memories steadily decreasing in price, their use is becoming a cost-effective means of interfacing a microprocessor with its memory. Unfortunately, the user must replace the PROM each time the address-decoding space needs changing. However, by using a static random-access memory, this design allows dynamic-address relocation and translation, provides remote control and diagnostic capabilities, and enables the

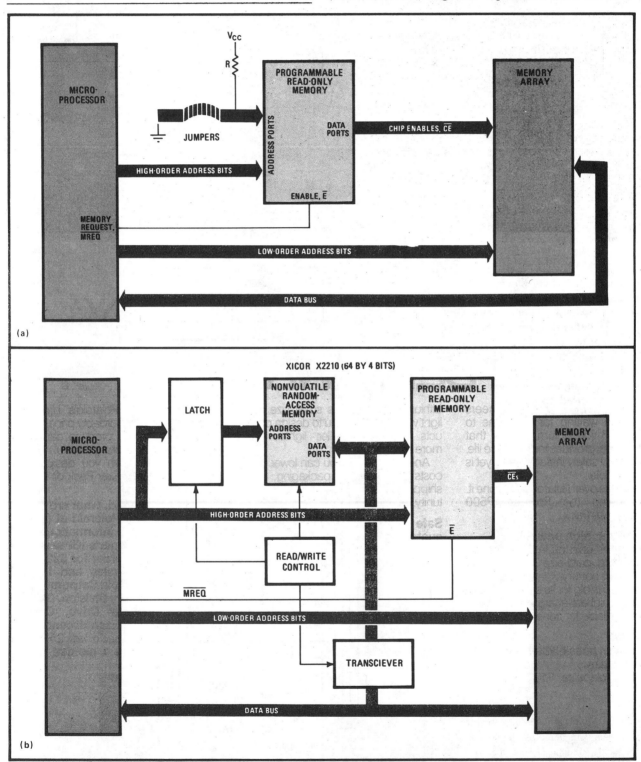

Relocating. Jumper wires select address location options for a system that uses a PROM to interface the microprocessor with its memory (a). When the jumpers are replaced with Xicor's 64-by-4-bit static RAM (b), dynamic address decoding under program control may be attained.

operator to select address locations easily.

To interface a processor with its memory, a high-order address from the microprocessor is usually fed into the PROM address inputs, and the PROM data-output lines enable chips in the memory array. The data pattern corresponding to the desired address space is then preprogrammed into the PROM that decodes the address. However, the PROM must be replaced whenever the address space has to be changed.

A better method is to preprogram the PROM with various address options and to use jumper wires to select the desired address location (a). But this solution is incomplete because the system must still be shut down to remove the board and latch up the right jumpers.

The shortcomings of earlier designs may be overcome by replacing the jumpers with a 64-by-4-bit RAM like the Xicor X2210 (b). This substitution allows the user to load the static RAM directly with a data pattern corresponding to the desired address space. As newer high-speed static RAMs are introduced, it may prove possible for just one of them to replace this latch, static RAM, and PROM combination. □

8X300 microcontroller performs fast 8-by-8-bit multiplication

by Sam Mallicoat
Tektronix Inc., Beaverton, Ore.

Although it is not geared to number crunching, the Signetics 8X300 bipolar microprocessor can execute its compact control-oriented instruction set rapidly and excels at bit testing and manipulation—qualities which make it easy to implement this high-speed algorithm for multiplying two 8-bit words. The multiplication routine can be performed in as little as 19 microseconds and in 24 microseconds in the worst-case situation.

Conventional routines perform an add and/or shift operation for each bit of the multiplier word to achieve fixed-point multiplication. This program, based on a technique described in detail by MacSorley,[1] halves the number of operations, and thus the execution time, by examining the multiplier's bits in groups. Depending on the group value, the multiplier is added to or subtracted from the previous partial product, either once or twice.

With the multiplicand and multiplier placed in registers R_3 and R_4, respectively, the 16-bit product appears at registers R_1 (high-order byte) and R_2 (low-order byte). All numbers are expressed in 2's complement format. □

References
1. O. L. MacSorley, "High-Speed Arithmetic in Binary Computers," Proceedings of the IRE, Jan. 1961, pp. 67–91.

MULTIPLICATION ROUTINE FOR 8X300 MICROCOMPUTER					
Label	Source statement	Comment	Label	Source statement	Comment
	XMIT 0, R1	clear product high byte		NZT OVF, LOOP	test for done
	XMIT 0, R2	clear product low byte		JMP EXIT	exit routine
	XMIT 377, AUX	place 2's complement of	TAB1:	XMIT 363, R6	
	XOR R3, R5	multiplicand in R5		XMIT 374, R6	
	XMIT 1, AUX		TAB2:	JMP SHIFT	
	ADD R5, R5			JMP P1	
	AND R3(7), AUX	test the sign of multiplicand		JMP P1	
	XEC AUX, TAB1			JMP P2	
	XMIT 0, R11	prepare multiplier		JMP M2	
	MOV R4(7), R4	mask for first pass		JMP M1	
	XMIT 6, AUX			JMP M1	
	JMP ENT			JMP SHIFT	
LOOP:	MOV R4(2), R4	rotate multiplier right twice	P1:	MOV R6(4), R11	add multiplicand once
	XMIT 7, AUX			MOV R3, AUX	
ENT:	AND R4, AUX	mask lower 3 bits		JMP SUM	
	XEC AUX, TAB2	eight-way conditional branch	P2:	MOV R6(4), R11	add multiplicand twice
SUM:	ADD R1, R1			MOV R3, AUX	
SHIFT:	XMIT 77, AUX	shift low product byte		ADD R1, R1	
	AND R2(2), R2	right twice		JMP SUM	
	XMIT 300, AUX		M2:	MOV R6(2), R11	subtract multiplicand twice
	ADD R6, R6	increment loop counter		MOV R5, AUX	
	AND R11, R11			ADD R1, R1	
	AND R1(2), AUX	shift 2 least significant bits of high		JMP SUM	
	XOR R2, R2	product byte to low product	M1:	MOV R6(2), R11	subtract multiplicand once
	XMIT 77, AUX	shift high product byte		MOV R5, AUX	
	AND R1(2), AUX	right twice		JMP SUM	
	XOR R11, R1	correct for product sign			

Octal latches extend bus hold time

by Jim Handy
Intel Corp., Santa Clara, Calif.

Although the timing of microprocessor-generated data and control signals is adequate for most applications, occasions do arise when the data hold time needs to be extended. Logic-analyzer designs and systems using slow memory or several stages of bus buffering are only two examples of the need for this retention. The simple circuit shown here uses just a chip and a spare gate to increase the data retention time to the entire duration of a bus cycle.

The outputs of the 74LS373 octal latch are each tied to appropriate data bus lines (a). Control signals are derived by combining the read and write signals through OR gate U_2, a function that is required only when the circuit is to be used for logic analysis. In other systems where a slower microprocessor is needed, the write signal alone needs to be connected to the clock and output control pins of the latch.

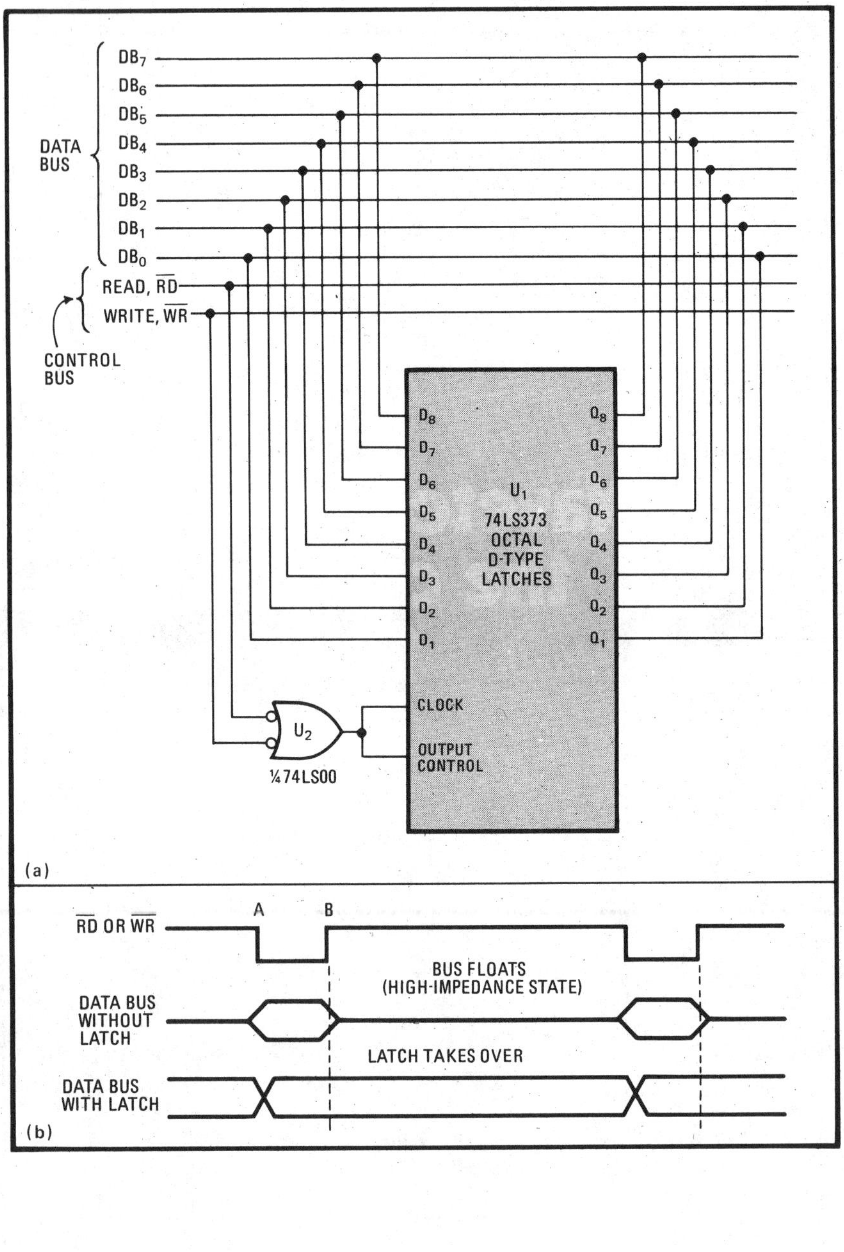

At reference point A in the timing diagram (b), the read or write signal disables the latch outputs, thereby allowing the data bus to assume the normal mode. The microprocessor or memory device drives the bus, and the clock input is simultaneously actuated. This allows the individual flow-through latches within U_1 to follow the data on the bus.

At time B, when the control signal disappears, the latch's clock signal is removed and the data is held at its last state. Also, U_1's outputs are activated within the delay time of the latch outputs. The total maximum delay of 35 nanoseconds is much smaller than the hold time of most available microprocessors and memory devices. As a result, there are no glitches on the data bus. Data is retained for the entire duration of the bus cycle and is dropped at the occurrence of the leading edge of the next control signal. □

Hold on. Using eight latches of the chip 74LS373 and a spare NAND gate wired as a negative-input OR gate, this simple circuit (a) extends the data retention time to the entire duration of a bus cycle. The timing diagram (b) shows that the data is retained until the leading edge of the next control signal is reached.

Simple patch reconciles parity flags in Z80, 8080

by Zvi Herman
Elbit Computers Ltd., Haifa, Israel

Since the Z80's instruction set is a superset of the 8080's, a program written for the latter can be run on the former. However, there are subtle incompatibilities that may cause unexpected behavior when such a program is run on the Z80—one of them being the definition of the parity flag.

In the 8080, this flag indicates the parity of the result, whether the operation is arithmetic or logical. But the meaning of the flag has been modified in the Z80 so that it indicates the parity of the result after a logical operation only.

When an arithmetic operation is executed, the parity flag (called P/V in the Z80) instead indicates an overflow condition. Hence, when an 8080 program relies on the parity resulting from an arithmetic operation, that piece of code could produce erroneous results when run on the Z80.

For instance, suppose the content of the accumulator is 55_{16} and the following instructions are executed:

ADI 11H (add 11_{16} to the contents of the accumulator)
JPE NEXT (if parity is even, jump to NEXT)

The 8080 adds 11_{16} to 55_{16}, the result is even, and the condition for the jump is true. But when the Z80 adds 11_{16} to 55_{16}, the result causes no overflow and the P/V flag stays off. Now the condition of the jump is false, and the Z80 continues right through the next instruction, rather than jumping to NEXT.

It is obvious that a patch is needed here. This patch (between the ADD and the JUMP) should leave the accumulator and the flags intact. The only change required is modification of the P/V flag to reflect the parity result from the arithmetic operation.

There is no single instruction in the Z80, however, that can do all that. All instructions that modify the P/V flag, in the sense of parity, change either the accumulator or other flags; a series of instructions, however, will produce the desired result.

The following sequence provides the solution:

```
ADD  A,11H
RLD       (rotate left one digit [nibble] to the left)
RLD
RLD
JP   PE,NEXT
```

The RLD instruction rotates the least significant (4-bit) digit of the accumulator through the memory location pointed to by register HL (see figure). It modifies the P/V flag to reflect the parity of the accumulator and does not affect the carry (CY) flag.

If the instruction is used three times, the accumulator is returned to its initial value and the CY, sign (S), and zero (Z) flags are restored. Note that RRD may be used to rotate the digit right with the same success.

This patch, however, does contain a few pitfalls. For one, the half-carry flag is zeroed by the RRD/RLD instruction. When the parity of the result of an arithmetic instruction is desired, the chance that the half carry is needed somewhat later is small, but it does exist.

When HL points to locations in read-only memory, this patch is not suitable, since ROM locations cannot be written into and thus cannot be used for temporary storage. Finally, the patch adds execution time to the program, and that factor should be considered when software timing is important. ☐

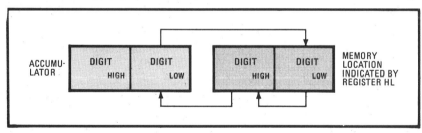

Bit brigade. Parity is flagged for arithmetic operations in the Z80 by rotating the least significant nibble in the accumulator through memory and back. Three executions of the RLD or RRD instructions perform this task.

Mapping an alterable reset vector for the MC68000

by Ron L. Cates
Motorola Semiconductor Products Sector, Phoenix, Ariz.

Microprocessor-based system design requires interrupt vectors to be resident in the random-access memory, so that the operating system may control the interrupt routines efficiently. However, many of the 16-bit microprocessors available today use the interrupt structure to handle the power-on-reset function. But since the RAM is in an unknown state as soon as the power is turned on, the reset vector must reside in nonvolatile storage.

The classic solution, implemented here for the MC68000 microprocessor, is a circuit that maps the reset vector for the processor when the power is turned on and then helps the processor relocate the vector from read-only memory to RAM. But this circuit in fact does more—it also lets the operating system change the location of the reset vector if necessary.

The processor uses addresses 000000 to 000007 to fetch the initial values for the stack pointer and the program counter. The interrupt vectors reside at hexadecimal addresses 000008 to 0003FF. The B part of dual timing circuit U_2 generates a 0.110-second reset pulse when the power supply stabilizes. This positive pulse is inverted and is used to drive the halt and reset inputs of processor U_1 (see figure). When the reset pulse is removed, U_1 reads the first four locations in memory.

Shift register U_3 is cleared only by the power-on-reset condition and is clocked by the address strobe of the processor. The first four memory cycles are mapped into the read-only area of memory. During these cycles, U_3 provides chip-select inputs for both ROM and RAM. The read-write pulse to RAM is inverted by U_4. This lets the processor read the data from ROM while simultaneously

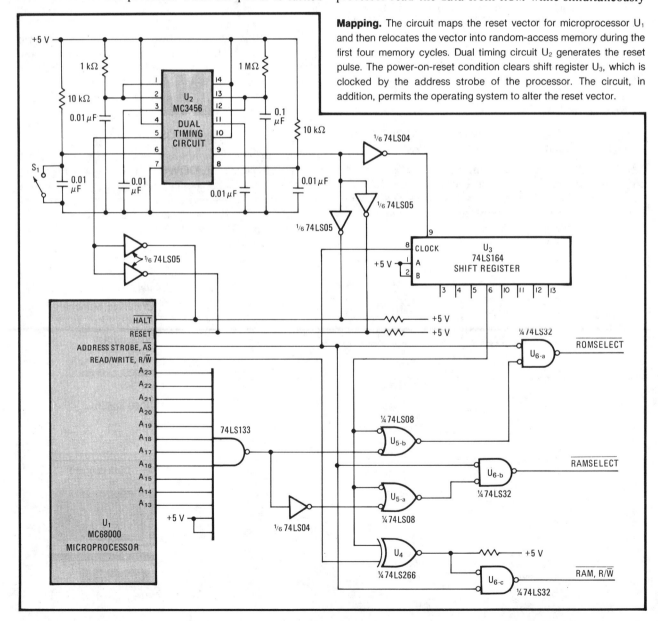

Mapping. The circuit maps the reset vector for microprocessor U_1 and then relocates the vector into random-access memory during the first four memory cycles. Dual timing circuit U_2 generates the reset pulse. The power-on-reset condition clears shift register U_3, which is clocked by the address strobe of the processor. The circuit, in addition, permits the operating system to alter the reset vector.

writing it into RAM locations 000000 to 000007.

When the first four memory cycles are over, U_3 goes into the high state and the decoding occurs. The other half of U_2 furnishes a switch-reset pulse when the system has stabilized. This 11-microsecond pulse sets the processor but does not clear the register. Thus, for all reset conditions set by the U_2-based switch, the vectors must be fetched from the RAM, thereby allowing the operating system to alter them. □

Reset circuit reacts to fast line hits

by Edward R. Miller
Norand Corp., Cedar Rapids, Iowa

This circuit generates a start-up signal on power-up or a reset pulse of specified width whenever the supply voltage to any device momentarily drops below a preset value. It is therefore ideal for use in initializing microprocessor-based systems. Much faster and more reliable than most RC-type reset circuits, it will respond to power glitches of microsecond duration.

As shown, open-collector operational amplifier A_1, working as a Schmitt trigger, compares the preset trip voltage, V_{ref}, to a preset fraction of the supply voltage, V_s. On power-up, voltage V_s' will come up before voltage V_{ref}, and A_1 will fire to enable capacitor C_1 to charge to $V_1 = [V_s' - 0.3] [R_1/(R_1+R_2)]$, with resistor R_1 selected to keep A_2 in its linear operating range and resistor R_3 chosen to limit the peak output current of A_1. A_3 turns on when V_{C1} exceeds its threshold voltage, which is $V_{ref}R_5(1/R_5 + 1/R_6 + 1/R_4)$.

During a power glitch, V_s' will drop below V_{ref}, causing A_1 to go low and bringing the noninverting input of A_2 to about 0.3 volt. Open-collector comparator A_2 then goes low, in effect latching (clamping) A_1 until C_1 discharges through R_2, whereupon the reset output at A_3 goes (active) low for a time equal to R_2C_1.

At this time, V_{C1} drops below V_Z, and A_2 once again becomes transparent to circuit operation, thus enabling the V_s'-to-V_{ref} comparison. A_1 will then allow C_1 to recharge and bring the reset output high until the next power glitch, whereupon a reset pulse is again generated. Note that if on power-up there is any voltage on C_1, the circuit will latch until the capacitor is completely discharged and then will allow it to charge through R_2 for the full reset time. □

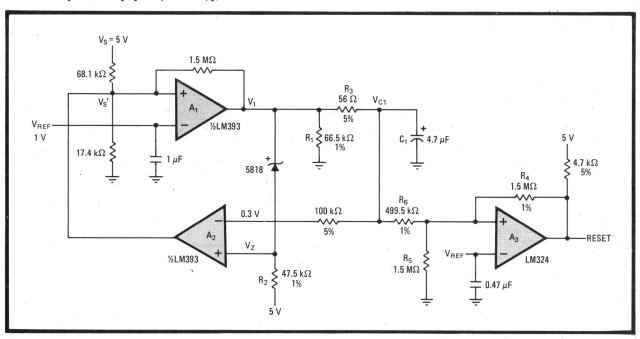

Fail-safe. Power-reset circuit provides full initialization of microprocessor-based systems for glitches lasting only microseconds. Circuit generates start-up signal on power up, then goes low for a time given by T = R_2C_1 when V_s' drops below user-set reference V_{ref}.

Z80B controller waits for slower memory

by Robert E. Turner
Martian Technologies, Spring Valley, Calif.

In order to run at full speed, the Z80B microprocessor requires memories that have an access time of 140 to 180 nanoseconds. However in small systems where memory costs must be kept to a minimum, problems may occur when data is pumped through during each cycle of the Z80B's 6-megahertz clock because of the system's generally slower (less costly) memories.

This programmable wait-state generator can tailor system speed to the memory type used, permitting fast and slow memories to be intermixed within the address space of the Z80. For example, a 2-K-by-8-bit random-access memory having a 120-nanosecond access time, such as the 6116, and a 2-K-by-8-bit erasable programable read-only memory having an access time of 450 ns, such as the 2716, may be used together.

Because the wait generator slows down the Z80's operation only when slower memories are addressed, a high throughput rate can still be achieved when fast RAMs are used. The advantages to be gained in terms of speed outweigh those achieved in systems using a slow clock rate to accommodate both memory types. The most memory-intensive Z80 operations involve the stack, and the operations depend on data variables.

Subroutines are a good example. A CALL has two stack operations; a RETURN has two more. If registers are saved on the stack, even more operations will occur during the operation of the subroutines. Thus, by using fast RAMs and the Z80B, system performance may be sped up by as much as 40% when compared with the Z80A arrangement and its 4-MHz clock.

As for circuit operation, a 74LS164 shift register is used to add one to seven clock periods to a given Z80 memory cycle. This addition allows the Z80B to drive memories that have 166-ns-to-1.166-ms access times.

The shift register is held in a cleared state until the Z80's memory-request output becomes active low. Then the register clocks the wait request through, from the Q_A output toward the Q_H output on each rising clock cycle.

When slow memory is addressed by the Z80, the wait-request line is pulled low by the memory addressed. This in turn sets Q_A of the register high on the next clock. The Z80's ready input then goes low until a logic 1 is detected by the 74LS04 inverter that is connected to the 74LS164. The output of the 74LS00 gate goes low when the wait line goes low and remains there until the selected register output goes high. The number of wait clocks is selected by connecting the input of the 74LS04 to the appropriate output of the register.

This generator will not alter the RAM's refresh cycle. The Z80 address bus has all 0s for the top nine output bits during such a cycle. These 0s could create a problem in using the address bus for wait selection, so the refresh line is used to disable the external wait line. In this way, the generation of a wait state is prevented. □

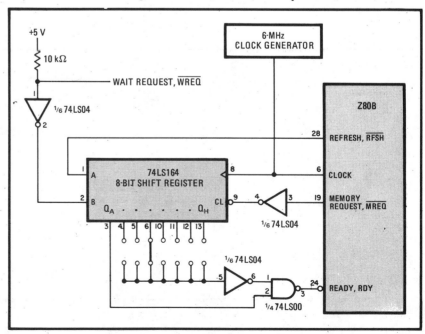

Patience at the PROM. A programmable wait generator for the Z80B microprocessor allows intermixing of fast and slow memories for maximum throughput. Precautions are taken to avoid generating a wait state during memory refresh cycles by using the refresh line to disable the external wait input. This circuit may be used without any drawbacks with any programmable-I/O, serial-I/O, or peripheral chip in the Z80 line.

Enabling a processor to interact with peripherals using DMA

by Trung D. Nguyen
Litton Data Systems Division, Van Nuys, Calif.

Most peripheral devices connected to a microcomputer by the peripheral interface bus can request services from the computer over the interrupt line. This facility, though, is lacking on those input/output devices that use direct memory access to transfer data at high speed between themselves and the processor's main memory. However, a simple interrupt logic circuit will enable such I/O devices to notify the computer when a DMA operation is complete.

As shown in (a), the interrupt-enabling flip-flop U_{1-1} is initially enabled by the enable-service-interrupt ($\overline{ESI}$) command provided by the computer. The computer next sends the external-function-enable ($\overline{EFE}$) command to an addressed I/O device to indicate its readiness for data transfer. If the addressed I/O device chooses to respond to this code, it generates an I/O interrupt request. Interrupt flip-flop U_{2-1} holds this request until the computer

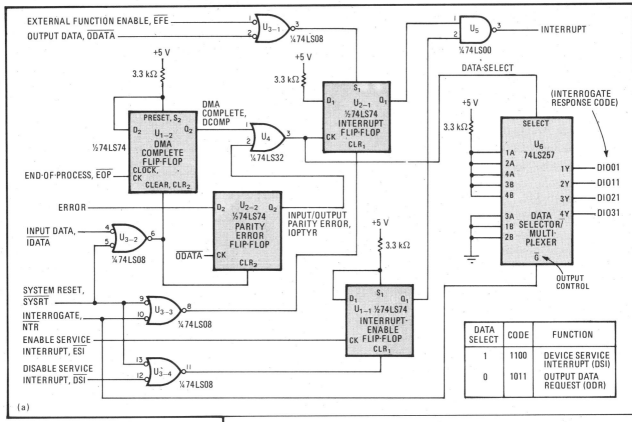

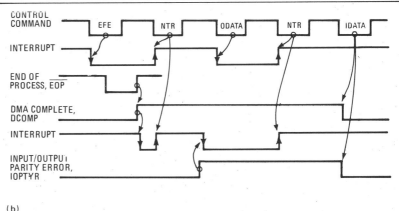

Interrupt. This interrupt logic circuit (a) allows a microcomputer to respond to input/output devices that access its memory directly. The timing diagram (b) of the logic shows that the interrupt line is set when the peripheral device detects an EFE or an output-data command and is reset by the NTR signal. The line is also set by a signal that a direct memory access is complete or that a parity error has been detected.

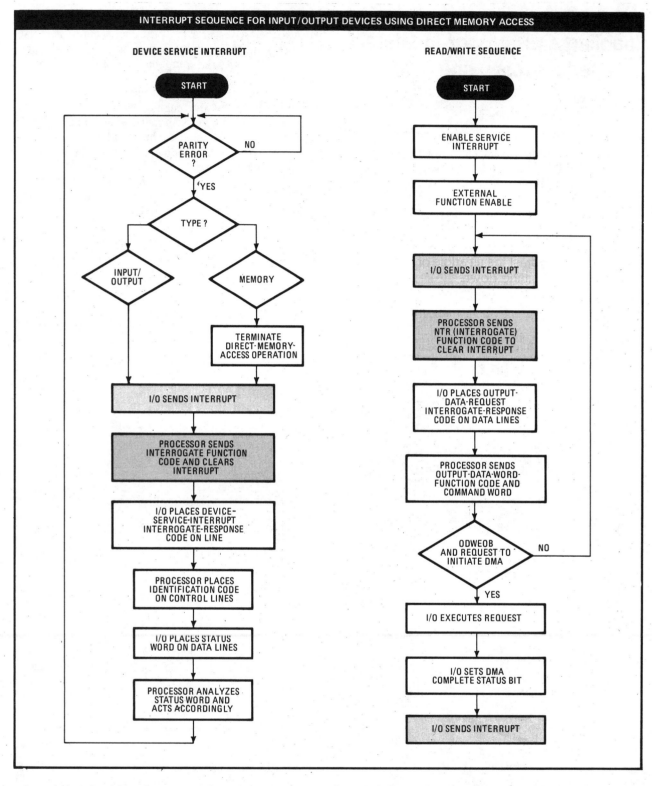

acknowledges it with an interrogate function code.

The I/O device next places the output-data-request interrogate-response code (ODR IRC) on the data lines, whereupon the processor places the output-data-word (ODW) function code and command word on the data lines. Then the I/O device interrupts the processor again for more ODW code, and this procedure continues until all parameters in the command word are sent. The processor indicates completion of the task by sending an output data word with an end-of-block (ODWEOB) signal on the control lines and by requesting the initiation of a DMA operation on the data lines.

Subsequently, the end-of-process (EOP) signal from the DMA indicates that the DMA operation is complete. The I/O device now sets the DMA-complete status bit and issues an interrupt to the processor.

If a memory parity error occurs during the memory-to-memory transfer cycle, the external end-of-process signal (not shown) causes the abrupt completion of the DMA operation and makes the I/O device issue an interrupt. The I/O parity error sets the I/O parity error flip-flop U_{2-2}, which enables the I/O device to generate an interrupt signal (b). The I/O device responds to this parity error interrupt by placing the device-service-interrupt interrogate-response code (DSI IRC) on the data lines. In addition, the processor sends the interrogate command code over the control lines and clears the interrupt. Next, the processor sends the input command code, and the I/O device places the status words on data lines and clears the parity flip-flop. The processor then analyzes the status word and acts accordingly. The table shows the step-by-step interrupt sequence.

The interrupt response codes are generated by data selector–multiplexer U_6. When the processor sends the interrogate ($\overline{\text{NTR}}$) command, U_6 selects either device-service-interrupt (DSI) or output-data-request (ODR) on the basis of the state of the data-select line. If a parity error or DMA completion occurs, the data-select line goes high and DSI is selected. If nothing happens, the data-select line remains low and ODR is selected. The codes for each function are placed on the data line bus, and the processor is programmed to recognize them. □

Power-fail detector uses chip's standby mode

by Jerry Winfield
Mostek Corp., Carrollton, Texas

As microprocessors are used more frequently for consumer and industrial applications, the need to preserve data and program status during power outages has become increasingly important. Until now, this function has been implemented with a small, outboard complementary-MOS random-access–memory battery-charging circuitry, power-fail circuitry, and a battery.

The MK3875, one of the MK387X family of single-chip microcomputers, simplifies the job of providing a battery backup system by incorporating 64 bytes of standby RAM and the battery-charging circuit onto the microprocessor chip. The only external components required are a battery and the power-fail circuit.

Figure 1 shows a simple, low-cost power-fail–detection circuit that can be designed with readily available parts. Figure 2 details the timing relationship between $\overline{\text{RESET}}$ and V_{cc} for enabling and disabling the standby RAM function. Simply stated, $\overline{\text{RESET}}$ must be low when V_{cc} is below its specified voltage, which is 4.75 volts for a 5% part and 4.5 v for a 10% part.

The circuit shown in Fig. 1 detects power failures and resets the device automatically at power-on and manually during operation. The circuit monitors the unregulated voltage that feeds the V_{cc} regulator and compares this voltage against a voltage reference. When the voltage drops below the reference, the $\overline{\text{RESET}}$ line is pulled low. Hysteresis is designed into the comparator to prevent the oscillation caused by slow rise and fall times.

The trimming potentiometer, R_{11}, should be adjusted so the negative threshold voltage V_{th-} is greater than the minimum input voltage of the V_{cc} regulator. Adjusting the threshold voltage above the minimum will yield additional time before the $\overline{\text{RESET}}$ line goes low should the external interrupt be used for saving variables.

As mentioned earlier, the circuit also functions as a power-on or manual reset. The power-on reset is created by the addition of C_2 and D_1, and the manual reset is created by the addition of R_2 and SW_1.

The power-fail circuit can also be configured to generate an external interrupt to the MK3875 to save variables before the $\overline{\text{RESET}}$ line is activated. Adding capacitor C_3 allows time for executing the save routine before the $\overline{\text{RESET}}$ line is pulled low; the external interrupt should also be programmed as an active-high input.

The MK3875 was designed primarily for use with a small 3.6-v nickel-cadmium battery and will automatically supply a maximum charging current of 19 mA at

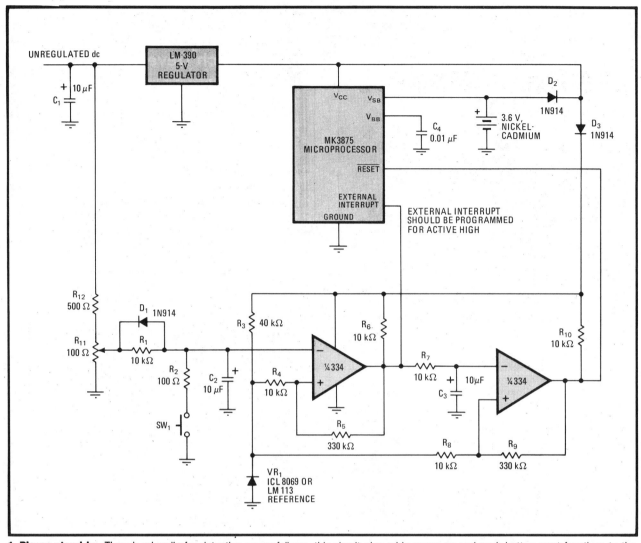

1. Please stand by. Though primarily for detecting power failures, this circuit also adds power-on and push-button reset functions to the MK3875 microcomputer, which has 64 bytes of standby RAM and a battery-charging circuit for implementing a standby power mode.

2. Wave goodbye. Where a save routine is needed to preserve critical data before power is lost, the external interrupt must be toggled before the reset line, during which time the routine is executed. The time allowed to save the data is determined by the value of C_3.

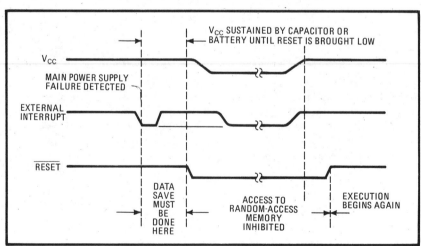

3.2 v and a minimum trickle charge of 0.8 mA at 3.8 v. The size of the battery will depend on the length of required standby time. (With a V_{cc} variation of 4.75 to 5.25 v, the standby current for the RAM will vary between 3.7 mA to 6 mA.)

Among those battery manufacturers whose small nickel-cadmium batteries could be used with the MK3875 are Varta, Yuasa, General Electric, Gould, Sanyo, and Panasonic. □

14. MODULATORS/DEMODULATORS

Switching regulator performs multiple analog division

by Orin Laney
Avocado Computer, Buena Park, Calif.

A simple two-quadrant analog divider that can divide many inputs simultaneously is useful in special-purpose modulators and demodulators. It needs only a monolithic switching regulator, a few passive components, and single-pole double-throw analog switches. However, a two-input example using just one switch is enough to explain the circuit's operation.

The division function is obtained from the switching voltage regulator TL497 (see figure) when multiple analog switch CD4053 is added to it. The input voltages at terminals Y and Q of the regulator are chopped by analog switch CD4053 according to the circuit's duty factor (DF), which is determined by the TL497.

These combined voltages are then filtered by R_2 and C_2 and fed back to the regulator. The duty factor is accordingly adjusted between 0% and 100% to equalize the feedback and the switching regulator's internal reference voltage (about 1.2 volts). The feedback voltage is:

$$V_{FB} = DF \cdot V_Y + (1 - DF)V_Q.$$ Solving for DF:

$$DF = (V_{ref} - V_Q)/(V_Y - V_Q)$$

The input voltages at terminals X and P, if the second switch is a slave of the first, are multiplied by the duty factor, yielding:

$$V_O = DF \cdot V_X + (1 - DF)V_P$$

where V_O is the output voltage and V_X and V_P are the voltages at terminals X and P. Eliminating DF yields:

$$V_O = \left[\frac{(V_{ref} - V_Q)(V_X - V_P)}{V_Y - V_Q}\right] + V_P$$

However, if terminals P and Q are grounded, then the desired output $V_O = V_{ref}V_X/V_Y$, where $V_Y \geq V_{ref}$.

Pull-down resistor R_1 and the emitter of the pass transistor contained in the voltage regulator together generate the logic drive for CD4053. Capacitor C_1 of TL497 sets the maximum frequency of oscillation—about 80 kilohertz for a 100-picofarad capacitor. In addition, time constant $R_3 \cdot C_3$ determines the allowable ripple for the output signal. □

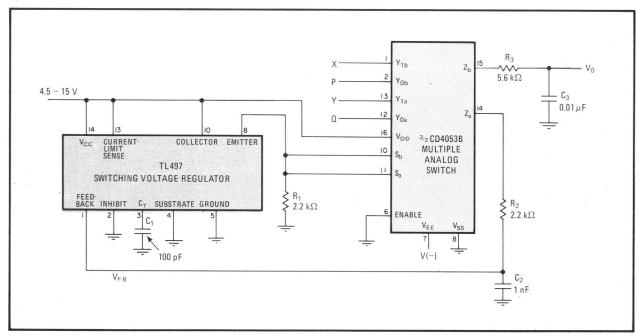

Divider. Switching voltage regulator TL497 generates a duty factor that is proportional to the reciprocal of the voltage V_Y. This factor enables the single-pole double-throw analog switch CD4053 to divide the input voltage V_X by V_Y. The output voltage is filtered by R_3 and C_3.

Tracking filters demodulate two audio-band fm signals

by Stephen Barnes
Center for Bioengineering, University of Washington, Seattle

Because of the way in which they retrieve recorded data, many systems designed for monitoring biomedical functions have to demodulate two closely spaced fm carrier signals in the audio-frequency band. The original data signals could be recovered with low-pass filters, but they are an expensive solution since their cutoff must be sharp to prevent cross modulation, signal blocking, or undue limiting of the bandwidth needed by one or both signals. The low-cost solution shown here, however, uses a dividing phase-locked loop to demodulate one signal and to provide the clock signal for a tracking notch filter that recovers the other channel of data.

The advantages of the circuit may be seen for a typical monitoring case in which a 30-hertz electrocardiogram signal (having a frequency too low to be recorded directly on cassette tape) is placed on a 9-kilohertz carrier. This signal is applied to the LF356 amplifier along with a 0-to-6-kHz signal from an ultrasonic doppler flowmeter that provides data on blood circulation.

Block A, the dividing phase-locked loop module, oscillates at a free-running frequency equal to $8f_m$. It is here that the 9-kHz carrier is directly demodulated. Also included in block A is a 74C193 divide-by-8 counter in the PLL's feedback loop, which provides the driving signals for the CD4051 multiplexer in block B. This block contains the sampling filter which passes frequencies equal to ⅛ the sampling frequency, $8f_m$, and its harmonics. Thus the doppler data is notched out by the sampling filter. But signal f_m is subtracted from the original input signal $f_o + f_m$ by the differential amplifier in block C. Therefore, only signal f_o will appear at the output.

Resistor R_Q, which is in the PLL's feedback loop, controls the width of the notch, which is given by $B = 1/8\pi R_Q C_Q$, where B is the width defined by the filter's upper 3-decibel frequency minus its lower 3-dB frequency. Potentiometer R_B is used for balancing the differential amplifier by nulling the 9-kHz feedthrough signal.

When the circuit is in the locked state, the minimum attenuation of the fm signal will be approximately 33 dB for input signals ranging from 540 millivolts to 8 volts peak to peak. Below signal levels of 540 mV, feedthrough from the multiplexer will reduce the attenuation.

Because the filter is a sampling device, it is subject to aliasing if any input-frequency components approach the Nyquist limit of $4f_m$, so that precautions should be taken to prevent this. Spurious output components that are higher harmonics of the 9-kHz fm signal or the clock signal can be removed easily.

Phase jitter in the PLL should also be minimized, for it causes narrow noise sidebands centered about the 9-kHz fm signal and separated from it by a frequency equal to the loop bandwidth of the PLL. The amplitude of these noise sidebands in a properly adjusted circuit should be down at least 33 dB from the level of the fm signals at the input. □

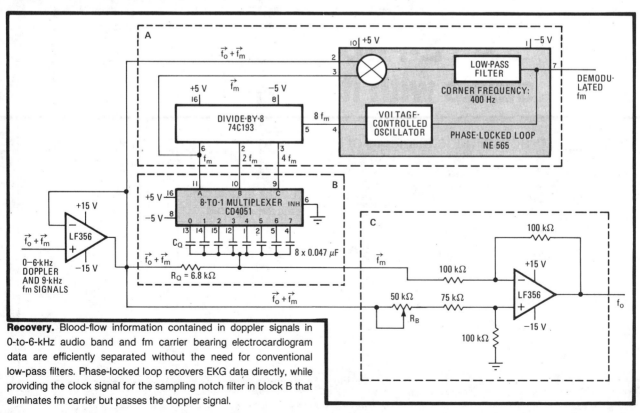

Recovery. Blood-flow information contained in doppler signals in 0-to-6-kHz audio band and fm carrier bearing electrocardiogram data are efficiently separated without the need for conventional low-pass filters. Phase-locked loop recovers EKG data directly, while providing the clock signal for the sampling notch filter in block B that eliminates fm carrier but passes the doppler signal.

Integrator improves 555 pulse-width modulator

by Larry Korba
Ottawa, Ont., Canada

In one method of providing linear pulse-width modulation with the 555 timer, a current source charges a timing capacitor, creating a ramp signal that drives the modulation input of the 555. Unfortunately, the circuit offers only a limited dynamic range of pulse widths and is highly sensitive to temperature. A better way is to use a resettable integrator as the timing element.

Charging with a constant current source (a) at best yields a 2:1 dynamic range for a supply of 5 volts—the linear operating range for voltage–to–pulse-width conversion is approximately 2.1 to 4.1 v, and the timing capacitor is totally discharged every timing cycle. Furthermore, the circuit requires temperature compensation to eliminate any timing fluctuation due to the temperature sensitivity of Q_1, since the base-emitter voltage varies at the relatively high rate of -5 millivolts per °C. And, to add to the circuit's woes, I_{cbo} varies with temperature as well.

The resettable integrator (b) made up of A_2, Q_1, C, and R applies a trigger pulse to the 555, causing Q_1 to turn off. Integrator A_2 then ramps up until the voltage level at the modulation input of the timer equals that at pin 5. When that happens, Q_1 is turned on again, resetting the integrator and turning off the 555.

The voltage applied to the integrator, V_c, is set to 2.1 v. This makes the shortest linearly modulated pulse width equal to the trigger pulse width—2 microseconds. With the timing values shown, the maximum pulse width is 6 milliseconds, producing a dynamic range of more than 3,000:1 over the linear operating region.

The active components affecting the timing circuit are A_2, Q_1, A_1, and V_{cc}. Since the average temperature coefficient for the offset voltage of A_2 is a very low 5 microvolts/°C (affecting the timing by only 2.5 parts per million/°C), the circuit's almost negligible adverse temperature effects are largely due to the variation with temperature of the off current of Q_1, I_{dss}. I_{dss} doubles every 10°C; for the 2N4360, it is about 10 nanoamperes at room temperature.

It is important to note that for both circuits, the effects of V_{cc} and the 555 on timing stability are the same. As a bonus, however, the new circuit provides a linear ramp output that can be loaded fairly heavily without seriously affecting circuit timing. □

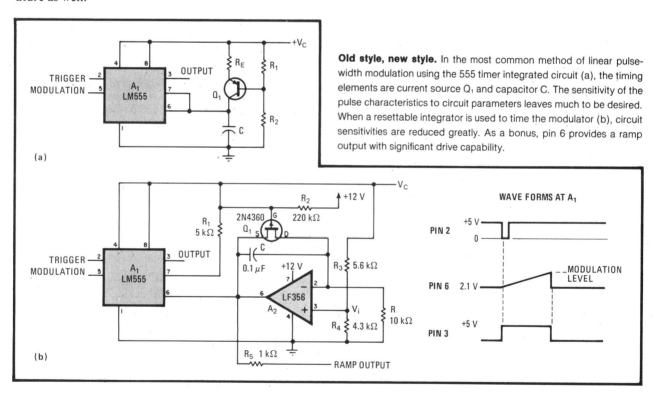

Old style, new style. In the most common method of linear pulse-width modulation using the 555 timer integrated circuit (a), the timing elements are current source Q_1 and capacitor C. The sensitivity of the pulse characteristics to circuit parameters leaves much to be desired. When a resettable integrator is used to time the modulator (b), circuit sensitivities are reduced greatly. As a bonus, pin 6 provides a ramp output with significant drive capability.

Linear one-chip modulator eases TV circuit design

by Ben Scott and Marty Bergan
Motorola Semiconductor Products Sector, Phoenix, Ariz.

The fact that Motorola's MC1374 chip has both frequency- and amplitude-modulation and oscillator functions simplifies the design of a television modulator. The device is ideally suited for applications using separate audio and composite video signals that need to be converted into a high-quality very high-frequency TV signal.

The a-m system (Fig.1a) of the 1374 is a basic multiplier combined with an integral balanced oscillator that is capable of operating at a frequency of over 100 megahertz. The fm oscillator-modulator (Fig.1b) is a voltage-controlled oscillator that exhibits a nearly linear output-frequency versus input-voltage characteristic.

This characteristic provides a good fm source with only a few inexpensive parts (no varactor is necessary). The system has a frequency range of 1.4 to 14 MHz and can produce a ±25-kilohertz–modulated 4.5-MHz signal with only 0.6% total harmonic distortion.

The a-m output for a complete TV-modulator circuit (Fig.1c) is taken at pin 9 through a double-pi low-pass filter with the load (R_7 = 75 ohms) connected across pins 8 and 9. Access to both (video and audio) inputs enables the designer to separate video and intercarrier sound sources.

The gain-adjustment resistor (R_8) is chosen in accordance with the available video amplitude. By making R_8

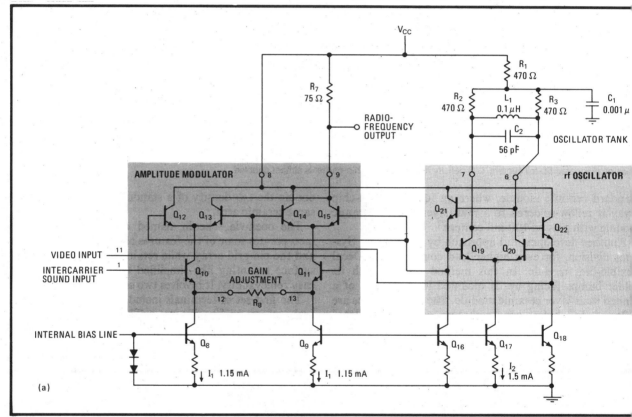

1. Modulator. An a-m (a) and fm (b) modulator and oscillator are incorporated in the design of MC1374. The complete TV modulator circuit (c) uses a simple low-loss second harmonic output filter. Gain resistor R_8 is 2.2 kΩ for an intended video input of about 1 V peak at sync tip. With a 12-V regulated supply there is less than a 10-kHz shift of rf carrier frequency from 0° to 50°C for any video input level.

2. Performance. The IRE test signal (shown at left) is used to evaluate the video modulator. The resulting modulated rf output (shown at right) from the MC1374 has a differential phase error of less than 2° and a differential gain distortion of 5% to 7%.

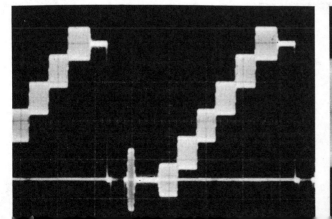
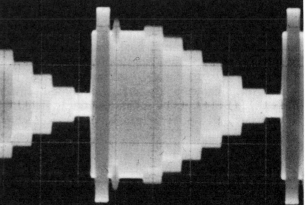

≅ 2 (peak video) volts/1.15 milliamperes, a low-level 920-kHz beat is ensured. To keep the background noise at least 60 decibels below the standard white carrier level, the minimum peak video should be at least 0.25 v.

An oscillator for channels 3 and 4 (61.25 and 67.25 MHz) uses a parallel LC combination from pin 6 to pin 7. For this configuration, the value of coil L_1 is kept small and the capacitance large to minimize the variation due to the capacitance of the MC1374 chip (approximately 4 picofarads).

Sloppy wiring and poor component placement around pins 1 and 11 may cause as much as a 300-kHz shift in carrier frequency (at 67 MHz) over the video input range. This frequency shift is due to transmission of the output radio frequency to components and wiring on the input pins. A careful layout keeps this shift below 10 kHz. The video signal and the corresponding modulated rf output (Fig. 2) have a differential phase error of less than 2° and differential gain distortions of 5% to 7%.

The fm system is designed specifically for the TV intercarrier frequency of 4.5 MHz for the U.S. and 5.5 MHz for Europe. The fm system's output from pin 2 is high in harmonic content, so instead is taken from pin 3. This choice sacrifices some source impedance but produces a clean fundamental output, with harmonics decreased by more than 40 dB.

The center frequency of the oscillator has approximately the same resonance as L_2 and the effective capacitance from pin 3 to ground. In addition, by keeping the reactance of the inductor at a point between 300 and 1,000 Ω, the overall stability of the oscillator is ensured.

Optional biasing of the audio-input pin (14) at 2.6 to

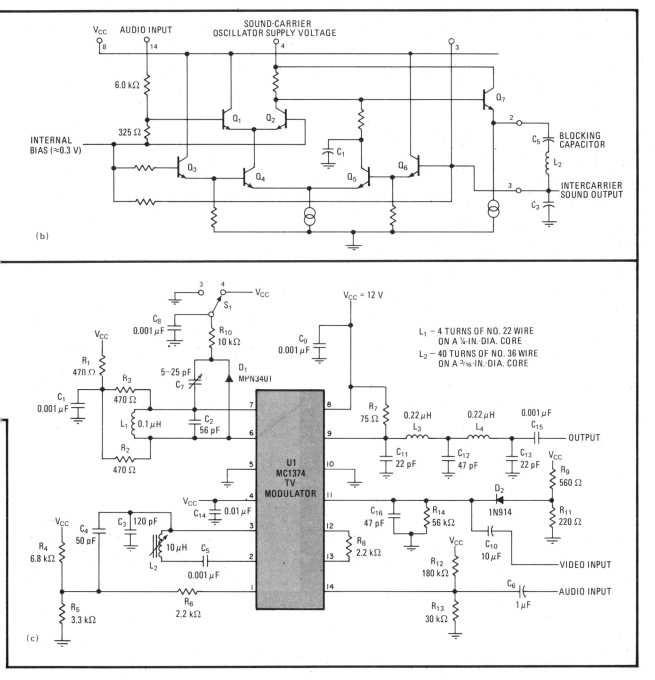

Achieving linear control of a 555 timer's frequency

by Arturo G. Sancholuz
Caracas, Venezuela

Though widely used as an astable multivibrator, the ever-popular 555 integrated-circuit timer is cursed with a nonlinear frequency response. However, in this configuration (a), the 555 is made to provide an easy, linear means for frequency calibration. In addition, it is unaffected by power-supply variations.

Capacitor C is charged by transistor Q_2, which is connected as a constant current source, and then discharged through pin 7 of timer U_1. Charging current $I = (V_1 - V_{be1} - V_{eb2})/R_4$, where V_{be1} and V_{eb2} are the base-to-emitter and emitter-to-base voltages of Q_1 and Q_2, respectively. When V_c is the common source voltage, K is a constant, and V_{be3} and V_{eb4} are the base-to-emitter and emitter-to-base voltages of Q_3 and Q_4, respectively, voltage V_1 can be expressed as:

$$V_{eb4} + V_{be3} + \frac{(V_c - V_{eb4} - V_{be3})KR_2}{R_1 + R_2 + R_3}$$

If the previous two equations for I and V_1 are combined and if it is assumed that voltages $V_{be1} = V_{be3}$ and $V_{eb2} = V_{eb4}$, when matched transistors are used, the charging current reduces to

$$\frac{(V_c - V_{eb4} - V_{be3})KR_2}{R_4(R_1 + R_2 + R_3)}$$

Because a 10-kilohm resistor is connected to the control voltage at pin 5, C charges from one quarter to half of supply voltage V_d instead of the normal third to two thirds. This change reduces the maximum voltage across C, which in turn provides a sufficient voltage drop across R_4 for Q_1 and Q_2 operation. The oscillation period can be obtained from the relationship $\tau = CV_d/4I$ where $V_d = V_c - V_{eb5} - V_{be6}$ and where V_{eb5} and V_{be6} are the emitter-to-base and base-to-emitter voltages of Q_5 and Q_6, respectively.

Rearranging the equations and assuming that $V_{eb4} = V_{eb5}$ and $V_{be3} = V_{be6}$ yields:

$$f = 1/\tau = \frac{4KR_2}{CR_4(R_1 + R_2 + R_3)}$$

where f is the oscillation frequency. The above relationship shows that the output frequency can be linearly

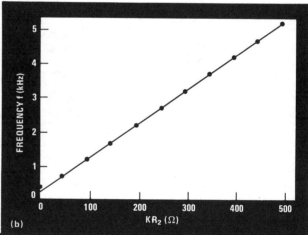

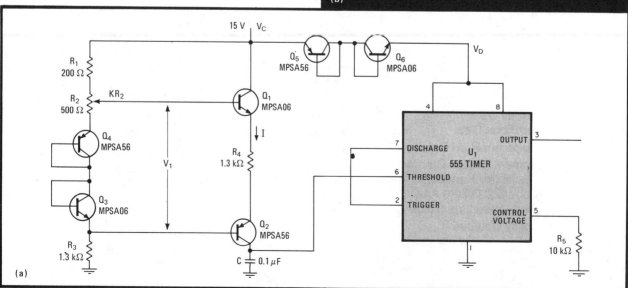

Linearity. Connected as an astable multivibrator (a), the 555 timer has an output frequency that is linearly controlled with potentiometer R_2. Capacitor C is charged by Q_2 at constant current I. Imperfect matching of transistors Q_1 to Q_4 and Q_2 to Q_3 may introduce nonlinearity at low frequencies. The calibration curve (b) shows that frequency can be linearly adjusted from a few hertz to 5 kHz.

Ultrafast hybrid counter converts BCD into binary

by L. J. Herbst
Teesside Polytechnic, Middlesbrough, England

Most ultrafast integrated-circuit counters that are capable of operating at 500 megahertz and above with a reset facility employ a binary-coded–decimal format. However, this ultrafast hybrid counter with a binary output and external reset control contains a BCD input stage followed by binary counters—it has a special conversion for a binary output. Because a conventional BCD-to-binary counter uses more integrated circuits, money is saved and conversion time shortened for this hybrid circuit. In addition, the method uses a novel decoding technique for faster conversions.

The counter output of the circuit (see figure) is $10N_B + N_D$, where N_B and N_D are the contents of the binary and decade counters respectively. The output is rewritten in the form $8N_B + 2N_B + N_D$, for code conversion. It is expressed through the following summation when the weight of 2^3 is assigned to the least significant bit of the binary conversion:

			D_3	D_2	D_1	D_0	N_D Term
B_n	B_4	B_3					$8N_B$ Term
B_{n+2}	B_6	B_5	B_4	B_3			$2N_B$ Term
O_n	O_4	O_3	O_2	O_1	O_0		Output

If standard full adders are used, the code conversion that is needed to implement the above expression is achieved with a normal two-word addition—except when output O_3 is selected. In this case, the input consists of D_3, B_3, B_5, and C_f (the carry forward from the 2-bit adder)—$D_3 + C_f$ is implemented with an OR gate. Because D_3 is 1 only when both D_2 and D_1 are 0, the input to yield O_3 is reduced to B_3, B_5, and $D_3 + C_f$.

The method is superior in cost and speed to an all-decade counter or the standard BCD-to-binary conversion technique. A hybrid counter with a 12-bit output, that uses the schematic shown, sports a Plessey Semiconductor BCD counter (SP8636B) and Texas Instruments' binary counters (SN74197) and adders (SN7483, SN7482). A BCD counter would have required four decades and 11 SN74184 BCD-to-binary decoders, giving a conversion time of 196 nanoseconds. However, the add time for two 16-bit words, using the SN7483s for example, is typically 43 ns. □

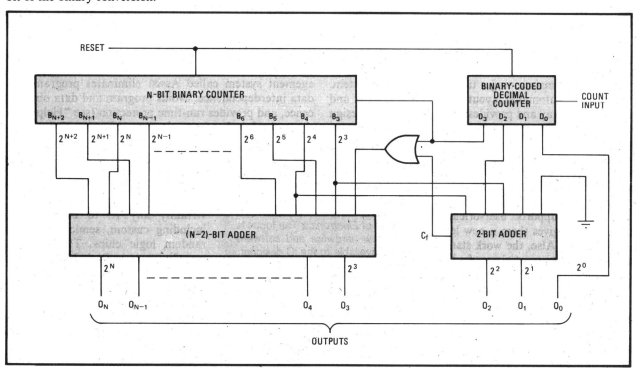

Hybrid counter. This ultrafast hybrid counter provides BCD-to-binary conversion. It consists of a BCD input stage followed by binary counters and full adders to achieve code conversion. The OR gate is used to reduce the inputs that yield O_3.

Pulse modulator provides switched-mode amplification

by P. H. Pazov
Polytechnic of Central London, England

Using pulse-width modulation to provide switched-mode amplification, this circuit affords the same advantages in the analog world as does its switching power supply counterpart—simplicity, efficiency, and low power consumption when a complementary-MOS logic family is employed. Noise rejection and distortion characteristics also are better than can be achieved with a conventional analog arrangement.

As shown in the general function diagram (a), a feedback current, I_{mod}, is derived from the audio input signal for the purpose of varying the pulse width of a free-running oscillator, which itself is formed by a digital integrated-circuit integrator and a hysteresis/power stage. C_2 and R_4 develop a dc feedback signal from comparison of V_{out} with the audio input, from which an error current is created. The ratio of R_4/R_5 sets the amplification factor. C_2 and C_3 act to integrate V_{out}, so that any long-term imbalance appears as an error and can be corrected.

The active digital elements thus switch in the linear mode, ensuring that the output pulse width is proportional to the amplitude of the input signal. The output, as taken at the junction of L and C_o, represents the instantaneous change in the oscillator's pulse width. With such an arrangement, the amplifier's efficiency is better than 90% at any signal level, as would be expected with any linear switched circuit.

The practical implementation is shown in (b). One 4049 inverter is used for all switching functions. C_2 and R_4 set the total bandwidth at about 20 kilohertz, with C_3 determining the lower cutoff frequency of 100 hertz. The circuit has an amplification factor of 15. It will drive a load as low as 2 ohms at a noise level that appears low enough for headphone-monitoring applications. With a supply voltage of 6 volts, the current consumption will be a mere 5.6 milliamperes.

As mentioned, the circuit's noise performance is very good. This is partly due to the fact that, as the modulation current increases toward the maximum charging current required to attain the maximum pulse width in the oscillator, a form of frequency modulation occurs, giving rise to an S-shaped transfer curve that is characteristic of an fm discriminator. This characteristic results in a lower noise and distortion factor than can be achieved using a purely linear approach.

As with all high-frequency circuits, a good earth ground and no ground loops are essential to proper operation. Otherwise, all sorts of oscillations appear and add to the noise level. □

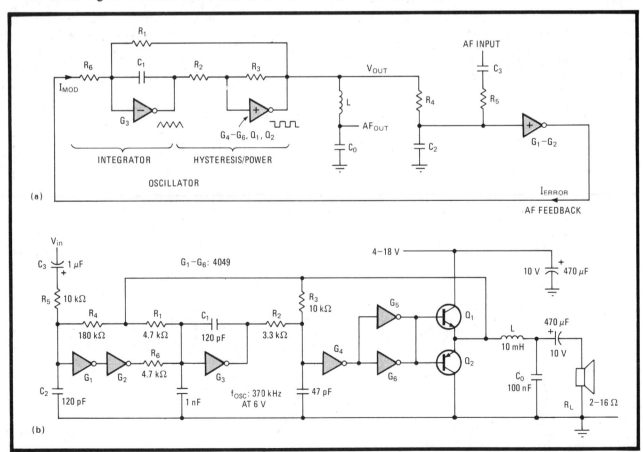

Linear logic. This digitally switched network (a) amplifies audio signals linearly by modulating the pulse width of an RC oscillator. The practical circuit (b) is uncomplicated. Excellent noise-rejection performance is achieved by the modulator's S-shaped transfer curve.

15. MULTIPLEXERS

HP64000 emulates MC6801/6803 using bidirectional multiplexer

by M. F. Smith
Department of Computer Science, University of Reading, Reading, England

Because Motorola's MC6801 and 6803 single-chip, 8-bit microcomputers have a multiplexed address and data bus, they are not compatible with MC6800 hardware and therefore cannot exploit the 6800 in-circuit emulation available on the HP64000 development system. However, with the aid of a two-chip bidirectional multiplexer, the HP64000 can be made to emulate the processor portion of these nearly identical chips.

The devices are attractive for a wide range of applica-

Emulation circuit. The circuit (a) allows the HP64000 to emulate the microprocessor portion of both the MC6801 and 6803 by using a converter board that is connected to the HP64212A emulator probe. The bidirectional multiplexer (b) combines the nonmultiplexed address and data lines of the MC6800 emulator.

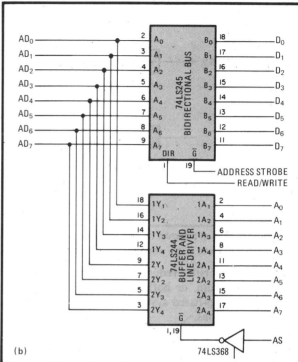

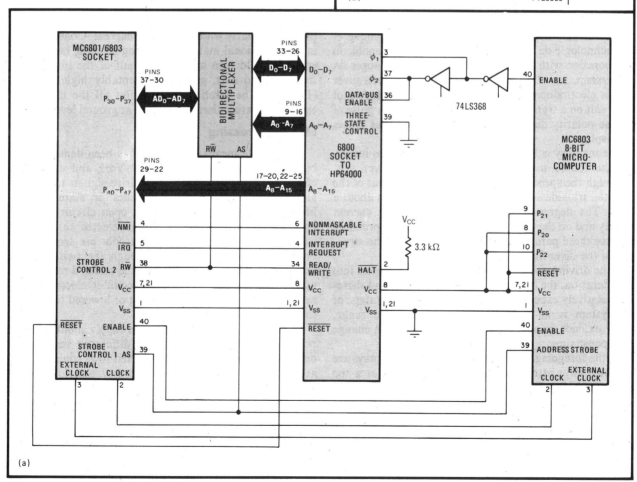

tions, particularly with Hitachi's recent introduction of a complementary-MOS version of the 6801. Besides having complete source and object compatibility with the 6800 software, they have extra instructions for 16-bit operations, an 8-bit multiply, and an X: = X+B instruction, not to mention PSHX/PULX. The units are about 20% faster than the 6800 at the same clock speed.

The circuit (a) allows the HP64000 development system to emulate the microprocessor portion of both the 6801 and 6803 by using a converter board that is connected to the HP64212A emulator probe. The bidirectional multiplexer (b) combines the nonmultiplexed address and data lines of the MC6800 emulator. An address strobe is derived by using a dummy 6801 or 6803 as a clock generator. The circuit allows only the expanded multiplied mode to be emulated. Thus it is not possible to emulate the random-access memory or the input and output of the 6801 or 6803 with this circuit. The assembler of the HP64000 for the 6800 allows use of the extended instruction set of the MC6801/6803. □

Multiplexers compress data for logarithmic conversion

by Andrzej Piasecki
Warsaw, Poland

Cascaded multiplexers and a few gates are all that is needed to build this digital log converter, which compresses an 8-bit signal into a 5-bit number according to the transformation $2^n \rightarrow 4n$. Conversion to higher numbers is achieved by cascading additional multiplexers and appropriate gating circuitry.

As seen in the figure and the truth table for n extending from 0 to 28, the design of the circuit is simplified because each of the circuit's 74157 multiplexers can transfer without alteration 4 bits of the signal formed by a preceding multiplexer.

Alternatively, following multiplexing it can transport input bits that extend the second and third digits to the two least significant bits at the output.

As a result, the two most significant output bits of any multiplexer are fixed within a given input-number range. They are encoded by transferring the given 0 and 1 logic states into successive multiplexer inputs, with the most significant input bit (at logic 1) switching on whichever multiplexer is appropriate for transferring the desired number to the output.

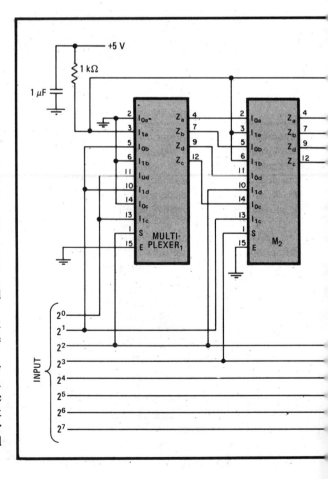

Multiplexer manipulation. Using digital multiplexers, this circuit converts 8-bit input numbers into their corresponding 5-bit logarithmic equivalents, performing the operation in 100 ns. The truth table illustrates the simplicity of the design technique that performs the conversion. Multiplexers may be cascaded for extending the range over which n may be transformed into its log value

As the input number decreases, the number of multiplexers required to transfer the desired data increases. NAND gates G_1 through G_5 derive the logic value of the most significant bit of the 5-bit number at the output. Note that the algorithm used will necessitate that the designer observe considerable care in wiring up the additional multiplexers that would be required to process larger numbers.

The propagation time of a digital logarithmic conversion is about 100 nanoseconds. The circuit draws no more than 120 milliamperes. □

| TRUTH TABLE: DIGITAL LOG CONVERTER ||||||||||||
| Number n | Input ||||| Number n | Input |||||
	2^4	2^3	2^2	2^1	2^0		2^4	2^3	2^2	2^1	2^0
0	0	0	0	0	0	0	0	0	0	0	0
1	0	0	0	0	1	2	0	0	0	1	0
2	0	0	0	1	0	4	0	0	1	0	0
3	0	0	0	1	1	6	0	0	1	1	0
4	0	0	1	0	0	8	0	1	0	0	0
5	0	0	1	0	1	9	0	1	0	0	1
6	0	0	1	1	0	10	0	1	0	1	0
7	0	0	1	1	1	11	0	1	0	1	1
8	0	1	0	0	0	12	0	1	1	0	0
10	0	1	0	1	0	13	0	1	1	0	1
12	0	1	1	0	0	14	0	1	1	1	0
14	0	1	1	1	0	15	0	1	1	1	1
16	1	0	0	0	0	16	1	0	0	0	0
20	1	0	1	0	0	17	1	0	0	0	1
24	1	1	0	0	0	18	1	0	0	1	0
28	1	1	1	0	0	19	1	0	0	1	1

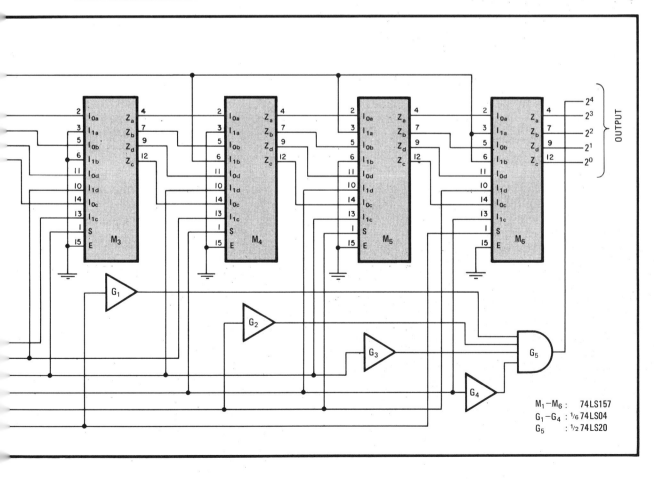

M_1–M_6 : 74LS157
G_1–G_4 : 1/6 74LS04
G_5 : 1/2 74LS20

Multiplexer does double duty as dual-edge shift register

by James A. Mears
Dallas, Texas

Useful and sometimes unusual logic functions may be found disguised as parts with other functional names, and they often offer circuit advantages not found in the part designed for a given task. A good example of this principle is the 74157 quad two-input multiplexer, which can serve in a pinch as a clocked shift register.

Unlike other registers that advance on one edge of the clock, it will shift on each successive transition. This attribute is particularly useful when it is necessary to process data at twice the clock frequency without using a frequency doubler.

A good application for this type of register would be in the timing chain of a dynamic memory controller, where it could give a fine timing resolution without the need for a high-frequency clock. The pin functions in the figure have been renamed to give a better understanding of the circuit operation of the modified multiplexer.

Signals applied to the serial input are routed to output 1 through the AND-OR gate combination, G_1 through G_3, with output 1 connected to the alternate input of G_1 and to the corresponding input of the next stage, G_4 through G_6. The last three stages are connected similarly to form a chain of D latches (another good use for the 74157).

Thus, when the clock is high, the input signal will pass to the output, back to the alternate input and to the next stage. On the low-going clock edge, the output of the first stage is latched, and the signal is passed to the next stage's output. This process is repeated for each clock transition, thereby shifting the signal through the register. At any time, the register can be disabled (all outputs low) by bringing the clear input high.

Variations of this circuit can be built with the 74257, which is equivalent to the 74157 but has three-state outputs. The circuit can also be built with the 74158 (74157 with inverted outputs) by adding inverters to the outputs and connecting them up as before. Registers may be cascaded by connecting the serial input to the preceding stage's output. □

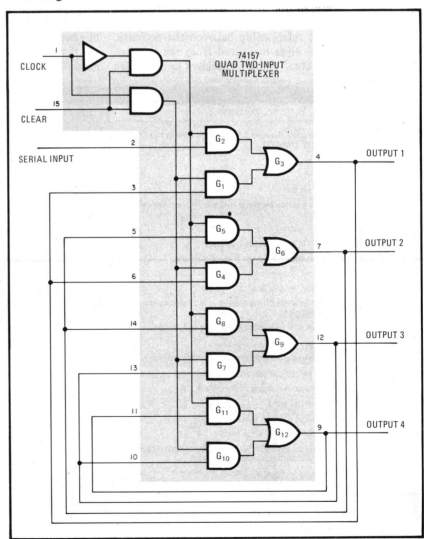

Disguise. Many logic chips can simulate other functions, such as this 74157 quad multiplexer, which performs well as a shift register. This part also gives the added advantage of dual-edge clocking, thus eliminating the need for frequency doublers in such circuits as the timing chain of high-resolution dynamic memory controllers.

Time-slot assigner chip cuts multiplexer parts count

by Henry Wurzburg
Motorola Inc., Semiconductor Group, Phoenix, Ariz.

In some communications systems, particularly digital telephony equipment, it is hard to examine the data from a given source after it has been time-division–multiplexed with other data for serial transmission over a common data line. Capturing the data from its time slot and converting it into parallel form for examination usually requires many integrated circuits, since the slot must be programmable.

A special-purpose IC, the MC14417 time-slot assigner carries out this serial-to-parallel function with the aid of only a few inverters and one other IC. What's more, the cost of implementing the circuit is only a few dollars.

The timing of a simple three-slot TDM system is shown in (a). In digital telephone systems, a data frame may consist of anywhere from 24 to 40 time slots, each containing 8 bits of data transmitted at rates of up to 2.56 megabits per second.

In the all–complementary-MOS capture circuit of (b), the MC14094 shift register acts as a serial-to-parallel converter, while the 14417 computes when the data is to be captured and converted. Just which time slot it captures is determined by the binary data present at inputs D_0–D_5 of the 14417. The circuit also provides a valid-data output signal. As for speed, the circuit works for clock rates of up to 2.56 MHz with systems having up to 40 time slots.

Implementing a parallel-to-serial converter for multiplexing data onto the TDM data line is equally simple if the 14417 is used as shown in (c). Here, a three-state buffer prevents the serial data bus from being loaded during idle time-slot periods. The frequency limitations of this second circuit are the same as for the capture circuit. □

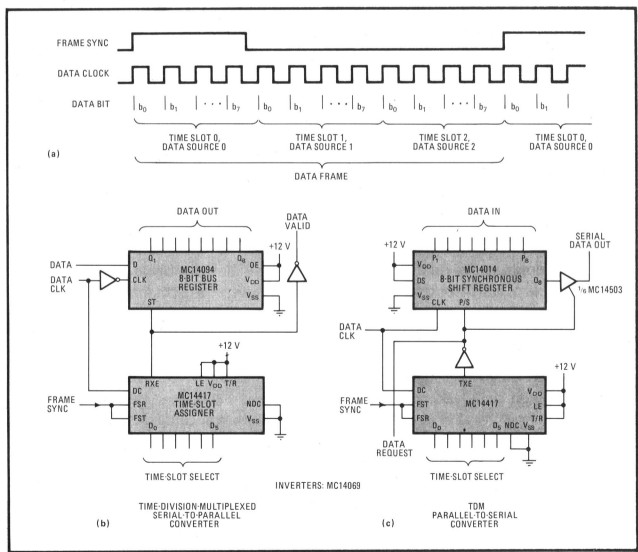

The right slot. Time-domain multiplexing (a) assigns to data from several sources specific time slots in a serial data stream. Capturing data from a specific slot is made easy with the MC14417 time-slot assigner (b), which works with the MC14094 shift register to provide data from the source dictated by the select inputs of the 14417. The versatile chip can also provide parallel-to-serial multiplexing (c).

16. MULTIPLIERS

Frequency multiplier uses digital technique

by Marian Stofka
Kosice, Czechoslovakia

Offering wide range and high accuracy, this digital frequency multiplier uses a few low-cost integrated circuits to produce an output that is an integral multiple of the input frequency. In addition, the output signal is in phase with the input signal. The correct frequency of the output signals is set within two periods of the input, and

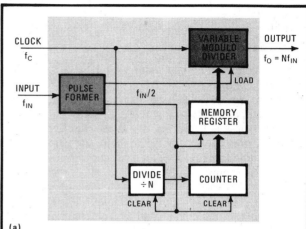

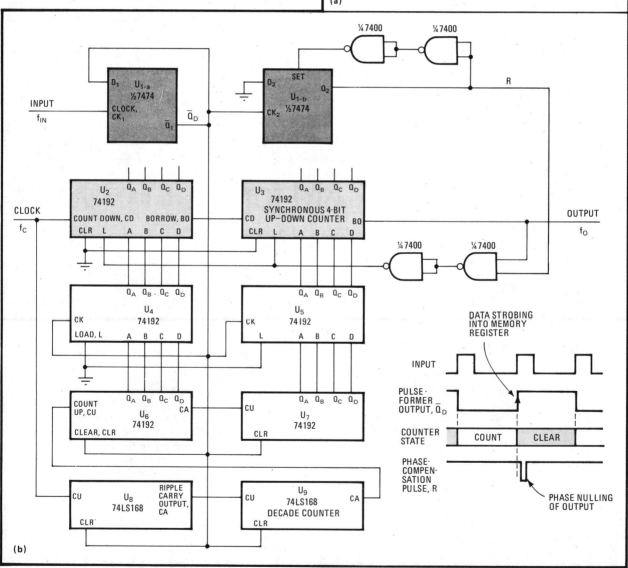

Multiply by N. The multiplier (a) generates an output whose frequency is N times the frequency of the input signal. The clock is divided by N, and the data is stored in a memory register that determines the modulus of the variable-modulo divider. The output of this divider is given by $f_o = Nf_{in}$. The hardware realization (b) uses a few chips to achieve an output that is 100 times f_{in}.

in fact the design could be integrated into one chip.

The fundamental schematic (a) comprises a pulse-former, divider, counter, memory register, and variable-modulo divider. The clock pulse is divided by N and then counted by the counter for a period corresponding to one cycle of the input signal. At the end of this period the counter output data is strobed into the memory register, which determines the modulus of the variable-modulo divider. This data is valid until new data is accepted. The variable-modulo divider is driven by the clock pulse, and the counter output data at the end of the counting period is $C_o = T_{in}f_c/N$, where T_{in} is the period of the input signal, f_c is the frequency of the clock, and N is the modulus of the divider. As a result, the output signal (f_o) obtained from the variable-modulo divider satisfies the relation $f_o = f_c/C_o = N/T_{in} = Nf_{in}$, where f_o is the frequency of the output signal.

The practical circuit (b) uses two decade counters, six synchronous 4-bit up-down counters, a flip-flop, and a NAND-gate chip to realize the multiplier design. The counter output is stored in the memory register consisting of U_4 and U_5. The clock frequency is determined from the equation $f_c = 0.8N10^k f_{in}$, where k is the number of decade counters connected in the multiplier counter. For the components shown, the output frequency is $f_o = 100f_{in}$. The worst-case relative-output phase error is $10^{-k}/0.6$. The output is compared with the input to ensure the output waveform has the proper phase. □

Joining a PLL and VCO forms fractional frequency multiplier

by S. K. Seth, S. K. Roy, R. Dattagupta, and D. K. Basu
Jadavpur University, Calcutta, India

Most frequency multipliers are hampered by the fact that they can multiply frequencies only in integer amounts. As a result, if a certain output frequency is desired, the input frequency must be carefully selected. Such exact choosing is no longer needed because this circuit can multiply pulse frequencies by any real number through the simple adjustment of two potentiometers. In addition, it operates over a wide input-frequency range and has a more stable output than do conventional multipliers.

This design combines a phase-locked–loop frequency-to-voltage converter and an external voltage-controlled oscillator for pulse-frequency multiplication. However, conventional multiplication circuits employing PLLs use either harmonic locking or a frequency divider between its VCO and phase comparator. Thus, the output is only an integer multiple of the input.

A PLL connected as a frequency demodulator, generates voltage V_d that is related to the input frequency by $V_d = kf_{in}$, where k is a constant and f_{in} is the frequency of the input signal. In addition, the input frequency of the internal VCO (contained in the PLL) is $f_{in} = V_d/VR_1C_1$, where R_1 and C_1 are the frequency-determining components of the internal VCO and V is the supply voltage.

Demodulated voltage V_d is fed to the control-voltage input of the external VCO whose output frequency is $f_{out} = V_d/VR_2C_2$, where R_2 and C_2 are frequency-determining components of the external VCO. Solving for the output frequency: $f_{out} = f_{in}R_1C_1/R_2C_2$ and thus $n = (R_1C_1)/(R_2C_2)$. The multiplication factor n is only determined by the externally connected resistors and capacitors and therefore can be chosen for any value.

The circuit (a) uses National Semiconductor's general-purpose PLL LM565 and VCO LM566. Operational amplifier μA741 serves as the buffer between the two.

The multiplication factor for this particular circuit is 6.15, and its input-frequency range is 2 to 6 kilohertz. The oscilloscope display (b) shows the input and output waveforms for an input frequency of 4 kHz and multiplication factor of 6.15. For stable circuit operation, R_1 and C_1 should be selected according to the input frequency, and R_2 and C_2 should be chosen to generate the desired multiplication factor. □

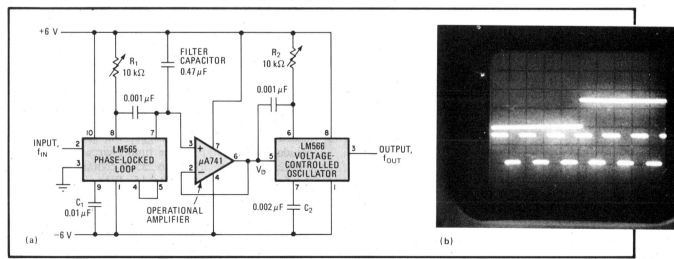

Multiplier. The circuit (a) uses a phase-locked loop, an external voltage-controlled oscillator, and a buffer. The oscilloscope display (b) shows the input and output waveform for an input frequency of 4 kHz and a multiplication factor of 6.15.

Simplified multiplier improves standard shaft encoder

by Michael M. Butler
Minneapolis, Minn.

The pulse multiplier proposed by Amthor [*Electronics*, Sept. 11, p. 139] for increasing the resolution of a standard shaft encoder may be simplified, and improved as well, with this low-power circuit. Using only three complementary-MOS chips to derive a proportionally greater number of pulses from the encoder's output for a positioning up/down counter, it is relatively insensitive to encoder phase errors, uses no temperamental one-shots, and can detect the occurrence of illegal transition states generated by the encoder or circuit. It will serve well in electrically noisy industrial environments.

As in Amthor's circuit, the multiplier (a) used to drive the counter produces four pulses for each square-wave input of the two-phase encoder, whose outputs are displaced 90° with respect to one another. In this circuit, however, two 4-bit shift registers, three exclusive-OR gates, and an eight-channel multiplexer (b) derive the pulses. Previously four one-shots and 16 logic gates were required for the same task.

As seen, the clocked shift registers generate a 4-bit code to the multiplexer for each clock cycle. To ensure that no shaft-encoder transitions are missed, the clock frequency should be at least 8 NS, where N is the number of pulses produced by the encoder for each shaft revolution and S is the maximum speed, in revolutions per second, to be expected. A clock frequency of 1 megahertz or less is recommended for optimum circuit operation.

The code is symmetrical, and so only three exclusive-OR gates are required to fold the data into 3 bits. These bits are then applied to the control ports of the three-input multiplexer, which will generate the truth table shown opposite. □

MULTIPLIER LOGIC					
A	B	Multiplexer code				Count
		Old A	New A	Old B	New B	
0	0	0	0	0	0	no change
0	↑	0	0	0	1	down
0	↓	0	0	1	0	up
0	1	0	0	1	1	no change
↑	0	0	1	0	0	up
↑	↑	0	1	0	1	error
↑	↓	0	1	1	0	error
↑	1	0	1	1	1	down
↓	0	1	0	0	0	down
↓	↑	1	0	0	1	error
↓	↓	1	0	1	0	error
↓	1	1	0	1	1	up
1	0	1	1	0	0	no change
1	↑	1	1	0	1	up
1	↓	1	1	1	0	down
1	1	1	1	1	1	no change

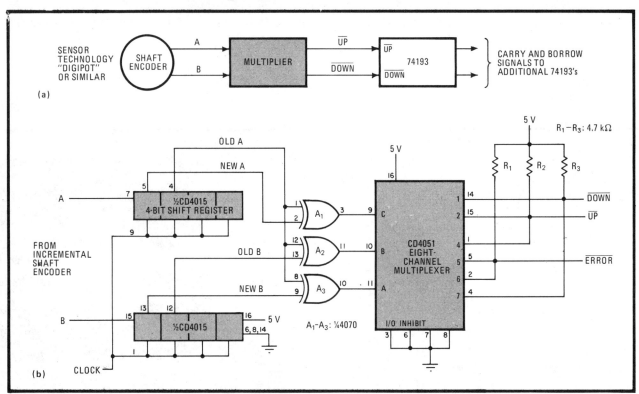

Accretion. Requiring only three chips, pulse multiplier gives better resolution to optical positioning systems that are driven by a shaft encoder. Low-power circuit is relatively insensitive to encoder phase errors. Stability is high because no one-shot multivibrators are used.

Enhanced multiplier cuts parts count

by Guy Ciancaglini
Fellows Corp., Springfield, Vt.

The encoder-pulse–feedback multiplier circuit proposed by Frank Amthor [*Electronics*, Sept. 11, 1980, p. 139] and later simplified by Michael M. Butler [*Electronics*, Nov. 20, 1980, p. 128] may be implemented in a higher-speed TTL version that costs less and has fewer parts.

Butler's circuit reduced the original four-latch outputs to three lines so an eight-channel analog-multiplexer can be used. But the price paid was a space-consuming exclusive-OR package.

The TTL design implements the multiplier's common edge detector with a 74379 quad D-latch connected as shown in Fig. 1a. Furthermore, the 74159 4-to-16-line decoder enables it to use all four latch outputs, eliminating the exclusive-OR package.

Another plus for the TTL design is its versatility, which

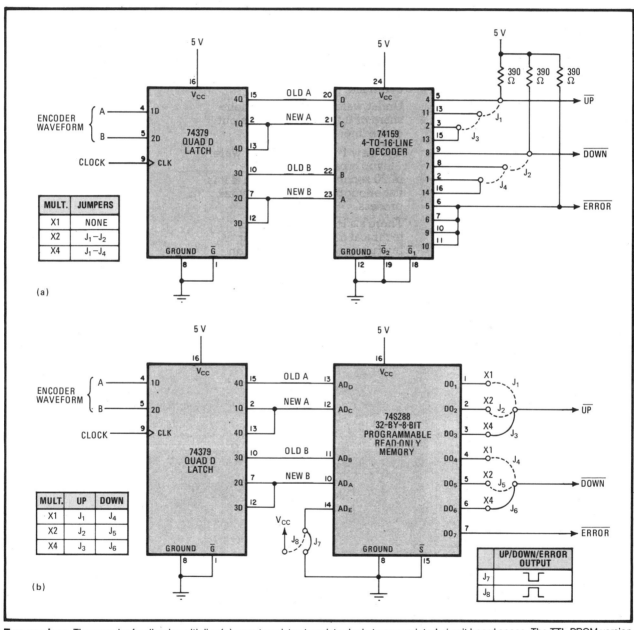

Two versions. The encoder feedback multiplier (a) uses transistor-transistor logic to save printed-circuit board space. The TTL PROM version (b) costs a bit more, handles positive or negative true logic, and needs even less board space. Both circuits are programmable.

A	B	Latch output				NO.	Up*			Down*			Error*
		Old A	New A	Old B	New B		X1	X2	X4	X1	X2	X4	
0	0	0	0	0	0	0	1	1	1	1	1	1	1
0	↑	0	0	0	1	1	1	1	1	1	1	0	1
0	↓	0	0	1	0	2	1	1	0	1	1	1	1
0	1	0	0	1	1	3	1	1	1	1	1	1	1
↑	0	0	1	0	0	4	0	0	0	1	1	1	1
↑	↑	0	1	0	1	5	1	1	1	1	1	1	0
↑	↓	0	1	1	0	6	1	1	1	1	1	1	0
↑	1	0	1	1	1	7	1	1	1	1	0	0	1
↓	0	1	0	0	0	8	1	1	1	0	0	0	1
↓	↑	1	0	0	1	9	1	1	1	1	1	1	0
↓	↓	1	0	1	0	10	1	1	1	1	1	1	0
↓	1	1	0	1	1	11	0	0	1	1	1	1	1
1	0	1	1	0	0	12	1	1	1	1	1	1	1
1	↑	1	1	0	1	13	1	1	0	1	1	1	1
1	↓	1	1	1	0	14	1	1	1	1	1	0	1
1	1	1	1	1	1	15	1	1	1	1	1	1	1

TRUTH TABLE FOR ENCODER MULTIPLIER

*For PROM version: when address E is tied high, all output states are inverted.

it owes to the fact that the collector outputs of the decoder are now open and may be hooked together in a wired-OR configuration. By including four jumpers or a dual–in-line–packaged switch, these outputs may be selected for one, two, or four times the original encoder feedback (see table). Since the decoder has a slower propagation delay than the latch frequency, the latch clock rate is limited to at most 25 megahertz.

Use of a PROM programmer can save even more space, but at a slightly higher cost. By replacing the decoder with a 32-by-8-bit TTL programmable read-only memory, the package is reduced in size and in pin count from 24 to 16 pins. The output pull-up resistors are also eliminated because of the PROM's three-state outputs. In this design variation, the ×1, ×2 and ×4 outputs are on individual lines. By making use of the fifth address input to the PROM, the output polarity may be set to positive or negative true logic using jumpers (Fig. 1b). □

17. OPERATIONAL AMPLIFIERS

Current booster drives low-impedance load

by Jeffrey L. Sharp
NAP Consumer Electronics Corp., Knoxville, Tenn.

Successfully interfacing operational-amplifier circuitry with a low-impedance load is a common problem encountered by analog design engineers. This current booster offers a fix by employing bipolar power transistors that provide a compromise between maximum output voltage capability and effective device protection. However, the use of power MOS field-effect transistors in the output stage of the circuit current-limits the circuit and delivers output voltage approaching the supply.

The conjugate symmetry output stage of the booster (see figure) is composed of bipolar transistors and complementary power MOS FETs to provide a current booster that has a low output impedance and a high bandwidth. The use of power MOS FETs in the output section of the circuit prevents cross conduction at frequencies above 100 kilohertz. Because power MOS FETs Q_5 and Q_6 act together as a nearly ideal transconductance amplifier, zener diodes D_3 and D_4 effectively limit output current by setting the output's gate-source voltage according to the relation: $V_z = V_{GSmax} = V_{GSth} + I_{Dmax}/g_{fs}$
where V_z = zener-diode voltage, V_{GSmax} = maximum gate-source voltage, V_{GSth} = gate-source threshold voltage, I_{Dmax} = maximum drain current and g_{fs} = common-source forward transconductance.

Careful device selection keeps the distortion at a minimum, especially in a class-B design. Because transconductance is a nonlinear function of drain current, the best selection method is a comparison of characteristic curves. Q_5 is composed of two p-channel MOS FETs that are connected in parallel. They effectively double this transistor's transconductance to 1 mho at a 1-ampere drain current and thereby match it to the n-channel MOS FET's transconductance of 0.9 mho at 2 A.

A 12.5-volt sine wave may be driven into an 8-ohm load from ±15-v supplies with the circuit, and a 20-Ω load may be driven to a 14-v peak. Its total harmonic distortion is 0.2% at 10 kHz and 0.03% at 1 kHz. In addition, its efficiency is 55% at full power and its output has a phase shift of about 10° at 100 kHz.

To ensure stability, the current-boosted amplifier must be operated at a gain that is greater than the output stage's voltage gain and must have careful component layout. Sufficient heat sinking is required to protect the amplifier from faulty loads that may result in a high output voltage at maximum current. Diodes D_1 and D_2 prevent op amp U_1 from saturating when the amp loop opens temporarily or the load is shorted. □

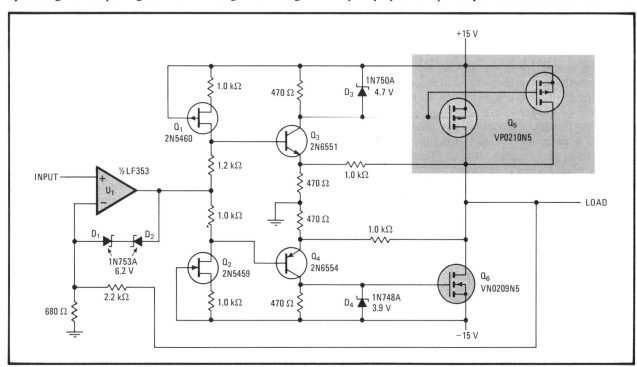

Low-impedance load. Using bipolar transistors and complementary power MOS FETs, this conjugate symmetry output circuit can drive a low-impedance load with high voltage and current over a wide range of frequencies. Matching FETs Q_5 and Q_6 reduce the distortion to less than 0.2% at 10 kHz and 0.03% at 1 kHz. Zener diodes D_3 and D_4 limit the output current.

Diode plus op amp provide double-threshold function

by Pavel Novak
Fraunhofer Institute for Solid State Technology, Munich, West Germany

Medical instruments that produce a signal or alarm when physiological measurements like heart rate or blood pressure reach undesirable levels need window comparators, which usually comprise two operational amplifiers each (a). But a new circuit cuts the op-amp requirement by half and achieves the comparators' double-threshold function by means of a simple diode.

When input voltage V_{in} reverse-biases diode D_1, V_{in} appears at the op amp's noninverting input (b). In addition, inverting input voltage V_n is the sum of supply voltage V_b and V_{in}. For this condition, $V_{in} = V_p$ (noninverting input voltage), and switching occurs when $V_p = V_n$. As a result, the upper threshold voltage V_u is now written as $V_u = V_b R_1/(R_1 + R_2)$.

On the other hand, when the input voltage forward biases D_1, the voltage at the noninverting input is $-V_d$, where V_d is the voltage across the diode under a forward-bias condition. The relationship for this second threshold voltage is now: $V_x = -V_b(R_3/R_2) - V_d[1 + R_3(1/R_1 + 1/R_2)]$ when $V_u \geq 0$ and $V_x \leq -V_d$.

This particular circuit leaves no room for threshold-voltage adjustment, but may be modified through the use of two potentiometers (c). Potentiometer R_4 sets D_1's reference potential, V_a. In addition, resistors R_1 and R_2 are replaced with pot R_5. The new set of threshold voltages are given by: $V_u = V_a$ and $V_x = V_a - V_d(1+r)$, where $r = R_3/R_1$. The window $V_u - V_x$ is valid when it is greater than or equal to V_d. Its characteristics are shown in (d). □

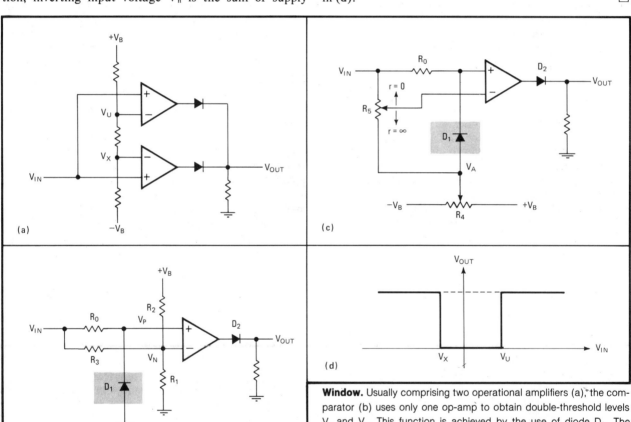

Window. Usually comprising two operational amplifiers (a), the comparator (b) uses only one op-amp to obtain double-threshold levels V_u and V_x. This function is achieved by the use of diode D_1. The threshold levels can be easily varied by employing pots R_4 and R_5 (c), and the window characteristic of the comparator is shown in (d).

Op-amp summer forms simple high-speed phase generator

by Dieter R. Lohrmann
Department of the Army, Harry Diamond Laboratories, Adelphi, Md.

A simple operational-amplifier summing circuit can generate an output voltage that is a rectangular- or sine-wave function of a dc source potential. For analog applications such as the direct phase modulation of a radio-frequency carrier, this circuit combines fast response with low throughput delay, making it superior to the often-used method of digitizing and reconverting an input signal with the aid of a microprocessor. In digital uses it can generate a pulse train having almost any pulse voltage–versus–dc input voltage characteristic.

In the circuit's general configuration, which for simplicity omits the stabilizing circuitry like bypass resistors and capacitors, resistors R_1–R_7 form a voltage divider whose taps are alternately connected to the inverting and noninverting inputs of the adjacent comparators A_1–A_6. The outputs of the comparators are simply summed via resistors R_8–R_{13}.

For a digital application (a), the incoming ramp is used to generate a variable pulse-width train. When the ramp voltage is below the node voltage at point 1, comparators A_1, A_3, and A_5 are low (−7 volts), whereas A_2, A_4, and A_6 are high (+7 v). Thus, the output voltage will be midway between the two supply voltages, or at zero.

As the ramp rises above the node voltage at point 1, comparator A_1 switches, activating a fourth comparator, while two remain off. The output voltage thus jumps to 2 v, assuming resistances R_8 to R_{13} are of equal value. This voltage remains constant while the ramp voltage increases, until it exceeds the potential at the divider's second node, causing comparator A_2 to switch off. At this time, A_2, A_3, and A_5 are low, and A_1, A_4, and A_6 are high, returning the output voltage to zero.

As the ramp voltage climbs past the potential at the third node, a similar operation moves the output to 2 v, because four comparators will be on and two will be off. The output drops to zero again when the ramp moves to a potential higher than at node 4.

The duration of each transition will be dependent upon the node and dc input voltages, which may be appropriately selected by the user. Consequently, a pulse train having almost any set of variable-width characteristics can be ordered.

In analog applications, a good approximation of a sine-wave function can also be generated if the voltage divider's switching intervals are made equal to ΔV, the voltage increment required to switch the output of each comparator from negative to positive saturation. Because the comparator's transitions are nonlinear, the steep sides of each comparator's rectangular transfer function are rounded off, and therefore a sine wave can be very nearly approximated.

Such a scheme is useful in phase or frequency modulators, as shown in (b), because it enables direct modulation at any arbitrarily large modulation index without introducing additional frequency offset. The function generator's output is simply introduced at the intermediate-frequency port of a single-sideband modulator. At

Speedy switching. Op-amp summing stage (a), wired in an inverting-to-noninverting input arrangement, provides sine- or rectangular-wave phase generation for fast response and low throughput delay. If the comparators' thresholds are selected so that the summer's switching profile approximates a sine wave, the circuit may be used to phase- or frequency-modulate an rf carrier directly (b).

the output of the modulator will appear the sum of the i-f and rf frequency, which is a cosine function applied to the modulator's rf port.

In this case, the function generator performs a phase modulation, because the phases of the generator and the rf carrier are added in the modulator. Thus, V_{in} causes a phase modulation of the rf carrier. The maximum modulation index is determined by the number of amplifiers in the phase generator. For a maximum modulation index of M, $2M/\pi$ amplifiers are necessary.

If the SSB modulator is of the phasing type (which cancels the lower sideband by phase-shift mixing), orthogonal signals will be required at its i-f and rf ports. This function generator can be easily modified to generate a cosine-function signal, instead of a sine-wave output. Thus, a sine- and cosine-function generator may be combined to provide the orthogonal signals required for the SSB mixer. □

Absolute-value amplifier uses just one op amp

by Larry Mitchell
Omnimedical, Paramount, Calif.

In many applications, certain functions within a circuit require the absolute value of the input voltage, and generally two or more operational amplifiers are needed to obtain this value. However, this design reduces the op-amp requirement to one. It inverts negative input voltages and leaves the polarity of positive input voltages unchanged.

As shown in the figure, op amp A_1 serves as the unity-gain amplifier. When the input is negative, $V_{in} = -V$, diode D_1 conducts, thus producing voltage $V_c = -(-V+I_{in}R_D)$ at point c where R_D is the forward-bias resistance of the diode and I_{in} is the input current. Because the amplifier gain is unity, I_{in} is equal to the output current I_{out}. The output voltage is the sum of the voltages at point c and the drop across forward-biased diode D_4. Thus, $V_{out} = -(-V+I_{out}R_D)+I_{out}R_D = V$.

For positive inputs, $V_{in} = V$, D_2 conducts and produces voltage $(V-I_{in}R_D)/2$ at point b. Because the potential difference between points a and b is zero, the voltage at point a is also $(V-I_{in}R_D)/2$. Since D_1 is reverse-biased, there is no current through resistor R_1. Therefore the voltage at point a is due to the two resistors, R_4 and R_5, acting as a divider. This condition produces a voltage $(V-I_{out}R_D)$ at point c. D_4 adds an additional voltage of $I_{out}R_D$ volts. The output is again

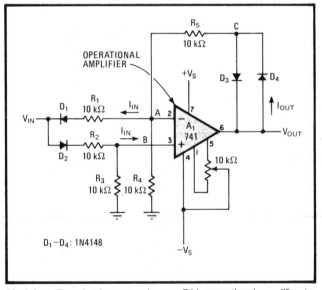

Modulus. The circuit uses only one 741 operational amplifier to achieve absolute-value amplification. Accuracy within 15 millivolts for low values of input, $V_{in} < <0.5$ V, is attained by using standard diodes and resistors in the circuit and having a gain of unity.

$V_{out} = V$. Diode D_3 eliminates latchup by allowing the output to drive the inverting input negative when V_{in} approaches 0 V.

When the circuit has standard diodes and unity gain, equal conduction currents are produced through the diodes, resulting in an accuracy of within 15 millivolts. Voltage gains other than unity may be obtained by altering the resistor values. The amplification factors are R_5/R_1 for the inverting port and, for the noninverting port, $(R_4+R_5)/(2R_4)$ when $R_2 = R_3$. □

Bi-FET op amps invade 741's general-purpose domain

by Jim Williams
National Semiconductor Corp., Santa Clara, Calif.

Thanks to their low-drift microampere supply currents and picoampere bias currents, recently introduced bipolar field-effect-transistor operational amplifiers like National's LF441 can be used in applications that general-purpose amplifiers like the 741 cannot address. A high-performance pH meter, logarithmic amplifiers, and a voltmeter-checker reference source may be inexpensively built with this bi-FET operational amplifier.

The low-bias input of the 441 provides an excellent nonloading port for a pH probe, which is used to measure the acidity or alkalinity of a solution (Fig. 1). This simple four-chip interface yields a linear 0-to-10-volt output corresponding directly to the value of the pH (0 to 10) being measured, a range that is more than adequate for many applications.

The output from buffer A_1 is applied to A_2, a tuned 60-hertz filter that removes power-line noise. A_2 also biases op amp A_3, which provides a compensation adjustment for the probe's temperature. A_4 allows the probe to be calibrated.

To calibrate the circuit, the probe is immersed in a solution having a pH of 7. The solution's temperature is normalized for the meter by R_1, a 10-turn 1,000-ohm potentiometer whose value may be set between 0 and 100 units. These values correspond directly to a solution temperature range of 0° to 100°C. Potentiometer R_2 is then adjusted for an output voltage of 7 V.

A conventional logarithmic amplifier (Fig. 2a) utilizes the well-known logarithmic relationship between the base-to-emitter voltage drop in a transistor and its collector current. Here, A_1 acts as a clamp, forcing the current through Q_1 to equal the input current, E_{in}/R_{in}. Q_2 provides feedback to A_2, forcing Q_2's collector current to equal A_2's input current, which is established by the LM185 zener-diode reference.

Because Q_2's collector current is constant, its emitter-to-base voltage is fixed. The base-to-emitter drop of Q_1, however, varies with the input current. The circuit's output voltage is therefore a function of the difference in the V_{be} voltages of Q_1 and Q_2 and is proportional to the logarithm of the input current. In this manner, the V_{be} drift is cancelled. The coefficient of this term will vary with temperature, however, and cause a drift in the output voltage. The 1,000-Ω thermistor compensates for this drift, stabilizing A_1's gain.

The 441's 50-pA bias current allows accurate logging down into the nanoampere region. With the values shown in the circuit, the scale factor for the amplifier is 1 V/decade.

A second type of logarithmic amplifier is shown in (b). This unconventional design completely eliminates the temperature-compensation problems of (a) by temperature-stabilizing logging transistor Q_1. This temperature problem is economically eliminated by utilizing the LM389 audio-amplifier-and-transistor array as an oven to control the logging transistor's environment.

Transistor Q_2 in the LM389 serves as a heater, and Q_3 functions as the chip's temperature sensor. The LM389 senses Q_3's V_{be}, which is temperature-dependent, and drives Q_2 to feed back the chip's temperature to the set point established by the 1-to-10-kilohm divider. The LM329 reference ensures that the power supply is independent of temperature changes.

Q_1, the logging transistor, operates in this tightly controlled thermal environment. When the circuit is first

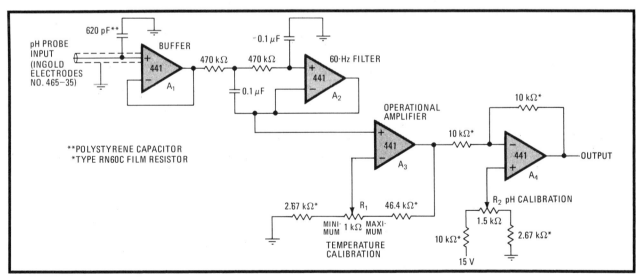

1. Acids and bases. This four-chip interface converts the output of a pH probe into direct readings of a solution's acidity and alkalinity. The circuit has a filter to reject the ac line noise that plagues instruments of this type. This unit can easily compensate for temperature variations.

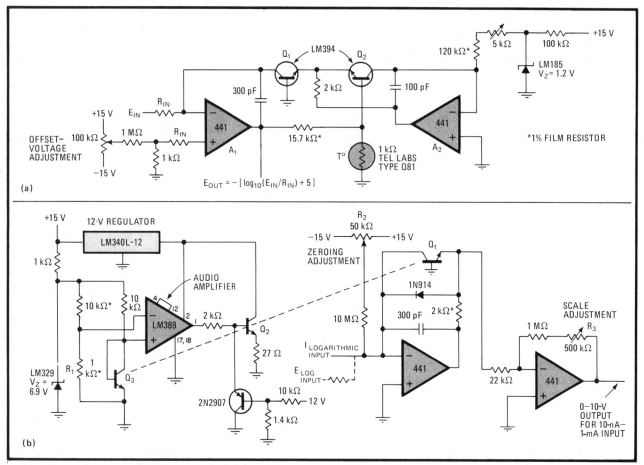

2. Low-power loggers. With amplifier A_1 clamping current through transistor Q_1 to input value E_{in}/R_{in} and with A_2 holding Q_2's current constant with LM185's reference, circuit (a) yields a logarithmic response by virtue of proportional differences between Q_1 and Q_2's well-known V_{be}-to-collector current relation. A more advanced version (b) uses an LM389 to eliminate temperature effects on output.

turned on, Q_2's current flow becomes 50 milliamperes forcing the transistor to dissipate about 0.5 watt, which raises the chip to its operating temperature rapidly. At this point, the thermal-feedback circuit takes control and adjusts the chip's power dissipation accordingly. The LM340L voltage regulator has only 3 v across it, so it never dissipates more than about 0.3 W. The pnp-transistor clamp at the base of Q_2 prevents feedback lock-up during circuit start-up.

To adjust this circuit, the base of Q_2 should be grounded, then the power applied to the circuit, and the collector voltage of Q_3 measured at room temperature. Next, Q_3's potential at 50°C is calculated, a drop of -2.2 millivolts/°C being assumed. The value of R_1 should be selected to yield a voltage close to the calculated potential at the LM389's negative input. After Q_2's base is removed from ground, the circuit will be operational.

A_1's low bias current allows values as low as 10 nanoamperes to be logged within 3%. Potentiometer R_2 provides zeroing for the amplifier. Potentiometer R_3 sets the overall gain of the circuit.

The low power consumption of the 441 is useful in a calibration checker for digital voltmeters that only draws 250 μA (Fig. 3). Here, the 441 is used as a noninverting amplifier. The LM385 is a low-power reference that provides 1.2 v to the input. This voltage is simply scaled by the feedback-resistor network to yield exactly 10 v at the circuit's output. The circuit will be accurate to within 0.1% for over a year, even with frequent use. □

3. Long-term accuracy. Using a single LF441 and a 1.2-volt reference, this circuit for calibrating digital voltmeters with a 10-V signal draws only 250 microamperes. Using a 15-v power source, the circuit has an output accuracy within 0.1% over a year's time.

Extending the range of a low-cost op amp

by Bob Darling
Department of Physics and Astronomy, Rutgers University, Piscataway, N. J.

Low-cost operational amplifiers often have limited power-supply and common-mode voltage ranges. This is the case with the recently introduced ICL7650CPD from Intersil Inc., a device with otherwise very good specifications that sells for under $3 in 100-piece quantities. The 7650 has a typical offset voltage of under 1 microvolt, drift of only a few nanovolts per degree Celsius, bias current in the low picoampere range, and a bandwidth of 2.5 megahertz.

However, its applicability is limited because its maximum supply voltage of 18 volts and its common-mode voltage range V^- to $V^+ - 2.7$ V yields a maximum allowable common-mode input of about 15 V. The circuit in (a) greatly extends the common-mode range in the follower mode.

The idea is to bootstrap the power supply of the 7650, limit the supply voltage on the integrated circuit with a zener diode, and float the complete package on a current source or another floating voltage source.

The 7650 has a maximum rated supply current of 4 milliamperes, so a current source capable of 5 mA was chosen. Transistors Q_2 and Q_3 (2N6718) provide a current mirror yielding the 5 mA current source. The 1N5242 zener diode limits the supply voltage to 12 V dc, and the 1N5228 — a 3.9-V zener diode — plus the gate-source voltage of the ITE4393 field-effect transistor is used to provide the greater-than-4-V bootstrap voltage needed for V^+ of the IC.

Bias for the 1N5228 comes through a 100-kilohm resistor, and the 10-kΩ, 150-picofarad filter stabilizes the circuit against high-frequency oscillations. Because the FET has a gate-source breakdown voltage (BV_{GSS}) of more than 60 V, the circuit has a common-mode range of −29 V to +25 V. In addition, the circuit can follow a 40-V peak-to-peak signal of over 20 kilohertz into a 50-kΩ load.

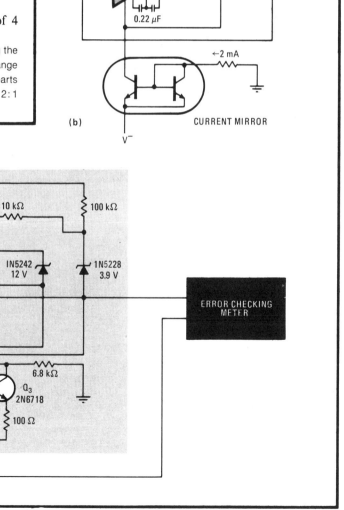

Ranging out. Bootstrapping the power supply, as well as limiting the supply voltage with a zener, helps extend the common-mode range of the inexpensive ICL7650 operational amplifier (a). Minimum parts version (b) of the extender circuit uses a commercially available 2:1 current mirror and a selected ITE4391 as the booster FET.

A minimum parts version of the first circuit is shown in (b). The current source is a 2:1 current mirror (from Texas Instruments), and the booster FET is a selected ITE4391 with a gate-source breakdown voltage (BV_{GSS}) greater than 60 v dc and a drain current greater than 5 mA when the gate-source voltage is equal to -3 v. The 100-kΩ and 10-pF stabilizing network is adjustable for maximum frequency response.

This type of circuit can be used to more than 100 v by substituting bipolar transistors or high-voltage FETs. And it can be built to be fully floating by replacing the current source with a pnp amplifier or p-channel FET □

Bi-FET op amps simplify AGC threshold design

by John H. Davis
Warm Springs, Ga.

Operational amplifiers with the bandwidth and input impedance available using bipolar–field-effect-transistor (bi-FET) technology are well suited for integrating the threshold detection and automatic-gain-control amplification functions in audio limiters or receiver AGC circuits. Generally, such circuits are implemented with discrete components. But this often entails component selection and critical trimming adjustments, or both. An op amp approach makes an AGC design more predictable, stable, and easier to troubleshoot.

The circuit of Fig. 1 requires only one adjustment, to zero the output of the TL071 op amp under no-signal conditions. In this circuit, a control voltage is required over the range from zero (at full gain) to the negative value corresponding to the FET's cutoff voltage. A zener diode supplies the reference voltage. It is connected in a way that makes use of the common-mode rejection properties of the op amp; thus, R_3 nulls the static output, which thereafter is quite stable.

The threshold is the voltage appearing at the junction of R_1 and R_2, plus the forward drop of the detector diodes, and can be readily computed for any desired

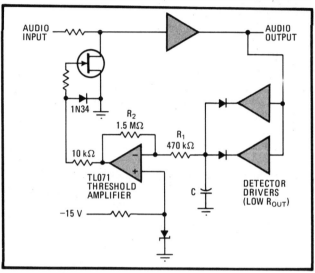

2. Simpler. Threshold detection, time constants, and amplification are consolidated in this single stage. For a receiver's i-f strip, an emitter follower is recommended, however. The control voltage here varies from a fixed negative value toward zero.

limiting level. For the detected peaks, the threshold detector has a voltage gain of:

$$A_{det} = (R_1 + R_2)/R_1$$

Not much gain is ordinarily required; too much imposes tighter tolerances on driver gain, diode properties, and

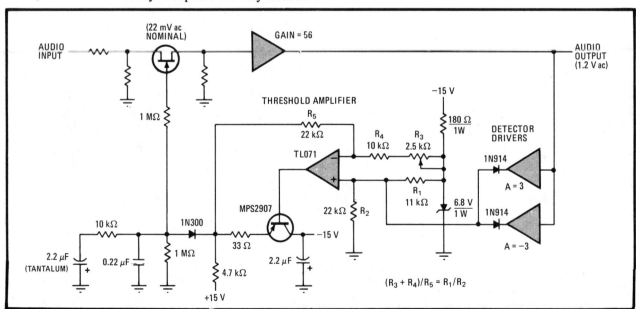

1. Easy play. Only one adjustment to zero the output of the TL071 op amp under no-signal conditions is needed in this AGC threshold amp. Good performance is achieved by using the amp's common-mode properties. The control voltage must vary from zero to a negative value.

trimmer adjustment. Driver gain can be adjusted, within output swing limits, to tailor limiting slopes.

The emitter follower improves the attack time of the time constant network. The dual set of time constants shown prevents short-duration peaks from depressing system gain longer than necessary.

Further simplification (Fig. 2) is possible if the driver amplifiers have low output impedance. Here the threshold detection, time constants, and amplification are consolidated in a single stage. This consolidation around one op amp means that little additional circuitry is needed when an FET is the voltage-controlled element.

The circuit assumes the control voltage must vary from a fixed negative value toward zero as gain reduction is needed. With no signal, the threshold op amp is referenced to the desired voltage by the zener diode. As long as no signal peaks are applied to time constant capacitor, C, the op amp acts as a voltage follower. Detected peaks charge C more negative than the reference, and the difference is amplified by a gain of $R_3 \div R_1$. This shifts the control voltage toward zero.

The release time constant is determined by C and R_1 and R_2. (Although only a single capacitor is shown, a dual arrangement as in Fig. 1 can be used.) The simplified circuit shown in Fig. 2 can also provide a fixed positive voltage that ranges toward zero for gain reduction if all the diodes and the reference-voltage polarity are reversed. □

Bi-FETs expand applications for general-purpose op amps

by Jim Williams
National Semiconductor Corp., Santa Clara, Calif.

With their excellent low-power consumption and low drift, bipolar field-effect-transistor operational amplifiers easily outperform general-purpose (741-type) op amps in a variety of applications [*Electronics*, Nov. 3, 1981, p. 134]. A low-power voltage-to-frequency converter, a battery-powered strip-chart preamplifier, and a high-efficiency crystal-oven controller can also benefit from those qualities of the 441 op amp.

The voltage-to-frequency converter (Fig. 1a) provides linearity to within 1% over the range of 1 hertz to 1 kilohertz. What is more, it does not need an integrator-resetting network using an FET switch, and its current drain is only 1 milliampere.

Integrator A_1 generates a ramp whose slope is proportional to the current into the amplifier's summing junc-

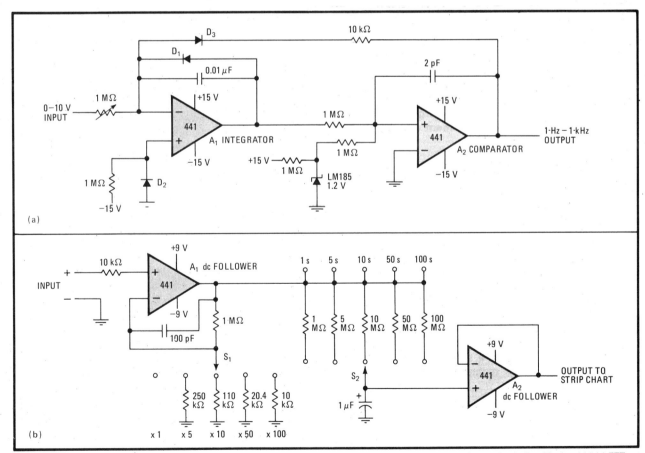

1. Low current, low cost. A voltage-to-frequency converter and preamplifier for strip-chart recorders may be built with the 441 bi-FET op amp. Converter (a), which is easily reset by a capacitor at A_2, provides linearity within 1% over a 0-to-1-kilohertz range and draws only 1 milliampere. A battery-powered preamplifier (b) has an adjustable gain and time constant. The circuit draws less than 500 microamperes.

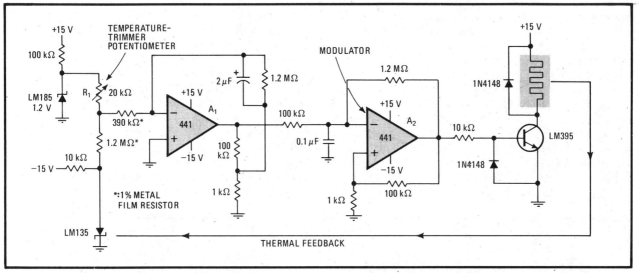

2. Heat switch. This feedback-type controller, using a switching modulator to conserve power, maintains the crystal temperature at about 75°C. Temperature, which may be trimmed over a 4°C range with potentiometer R_1, can be held to within ±0.1°C for a long time.

tion. The ramp's amplitude is then compared with the 1.2-volt reference at A_2, which serves as a current-summing comparator.

When the instantaneous amplitude of the ramp exceeds −1.2 v, A_2's output goes low, thereby pulling current from A_1's summing junction. This pulling, aided by diode D_1, causes A_1's output to drop quickly to zero. D_2 biases A_1's noninverting input, providing temperature compensation for the amplifier. These diodes and D_3 are 1N4148 parts.

The 2-picofarad capacitor at A_2 ensures that the output of the amplifier will remain high long enough to completely discharge the 0.01-microfarad capacitor at A_1, thus doing the job of the integrator-reset mechanism. As for calibration, the output is easily adjusted with the 1-megohm potentiometer for a 1-kHz output that is given an input voltage of 10 v.

The 441's low-bias current and its low-power consumption can also yield a simple and flexible preamplifier for strip-chart recorders (Fig. 1b). The circuit is powered by two standard 9-v batteries and may be plugged directly into the recorder's input. As a result, common-mode and ground-loop difficulties are minimized. The gain is variable from 1 to 100, and the time constant is adjustable from 1 to 100 seconds.

Input amplifier A_1 operates as a dc follower with gain. The gain has five ranges and is selected by S_1. The operational amplifier's input impedance is extremely high (10^{12} ohms) and consequently bias-current loading at the input is around 50 picoamperes. The 10-kilohm resistor in the input line provides current limiting under fault (overloaded input) conditions.

A_2, a second dc follower, buffers the RC filter composed of five resistors and a capacitor. The time constant is selected by switch S_2. This circuit draws less than 500 microamperes, ensuring long battery life.

The efficiency of the crystal-oven controller circuit (Fig. 2) is improved by having power switched across the heater element, instead of using a conventional linear-control arrangement. Oven temperature is sensed by the LM135 temperature sensor, whose output varies 10 millivolts/°C; thus its output will be 2.98 v at 25°C. This signal, converted into current as it flows through the 1.2-MΩ resistor, is then summed with a current derived from the LM185 voltage reference.

A_1 amplifies the difference between these two currents and drives A_2, a free-running duty-cycle modulator, over several kilohertz of frequency to power the output transistor and the heater.

Generally, when power is applied to the circuit, A_1 attains a negative saturation, forcing A_2's output to a positive one. The LM395 then turns on and the oven warms. When the oven is within 1°C of the desired setting, A_2 becomes unsaturated and runs at a duty cycle dependent upon A_1's output voltage. The duty cycle is determined by the temperature difference between the oven and the setpoint. For the given values, the circuit will maintain an oven temperature at 75°C, ±0.1°C. □

18. OPTOELECTRONIC CIRCUITS

Manchester decoder optimizes fiber-optic receiver

by Dwayne Yount
Measurex Corp., Cupertino, Calif.

Using a low-cost high-speed optical receiver and a simple Manchester II decoder circuit, this inexpensive fiber-optic receiver accepts data at rates of 10 megabits per second and higher and from a distance of at least 1 kilometer. Because the Manchester code minimizes any frequency drift in the receiver, the circuit works well for such fiber optic communication systems and in addition facilitates the use of an automatic gain control.

The optical signal is detected by the p-i-n photodiode D_1 (see figure), and the corresponding electrical current is then amplified by the optical receiver U_1. This receiver output is also converted to TTL levels by U_1. The Manchester-coded data is next fed into the delay line, which provides a maximum delay of 100 nanoseconds. The MCODE signal is delayed by 50 ns, 80 ns, and 100 ns to provide additional three signals M_{50}, M_{80}, and M_{100}.

MCODE and M_{50} are fed into the exclusive-OR gate U_2, whose output is inverted by U_3 to produce the DATA$_1$ signal. Gate U_4 combines M_{80} and M_{100} to produce a clean 20-ns-wide (clock 1) pulse that is used to clock the D-type flip-flop U_5. The clock 2 output of U_5 clocks U_6 only when the data makes a transition. D-latch U_6 reproduces the original data at the $\overline{Q}_2$ output.

The AGC and the receiver gain of 60 decibels optimize the signal-to-noise ratio. As a result, the circuit is capable of handling optical powers of the level of 1 microwatt. The self-clocking ability of the Manchester code makes it possible to recover the clock from the data. □

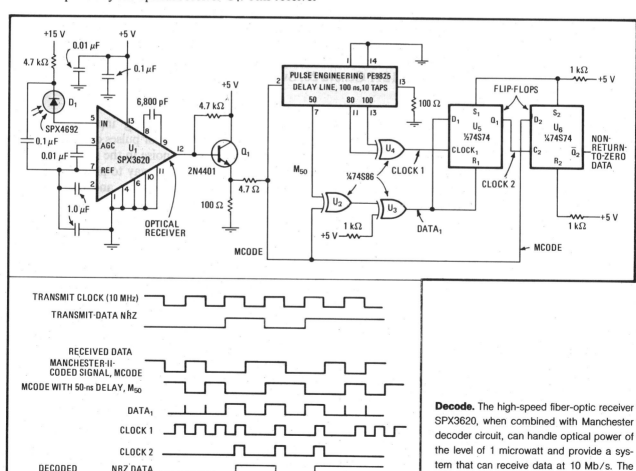

Decode. The high-speed fiber-optic receiver SPX3620, when combined with Manchester decoder circuit, can handle optical power of the level of 1 microwatt and provide a system that can receive data at 10 Mb/s. The output signal-to-noise ratio is optimized using AGC and a 60-dB receiver gain.

Fiber-optic link taps time-division multiplexing

by Mark Amarandos
Burr-Brown Research Corp., Tucson, Ariz.

Although fiber optics holds many advantages for process control and data collection, it is rarely used to transmit analog data because of the large cost per channel. However, through the use of time-division multiplexing, this circuit may carry up to 16 channels over one fiber and greatly increases the information-carrying capacity and cost effectiveness of a fiber-optic link. The design may be expanded to carry more than 16 channels and can transmit analog signals as weak as 10 millivolts. In addition, lines up to 9.7 kilometers long are attainable.

All 16 channels are scanned and updated once a second by the circuit (a). The input multiplexer examines the channels with respect to a free-running clock; the circuit requires no control from the receiving end of the link. The link's output, V_{out}, is a combination of a voltage that tracks input levels and a 4-bit binary word that represents the current channel address. This channel address is decoded by a phase-locked loop whose output through a 4-bit counter is used to track the state of the 4-bit counter at the transmitter.

The oscillator comprising exclusive-NOR gates U_{4-a} and U_{4-b} clocks 4-bit counter U_3 at 16 hertz. The output of U_3 sets the channel addresses of multiplexers U_1 and

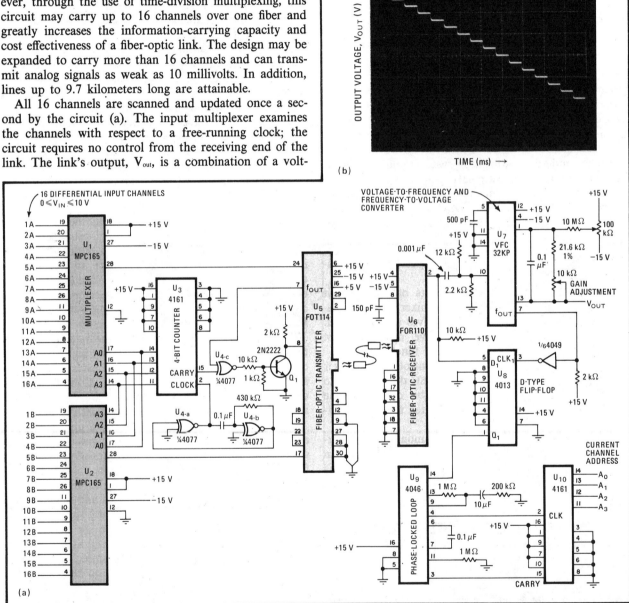

Optical link. Using a time-division multiplexer, 16 differential input channels can be transmitted over one fiber-optic cable. The channel address is generated at the transmitter by a free-running clock. When the VFC output is inverted on channel 16, address information is transmitted along with the data. Flip-flop U_8, following receiver FOR110, detects the inverted signal and creates a timing pulse that is used to recreate the channel address by PLL U_9. The trace (b) shows an output of the link.

U_2, which provide the 16 differential inputs to fiber-optic transmitter U_5. When channel 16 is selected, the carry output of the counter is a logic high. This output makes exclusive-NOR gate U_{4c} invert the VFC output of U_5 when channel 16 is selected, and this deviation in duty cycle is used when the current channel is 16.

In the receiver section, U_7 reconstructs the voltage applied to the VFC input of transmitter U_5. In addition, its one-shot output determines the input pulse train that is sampled for duty-cycle information. PLL U_9 locks onto this carry pulse signal from the flip-flop and reconstructs the input channel address. The 10-second time constant of the PLL filter causes a significant delay in locking. However, once the loop locks, the circuit remains stable.

For example, the trace (b) of the time-division-multiplexed link shows that the signal at V_{out} settles to within 0.1% in 30 milliseconds. The output of the link may be either loaded into an array of sample-and-hold circuits for a simultaneous readout of all analog voltages or converted to a digital form by a successive approximation analog-to-digital converter. Although the system is best suited for applications where the inputs operate over the same voltage range, an additional multiplexer can select different gain set resistors allowing compatibility with various input ranges. □

Optical coupler isolates comparator inputs

by Dennis J. Eichenberg
Cleveland, Ohio

Many dc-comparator applications need complete signal isolation. However, Motorola's optically isolated linear coupler MOC5010 eases this problem by eliminating the complex circuitry that is required with other techniques. The circuit's use of a single-ended power supply further simplifies the design.

The comparator circuit (see figure) compares two 0-to-12-volt signals that must be completely isolated. Resistor R_1, calculated for a current of 40 milliamperes, creates an acceptable current from V_{in} for the light-emitting diode of optocoupler A_1. Because there is an offset voltage at the output of A_1 (V_{in} = 0 v), the voltage at the inverting input of A_2 is made equal to the voltage at the noninverting input by adjusting the offset trimmer potentiometer R_4. This adjustment is done when V_{in} and V_{ref} are zero. Resistors R_5 and R_6 protect A_2 by limiting surge current.

Potentiometers R_2 and R_3 permit the slope of the input voltage for A_2 to be adjusted at the maximum V_{in} and V_{ref} by a desired ratio. When V_{in} exceeds V_{ref} by this ratio, the output goes high. Hysteresis may be provided by connecting an appropriate resistor from the output to the comparator's noninverting input. □

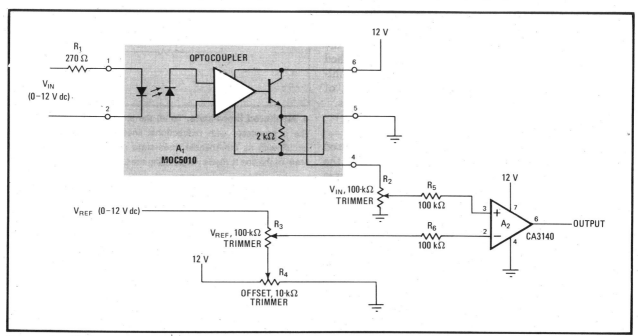

Comparator. The circuit compares two 0-to-12-V dc signals and provides complete isolation between the two signals. The circuit uses optocoupler MOC5010 to provide isolation and requires a single-ended power supply. Potentiometer R_4 balances the offset voltage.

Improving the LM395 for low-level switching

by Yehuda Gabay
Israel Atomic Energy Commission, Beersheba, Israel

The most significant drawback of a power transistor like National Semiconductor's LM395 is its relatively high quiescent current (10 milliamperes or so), which makes it impossible to use as a reliable switching device for small loads or loads that require dynamic currents ranging from zero to some high value. Adding a transistor-diode network and an optocoupler to the circuit, however, adapts the LM395 as a low-level (down to 0-mA) switch without sacrificing the current-handling capabilities of the power transistor and provides input-to-output isolation as well.

This circuit is configured as a normally-off switch whose quiescent load voltage is a maximum of 0.6 volt. Placing a logic 1 at the input of optocoupler U_1 causes transistor Q_1 to turn off. Thus the power transistor, Q_2, conducts and the desired current flows through the load, R_L.

If the input to the optocoupler goes to a logic 0, Q_2 cuts off and no current flows through the load. In this state, Q_1 conducts and the quiescent current of Q_2 that must flow is shunted through diode D_1 and through Q_1 to ground. It should be noted that D_2 bypasses transients to ground that are caused by an inductive load.

In the case where the user desires to implement a normally closed switch, it is only necessary to remove the circuitry centered around Q_1. Then, U_1's output transistor will serve to bypass Q_2's quiescent current to ground when necessary. □

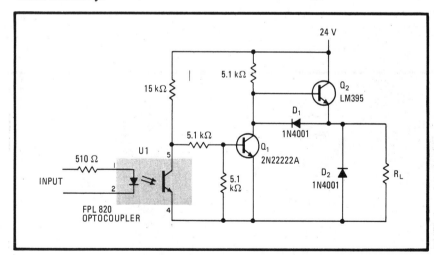

Bypass. A floating switch in the form of an optocoupler adapts power transistor Q_2 for handling small load currents, directing transistor Q_1 to bypass Q_2's high quiescent current (10 mA) when a logic 0 is applied to the circuit input. If a normally on switch is desired, Q_1 and its associated circuitry need only be removed. U_1's output transistor then will take Q_2's quiescent current to ground.

Twin optocouplers raise serial transmission speed

by Luis E. Murguis
Autotrol SA, Buenos Aires, Argentina

In a balanced 20-milliampere current loop for long-distance serial data transmission, optical couplers are a convenient way of connecting both receiver and transmitter to the transmission line, and provide isolation as well. However, an active pullup scheme employing an additional optical coupler at the receiver can improve transmission speed by an order of magnitude.

In the setup shown in (a), the fall time of the output voltage depends on the saturation current, I_i, of the coupler's input. However, the rise time of the output voltage, which determines the maximum transmission frequency, corresponds to the turn-off time of the coupler's output and is a function of load resistor R_L. Lowering the value of R_L raises the transmission rate, but only up to a limit set by the amount of current the optical coupler can handle.

Instead of trading off transmission speed and coupler loading, a second optical coupler produces a faster rise time and improves the transmission frequency almost 10 times over systems configured in the conventional way. The two optical couplers are connected as shown in (b) to produce an active pull-up and pull-down circuit at the output and thus speed up the output-voltage rise time. Both the rise and fall times are now a function of I_i, as the couplers alternate between their on and off states. Resistors R_1 and R_2 are optional and provide a fixed bias in case a circuit failure causes I_i to fall to zero. Another advantage of this circuit is that it improves fanout since a load resistor is no longer needed. □

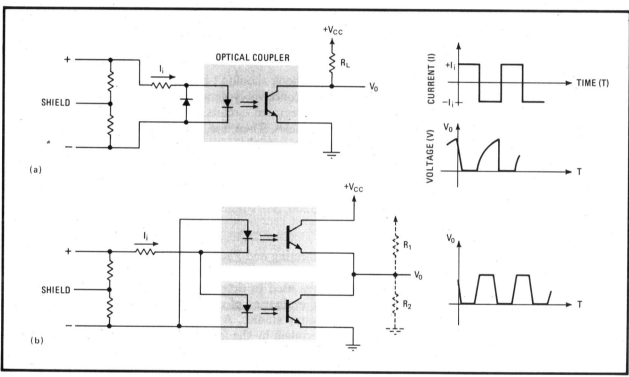

Active output. A conventional, single-coupler design for a 20-milliampere current loop (a) limits the transmission rate because the signal rise time is a function of resistor R_L. Using two couplers in an active pull-up output (b) forces faster rise times and hence higher transmission rates.

Programmed comparator finds loss in optical fiber

by J. T. Harvey and G. D. Sizer
AWA Research Laboratory, North Ryde, NSW, Australia

This circuit simplifies the job of finding an optical cable's quality for systems that use time-domain reflectometry (TDR) to measure transmission loss. The novelty of the circuit is that it is used as a programmable comparator to determine the times (and thus the cable length) corresponding to preset levels of the optical backscatter signal generated by the system, instead of to sample the amplitudes of the signal at preset times, as is normally the case. Consequently, circuit complexity is minimized and the measurement procedure is made significantly easier. Although the measurements will not yield results as accurate as could be achieved in the lab, they are adequate for most field applications—typically, to within 0.2 decibels per kilometer of the true cable attenuation.

In the TDR technique, the backscatter signal, which is derived from the fiber's reflection of an infrared pulse generated by the system's laser pulser, is detected by a photodiode amplifier and introduced to the vertical input

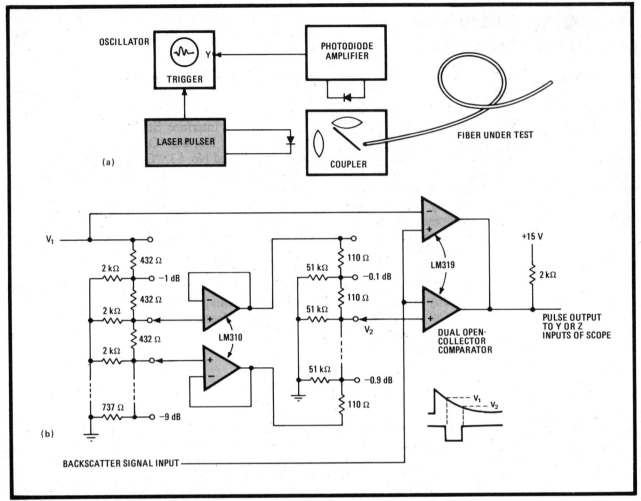

Simplified. Comparator determines the threshold crossing times corresponding to preset levels of an infrared backscatter signal so that loss in optical cable can be found directly. Accuracy of system is to within 0.2 dB of true value for fibers having 3 to 6 dB/km of loss.

of an oscilloscope. The laser pulser triggers the scope periodically as shown in (a) so that the profile of the backscatter signal can be sampled and the resulting cable loss calculated. Sampling, averaging, and computation are done in a variety of ways[1,2], often requiring a rather complex digital timekeeper to coordinate the measurement activities and a scope reading for estimating cable loss.

The slope of the backscatter signal (actually, the log of the slope, as it is exponential for a cable having constant loss per unit length) can be found easily with the circuit shown in (b), which uses a variant of a two-point sampling technique described in reference 2. There, the backscatter signal is sampled at two set times and the samples digitally processed to yield the loss between sample locations. With this circuit, however, the sample levels are set and the propagation times (and cable loss) corresponding to them are found.

Essentially, the circuit is a window detector that generates a pulse width corresponding to the time it takes for the backscatter signal to cross two preset thresholds. The first threshold is continuously variable, being set by voltage V_1. The second threshold is selectable in 0.1-dB steps from 0 to 9.9 dB by means of the resistive ladder, which is designed so that the voltages corresponding to those steps are given by $5 \log V_2/V_1$. This response matches that of the typical square-law photodetector and optic cable system.

The circuit can be used in several ways. One convenient way to measure cable loss is to set V_1 to initiate a pulse edge on the incident laser pulse and to adjust the second threshold so that the end of the pulse terminates 10 μs later. (This is the time taken for a double transition of a pulse through a 1-km fiber with a refractive index of 1.499, which is within 1% of the value expected in commonly available optical cables.) The cable loss, in dB, is then found from direct readout of the tapped resistive ladder.

The resolution of the measurement will be within 0.1 dB/km for a cable having a loss of from 3 to 6 dB/km. The estimated error in setting the pulse trigger time for a 1-km cable corresponds to a length deviation of about 30 meters, due largely to jitter on the laser pulse. The overall measurement error is thus 0.2 dB/km for cables having a low to medium loss—more than adequate for field use. □

References
1. M. K. Barnoski and S. M. Jensen, "Fiber Waveguides: A Novel Technique for Investigating Attenuation Characteristics," Applied Optics, Vol. 15, No. 9, Sept. 1976, pp. 2112-2115.
2. A. J. Conduit et al., "An Optimized Technique for Backscatter Attenuation Measurements in Optical Fibers," Optical and Quantum Electronics 12, 1980, pp. 169-178.

Optocouplers clamp spikes fast over wide range

by Alex Kisin
Digitus Corp., Baltimore, Md.

A series resistor and shunting diode are the cheapest way to protect the input stages of analog-measurement devices, at least partially, against overvoltage. But when high-voltage spikes come through, users find the large-value resistor often needed to limit surge currents may lead to an increased RC response time of the suppressor.

However, this increase cannot often be tolerated because of the propagation time of the analog data. By substituting an optocoupler for the shunting diodes, the device being protected may be isolated while ensuring a minimum delay time in eliminating glitches.

A common circuit for securing pulse protection is shown in (a). To increase protection, clamp voltages $+V_1$ and $-V_2$ may be raised, which is not always convenient. Alternatively, R may be increased but the response time of the device then becomes $\tau = R(C_L + 2C_D)$—a value that is often too high to be practical.

Adding opto-isolators (b) cures the problem. When the input voltage exceeds $+V_1$ or $-V_2$, the respective light-emitting diode will glow, transmitting light to the transistor optosensor. The flip-flop will be set and, in turn, the switch relay, which may be a small field-effect transistor, will be deactivated and the input signal electrically disconnected from the circuit.

The small activating and high surge current of the LEDs allow the input resistance (R_1) of the circuit to be reduced without sacrificing reliability.

This circuit, having $V_1 = 15$ volts, $V_2 = -15$ v and $R_1 = 470$ ohms, was used with 4N33 optocouplers to protect an LF198 sample-and-hold amplifier. The propagation delay was less than 50 nanoseconds. □

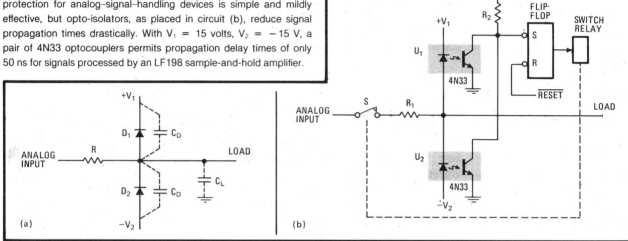

Short suppression. A rudimentary circuit (a) securing overvoltage protection for analog-signal-handling devices is simple and mildly effective, but opto-isolators, as placed in circuit (b), reduce signal propagation times drastically. With $V_1 = 15$ volts, $V_2 = -15$ V, a pair of 4N33 optocouplers permits propagation delay times of only 50 ns for signals processed by an LF198 sample-and-hold amplifier.

Swapable fiber-optic parts ease isolation problems

by Jim Herman
Motorola Semiconductor Sector, Phoenix, Ariz.

Assembling opto-isolators from their component parts to meet various high-voltage, high-frequency applications is now simplified with the introduction of interchangeable fiber-optic emitters and detectors. These devices can be built at a lower cost than conventional hybrids.

Therefore, systems such as a simple and effective 25-megahertz analog transmission channel and a 20-megabit-a-second emitter-coupled-logic data-handling system, which provide ac and dc isolation up to 50,000 volts, may be easily constructed.

A light-emitting diode and an optically-coupled photosensitive detector make up the basic optical isolator (Fig. la, p. 124). The plastic cable-splice bushing and plastic retainer caps housing the ferruled emitters and detectors are manufactured by AMP Inc., Harrisburg, Pa. When assembled, the components form an isolator that measures 0.75 inch long and 0.5 in. wide.

Characteristics of the isolator are determined by the selected emitter and detector. In particular, the interchangeable detectors provide the designer with several options that include interfacing with TTL or ECL loads, wide bandwidths, and analog or digital formats. The isolation voltage of the device is directly related to the separation between the LED and the detector, the package material, its size and shape, and the value of the parasitic capacity (C_C). This capacity determines the amount of ac protection.

The actual isolation (breakdown) voltage may easily be determined by measuring the voltage potential across the isolator from input to output at a prespecified leakage current such as 80 microamperes.

The coupler (lb) contains light pipes that provide efficient coupling while maintaining a large separation

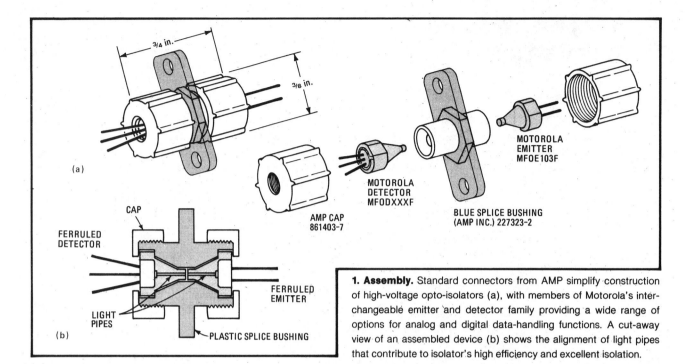

1. Assembly. Standard connectors from AMP simplify construction of high-voltage opto-isolators (a), with members of Motorola's interchangeable emitter and detector family providing a wide range of options for analog and digital data-handling functions. A cut-away view of an assembled device (b) shows the alignment of light pipes that contribute to isolator's high efficiency and excellent isolation.

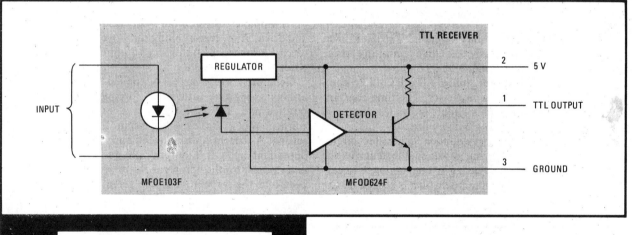

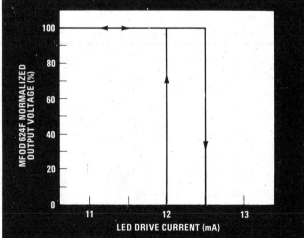

2. Applications. An isolator that can handle 500-kb/s data rates for TTL loads is configured with the MFOE103F emitter and MFOD 624F detector. The device's transfer function is also very sharp. Replacing the receiver with a 404F detector converts the unit to one that delivers a 10-Mb/s data rate. When the MFOC600 receiver is added, an isolator will provide 20-Mb/s data rates to ECL loads.

between the emitter and detector. As a result, the input and output have excellent ac and dc isolation. A wide selection of emitters and detectors are available.

Among the devices that may be readily assembled is a 25-MHz analog isolator built with the MFOE103F emitter and the MFOD104F pin detector. The transfer function of this device will be linear, providing a 0-to-20-μA pin-diode output for a 0-to-75-milliampere LED driving current.

Similarly, a 500-kilobit TTL isolator, Fig. 2, is easily built with the MFOE103F emitter and the MFOD624F integrated receiver. For wider bandwidths, the MFOD 404F integrated detector preamplifier and the MFOC 600 receiver circuit may be used. This combination will provide 10-Mb/s data rates for driving TTL loads or 20-Mb/s rates for driving ECL loads. In addition to high-voltage optical isolation, this isolator will also provide automatic gain control to stabilize output signals and an analog output port for status monitoring. □

Opto-isolated line monitor provides fail-safe control

by Eric G. Breeze and Earl V. Cole
General Instrument Corp., Optoelectronics Division, Palo Alto, Calif.

Because of the degree of isolation they provide, optically coupled ac-line monitors serve well as a small, reliable low-power interface in ac-to-dc control applications, where status information of the ac line is crucial. When a type is used that is TTL- and microprocessor-compatible—like the MID400, which performs the basic monitoring function on a single chip—a circuit can be built that ensures fail-safe control under the most difficult monitoring assignments. And when combined with a 555 timer, the monitor will have improved drive capability, more precise control of the turn-on and turn-off delay times, and better noise immunity than units that are typically available.

As shown in Fig. 1, two gallium arsenide infrared-light–emitting diodes connected back to back and optically coupled to an integrated photodiode and high-gain amplifier make up the MID400, which is encased in an eight-pin dual in-line package. In operation, each LED conducts on alternate half cycles of the ac-input waveform, together producing 120 pulses of light per second. Thus the photodiode periodically conducts, causing the amplifier to drive the npn transistor at the output to its on state. As the amplifier's switching time has been designed to be slow, it will not respond to an absence of input voltage lasting a few milliseconds; therefore it will not respond to the short zero-crossing period that occurs each half cycle.

The MID400 operates in one of two basic modes: saturated or unsaturated. It operates in the saturated mode when the input signal is above the minimum required current of 4 milliamperes root mean square and the photodiode pulses keep the output of the MID400 low. It operates in the unsaturated mode if the input current drops below 4 mA rms. Under these conditions, a train of pulses will appear at the output. In this way, a clean clock generator, devoid of power-line transients because it is isolated, may be realized.

If the input current drops below 0.15 mA, the device turns off. This causes the output of the MID400 to remain high.

Adding an external capacitor, C, to pin 7 of the device produces a time-delay circuit. The amount of delay on power-up is short because the photodiode has a low impedance when conducting. When ac is removed, however, the delay is long because the capacitor must discharge through the leakage resistance of the amplifier and photodiode. The larger the capacitor, the longer the delay.

In lieu of capacitor C, the use of a 555 timer also provides pulse shaping by yielding faster rise and fall times at the output. Here, the 555 is used as a Schmitt trigger with well-defined thresholds, the high state being $2/3\ V_{cc}$ and the low state being $1/3\ V_{cc}$. Besides providing

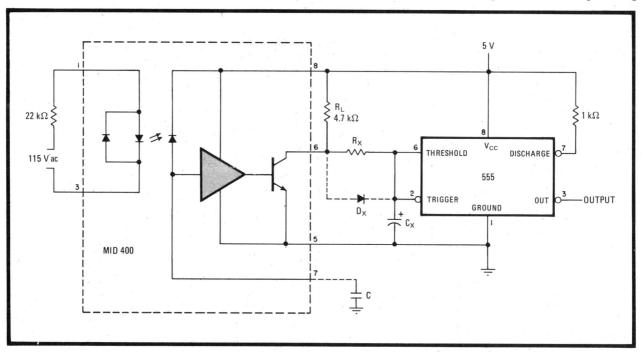

1. Control chip. Opto-isolated power-line monitor on single IC provides hash-free driving signals for ac-to-dc control applications. When united with 555 timer, circuit provides precise control of on and off delay times, improved driving capability, and good noise immunity.

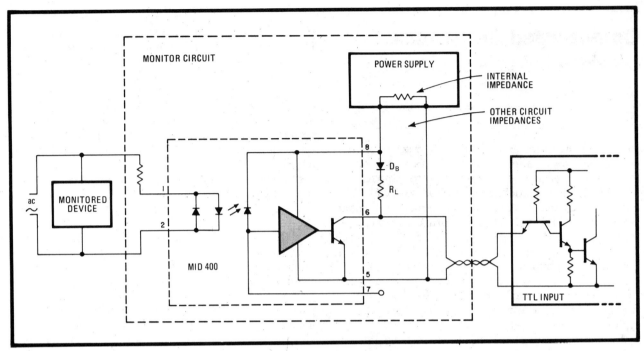

2. Immune. Monitor is easily modified for fail-safe control tasks, such as are required in industrial applications. Diode prevents component failures from creating output glitches. MID400's output circuit design inherently prevents generation of glitches caused by supply anomolies.

noise immunity, use of the 555 also minimizes oscillations that might occur if a TTL device were to drive a minicomputer.

Timing elements R_x and C_x set the delay. With appropriate choice of the time constant, the circuit can be made to respond to, or to ignore, one or more ac input cycles. Diode D_x permits the fast charge and slow discharge of C_x or, if the diode's polarity is reversed, the slow charge and fast discharge of the capacitor. The actual delay time will depend on the operating mode of the MID400.

For industrial, military, or medical applications in which fail-safe operation is important, the circuit's response must also be considered for the cases where either the ac input or the MID400 supply is removed. Fortunately, the MID400 has been designed so that its output transistor is on (low) only when both the ac input voltage and the supply voltage are present, thus simplifying the problem of providing a valid fail-safe control signal. Consider the case where the MID400 is powered by a separate 5-volt supply and the monitor drives a TTL-compatible interface through a twisted-pair line (Fig. 2). Normally, the inherent truth table of the device will prevent an erroneous output to the TTL interface circuit. Should R_L fall below 1 kilohm for one reason or another, however, and the power supply be off, the TTL input of the minicomputer will appear low because of excessive current flow through R_L. Diode D_B in series with R_L blocks any reverse current, eliminating the problem. □

Opto-isolated RS-232 interface achieves high data rate

by Vojin G. Oklobdzija
Xerox-Microelectronics, El Segundo, Calif.

When signals originating from isolated sources are transferred to a destination at a different voltage, coupling circuitry must be used to minimize signal distortion and interference. Unfortunately this circuitry slows down data-transmission rates. However, General Instrument's dual-phototransistor opto-isolator MCT66 may be used to isolate an RS-232 interface and still achieve a relatively high data rate of 9,600 bits per second.

The opto-isolated RS-232 interface (a) uses the MCT66, two diodes, an inverter, and a resistor. If pull-up resistors were used instead of transistors Q_1 and Q_3, the rising edge of the signal would be slow and thus would limit the transmission rate to below 1,200 b/s.

This limiting depends on the values of the resistors and the length of the RS-232 cable. However, if the pull-up resistor's value is reduced below 1 kilohm, intolerable power dissipation occurs.

This circuit not only achieves high data rates but also enables the polarity of the signal to be changed, without using an additional inverter, by altering the connections between transistors Q_1 and Q_2 or Q_3 and Q_4 (b). □

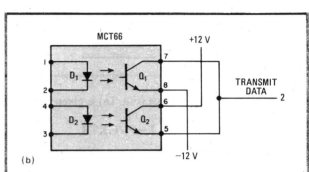

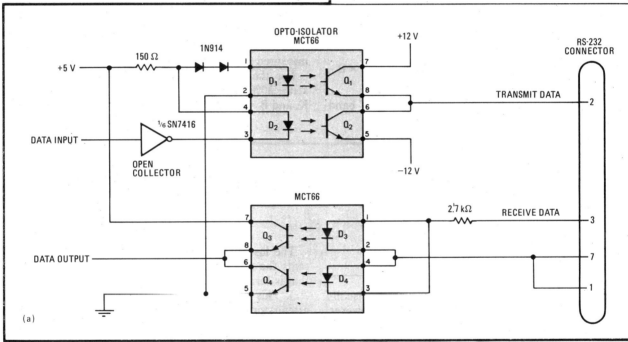

Isolation. General Instrument's dual-phototransistor opto-isolator MCT66 isolates the RS-232 interface (a) and achieves a high data rate of 9,600 b/s. The signal's polarity may be changed by altering the connections between Q_1 and Q_2 or Q_3 and Q_4 (b).

Optical agc minimizes video measurement errors

by D. Sporea and N. Miron
National Center of Physics, Magurele, Rumania

In an optical system, pulse width and peak amplitude are two information-bearing parameters that must be measured accurately for intensity-distribution and frequency-domain analysis—parameters that are difficult to determine when variations occur in the intensity of the signal's laser light source. This difficulty may be virtually eliminated by using the unmodulated laser signal to provide automatic gain control of the amplifier that processes the optical signal from a video camera. Such a scheme rejects the black level of the optical input signal, minimizes drift of low-level detector thresholds, and reduces noise caused by light scattering through optical components for near-zero level signals.

Signals are applied in a balanced fashion to the μA733 video amplifier, as shown. R_1C_1, R_2C_2, and the excellent common-mode characteristics of the operational amplifier reject most optical noise and bias the black level below the amplifier's active region.

The gain is controlled by a phototransistor operating in its linear region and two field-effect transistors whose drain-to-source resistance varies directly as a function of the reference laser signal. If the laser power increases, amplifier gain will be proportionally lowered, and vice versa, so that the output amplitude for a given video signal will be relatively independent of changes in the reference level. The signal is then amplified by the μA715 operational amplifier and presented to the μA710 high-speed comparator, where the switching threshold is set by the user.

In operation, the circuit provides excellent agc characteristics. Typically, the peak voltage (V_p) at the output of the 715 (see inset) will vary only 0.35 decibel for a given video signal and a change in laser input power of more than 10 dB. The roll-off response of filters R_1C_1 and R_2C_2 is such that the detector's threshold voltage, V_t, varies to a similar degree. Consequently, the ratio of V_p/V_b is virtually unchanged, a condition required for accurate measurements of width versus intensity. □

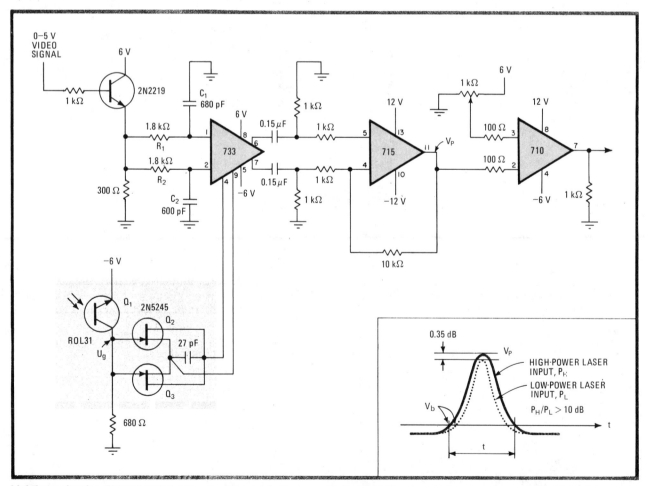

Lightbeam leveler. Phototransistor sets gain of input amplifier, ensuring that variations in output power of unmodulated laser light source have little effect on pulse-width measurements of optical input signal, which is derived from video camera. Change in output amplitude is only 0.35 dB for laser-power variations of more than 10 dB. Detector's threshold voltage varies to a similar degree.

Fiber-optic transmitter measures high voltages safely

by Larry Berkbigler and Greg Dallum
Lawrence Livermore Laboratory, Livermore, Calif.

In many industrial as well as research applications, there is a routine need to make accurate, high-voltage measurements in the region of 10 kilovolts or more. But unless sufficient isolation is provided between such a high voltage and the person measuring it, he (or she) will be in extreme danger.

This transmitter circuit converts a high voltage into an optical pulse train whose frequency corresponds to the voltage, providing a safe means of measuring high dc or low-frequency ac voltages. Since all power for the circuit comes from the measured source, faults cannot be conducted either through an ac power line or through the optical fiber that transmits the output pulse train.

The transmitter is basically a relaxation oscillator consisting of an oscillator circuit, an output-pulse-duration timer, a load and matching circuit, and a compensation network. In the oscillator circuit, capacitors C_1 and C_2 are charged to about 100 volts through R_1 at a rate proportional to the input voltage, V_{in}. When V_a reaches the value of the reference voltage V_g (see figure), the programmable unijunction transistor (PUT) fires and turns on the silicon controlled rectifier. C_1 and C_2 are thus discharged through the load, which also turns on the pulse-duration timer.

After about 8 microseconds, the output pulse turns itself off, which forces the SCR's biasing current to zero and switches it off. For high input voltages, where an 8-μs discharge time would cause significant errors in the measurement being made, the compensation circuit shortens the charging time.

The circuit shown is optimized for a maximum V_{in} of 30 kv and its output frequency is 0.2 hertz per volt ± 1% over the span of 6 kv to 30 kv; other ranges will require a number of component changes. The 8-μs output pulse is approximately half sinusoidal, with a peak power of 500 microwatts.

With inputs below 6 kv, the transmitter has two problems. First, it loses accuracy as V_{in} approaches the peak charge voltage of the capacitors, C_1 and C_2—the circuit has a typical error of ±2% with a 1-kv input. Second, the transmitter has a poor response time at low V_{in}; the output frequency is low, placing a burden on a frequency-to-voltage receiver.

The circuit is calibrated by first adjusting C_1 to provide a 2-kilohertz output with an input of 10 kv; at that input level, the compensation network has little effect. The compensation network is then calibrated by

High tension. Basically a high-voltage-to-frequency converter, this circuit safely measures voltages in the range of 1 to 30 kV. The output signal, which can be sent a distance of 20 yards or more over a fiber-optic link, is a TTL pulse train whose frequency is 0.2 Hz per measured volt. Above 1 kV, accuracy is to within 2%, and the circuit is powered only by the voltage that it is measuring. adjusting C_3 so that the output is 4.8 kHz with a 24-kV input. One iteration may be required in order to balance the system.

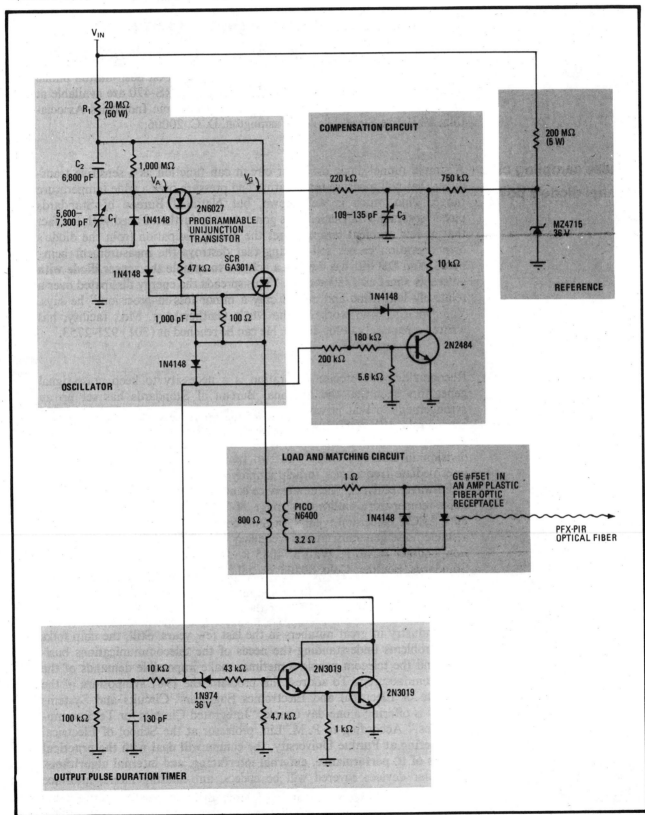

19. OSCILLATORS

Divider sets tuning limits of C-MOS oscillator

by Henno Normet
Diversified Electronics Inc., Leesburg, Fla.

Useful as it is, the square-wave RC oscillator implemented in complementary-MOS has one shortcoming—setting its maximum and minimum frequencies of oscillation independently while also maintaining accuracy is extremely difficult. By placing a voltage divider in the feedback loop of the conventional three-gate circuit, however, a one-time trimming adjustment can accurately set the maximum and minimum frequency excursion and will force the ratio of the upper to the lower limit of oscillation to approach a value virtually determined by the resistors used in the same divider.

The standard RC oscillator generates a frequency of $f \approx 0.482/R_1C$, where $R_1 = R_2$, as shown in (a). Generally, it is not practical or economical to use a variable capacitor for C. A potentiometer could be substituted for R_1 to tune the frequency, but slight differences in integrated-circuit parameters will preclude predicting the maximum and minimum frequencies of oscillation with any degree of accuracy for a particular chip. The only other method for setting the upper and lower frequency limits is to parallel several capacitors across C, a tedious procedure at best.

Alternatively, R_1 can be a potentiometer that is placed virtually in parallel with voltage divider R_4–R_5 through C (b). In this way, capacitor C is no longer charged from the fixed-voltage output of the middle gate in (a), but from the voltage divider across the output. R_1 is thus used to change the circuit's time constant without affecting the potential that is applied to C.

The upper and lower limits of oscillation are determined by the position of R_4's wiper arm and by the values of R_4 and R_5. With the tap at point A, the circuit will oscillate at a frequency given by $f = 1/2.2R_1C$. With the wiper at point B, the frequency will be $f = 1/1.39R_1C$. The frequency ratio to be expected is thus $2.2/1.39 = 1.6$. The actual frequency change measured with the particular chip used for breadboarding was 56%, which is thus very close to the intended value. The ratio will increase as R_4 is made larger with respect to R_5.

The circuit has only one small disadvantage—the load presented by R_4 and R_5 does increase the power-supply drain by approximately 0.5 milliampere. □

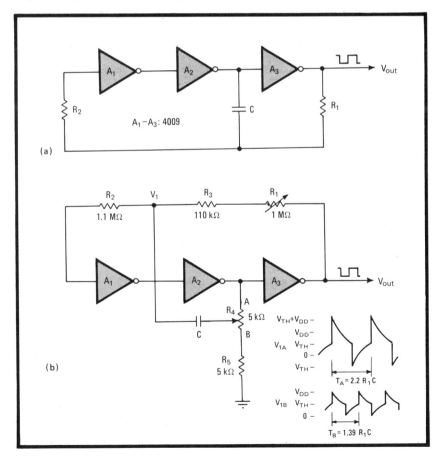

Calibrate. IC anomalies, inherent circuit imbalance, and the expense of making C variable preclude setting upper and lower oscillation limits of typical RC oscillator (a) with any accuracy. Placing R_1 virtually in parallel with voltage divider (b) through C gives circuit one-knob frequency control, with upper-to-lower oscillation ratio in effect determined by R_4 and R_5.

C-MOS IC achieves triggered phase-locked oscillations

by N. Miron, M. N. Ion, and D. Sporea
Central Institute of Physics, Romania

The application of dual vernier interpolation to time-interval measurements requires a triggered phase-locked oscillator that is phase-synchronized with the external trigger signal. This circuit (a) is a modification of the idea proposed by D. C. Chu[1] and uses a complementary-MOS medium-scale integrated circuit to make it simple and inexpensive.

Prior to the arrival of an external-phase–synchronization pulse, the oscillation frequency (f_1) of the voltage-controlled oscillator is in phase with the reference frequency (f_0). In the quiescent lock state, the positive transitions of the mixer and counter output occur at the same time, thereby satisfying $f_o - f_1 = f_1/N$ or in terms of period, $T_1 = (1 + 1/N)T_o$.

The arrival of the phase-synchronization pulse sets the latch whose output through the one-shot multivibrator U_2 inhibits the VCO (b) for a time that is determined by R_1C_1 ($\tau = 600$ ns). However, when the VCO starts again with a zero phase shift, the signal $\overline{INH}$ (inhibit) resets the divide-by-N counter.

This $\overline{INH}$ signal through NAND gates A and B also inhibits the three-state phase comparator (U_8) that will remain in a high-impedance state between the arrival of the phase-synchronization pulse and the first negative transition of mixer output $\overline{Q}_2$. The $\overline{INH}$ state is changed by this sequence, thereby enabling gates A and B and directing the mixer and counter outputs to the phase-comparator inputs.

The delay τ_1 introduced at the latch's output $\overline{Q}$ avoids locking a possible initial mixer negative transition that is not related to phase crossover. The circuit is set to the initial state with an external RESET signal. □

References
1. D. C. Chu, "The triggered phase-locked oscillator," Hewlett-Packard Journal, Vol. 29, Aug. 1978, p. 8.

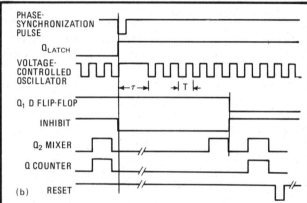

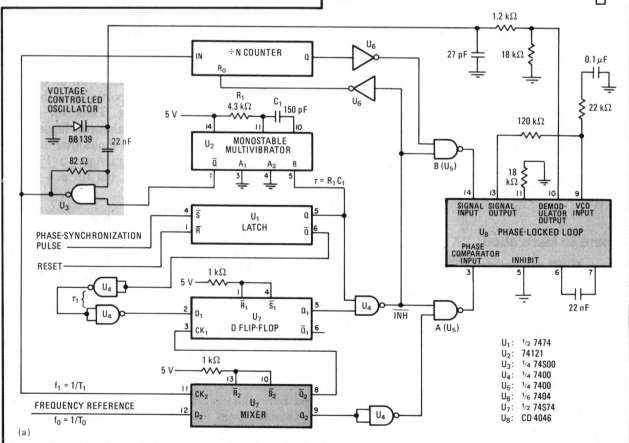

Phase-synchronized. This triggered phase-locked oscillator (a) uses a C-MOS phase-locked–loop CD4046 and achieves phase synchronization with an external pulse. The timing diagram (b) shows typical waveforms for different sections of the oscillator.

Stable sinusoidal oscillator has multiple phased outputs

by James J. Rede
Rede Electronics, St. Louis, Mo.

This sine-wave oscillator design is useful as, among other things, a pure signal source for calibration. It provides gain-independent operation, is easy to adjust, and has a wide frequency range and stable amplitude. What's more, the circuit is self-starting and provides multiple-phased outputs of equal amplitude.

As shown in Fig. 1, the feedback loop has two 90° phase shifters and a unity-gain inverter. Thus, the circuit meets the criteria for oscillation—a loop gain of 1 with a phase shift of 180°. Each 90° phase shift is provided by the delay equalizer circuits (A_2, A_3). These two circuits have a unity gain at all frequencies and a phase shift that is adjustable between 0° and $-180°$. It is these properties that make the oscillator's features notably superior to those of other designs.

The transfer function of the equalizer circuit is $T(s) = V_{out}/V_{in} = (s-a)/(s+a)$, where $a = 1/R_1C_1$. At any frequency or pole/zero value, the absolute magnitude is always unity: $|T(s)| = |s-a|/|s+a| = (a^2+\omega^2)^{1/2}/(a^2+\omega^2)^{1/2} = 1$. Phase, $B(\omega)$, which is plotted in Fig. 2 is given by $B(\omega) = -2\tan^{-1}(\omega/a)$, where $a = 1/R_1C_1$. At $\omega = a$, the zero contributes $-135°$ of phase shift and the pole contributes 45° of phase shift for a total phase shift of $-90°$.

The frequency of oscillation is completely determined by the two independent time constants, R_1C_1 and R_2C_2, and can be expressed exactly as $f = 1/[2\pi(R_1C_1R_2C_2)^{1/2}]$. If $R_1 = R_2$ and $C_1 = C_2$, then $f = 1/(2\pi R_1C_1)$.

Since the frequency range of oscillation is totally independent of any gain factor, the amplitude stability and the amplitude of oscillation are completely decoupled from the frequency-determining adjustment of R_4. A wide frequency range is assured for this oscillator. The amplitude of oscillation is determined by the maximum voltage swing of the op amp.

In the circuit of Fig. 1, C_1 and C_2 are equal; R_1 and R_2 are set equal; R_4 is adjusted for a total loop gain of 1. As the loop gain approaches unity, the pure sinusoidal oscillation begins. Further adjustment of R_4 permits the oscillation to be easily stabilized at its maximum amplitude and with no harmonic distortion. Adjustment of R_1 alters the phase relationship between the 0° and $-180°$ signals. It also changes the frequency of the oscillator without disturbing the amplitude, amplitude stability, or oscillation criteria. The placement of C_3 in the circuit prevents high-frequency parasitic oscillations from occurring in the operational amplifiers.

Oscillations over a large frequency range could be obtained by changing the value of C_1 and C_2 and varying R_1 and R_2 with a dual potentiometer. For single-element

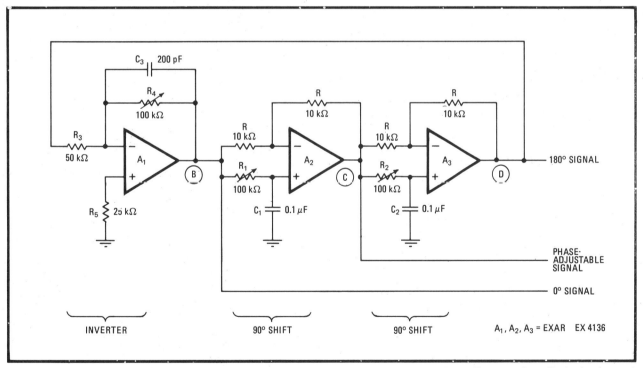

1. Phased out. An inverter (A_1) and two identical equalizer circuits (A_2, A_3) form a sinusoidal oscillator. Three equal-amplitude signals are at 0°, 90°, and 180° phase angles, respectively. The oscillator is self-starting and is capable of a wide range of frequencies.

2. Frequency equalizer. The oscillator circuit in Fig. 1 provides unity gain independent of frequency. The phase angle of the output varies from 0° to −180° with frequency, as shown above. At the frequency of oscillation, $\omega = a$, the phase angle is −90°.

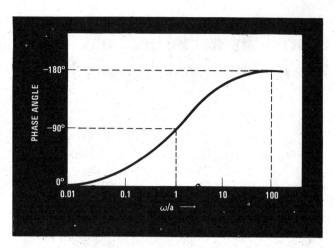

control, R_1 and R_2 could be voltage-controlled resistances. Over a narrower range, frequency can be adequately varied by just adjusting R_1.

The circuit is ideal for producing multiple signals with precise phase relationships of equal amplitude. By cascading increasing numbers of delay equalizer stages, signals of any phase can be easily obtained. □

Stable and fast PLL switches loop bandwidths

by Yekutiel Josefsberg
Israel Electronics Industries Ltd., Holon, Israel

Narrow phase-locked loops may be locked faster by starting them off with wider loop bandwidth, thus eliminating the usual "hunting" that occurs with just a single narrow bandwidth PLL. However, the switch back to the original bandwidth introduces transients into the loop that cause a loss of lock, and particularly when the ratio between the two bandwidths is high. This circuit eases such a switching problem by maintaining a stable lock for bandwidth ratios up to 1,000:1.

Two loop filters, a narrowband low-pass filter consisting of R_1, R_2, and C_1 and a wideband low-pass filter comprising R_3, R_4, and C_2, are connected in parallel (see figure). When the PLL is turned on, switches S_1 and S_2 close. This action widens the loop bandwidth, which subsequently charges capacitor C_2 to its peak voltage quickly. A fast lock results.

Once the circuit is locked, S_1 and S_2 open to narrow the loop bandwidth. As C_1 is already charged to the correct voltage, no transient occurs, and thus there is no loss of lock. The voltage on C_1 follows that of C_2 with a delay determined by C_1, the resistance of S_2, and the output resistance of buffer A. □

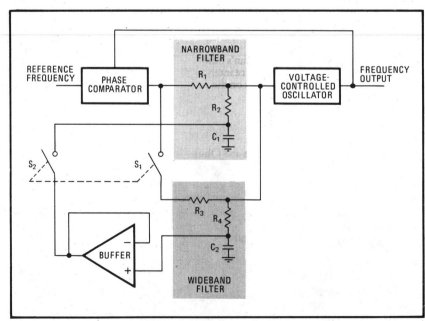

Acquisition. This phase-locked loop achieves a faster acquisition rate by switching from wide loop bandwidth to a narrow one. The circuit can be designed to switch a bandwidth ratio of 1,000:1 without the loss of lock.

Phase shifters simplify frequency-multiplier design

by Fred Brown
Lake San Marcos, Calif.

Phase-shift frequency multipliers, unlike conventional multipliers, can produce a spectrally pure output without filtering. However, by using wideband phase-difference networks for phase splitting, frequency-independent multipliers over many octaves may be obtained.

The principle of this type of multiplier is shown in Fig.1a. A sine-wave frequency is multiplied N times by dividing the input into N different phases that are equal-

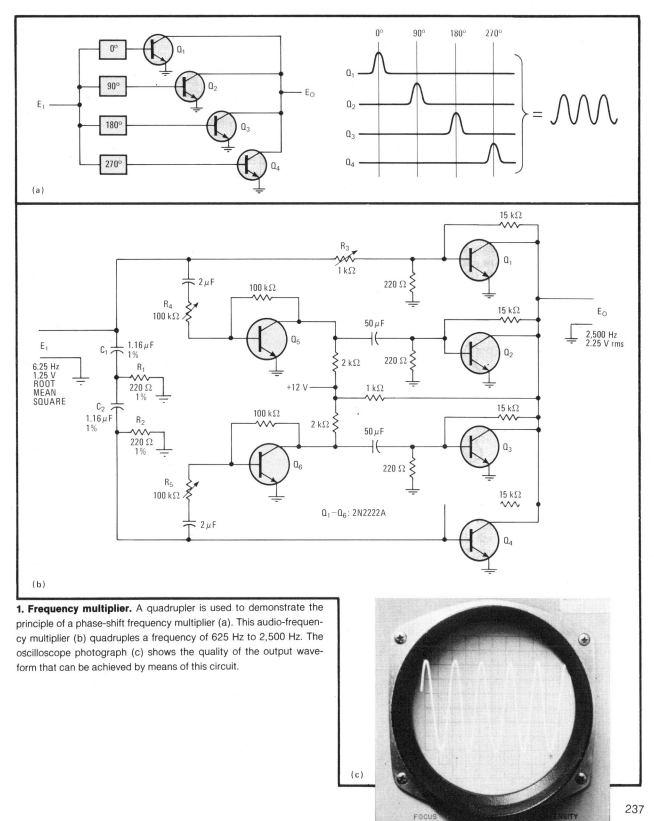

1. Frequency multiplier. A quadrupler is used to demonstrate the principle of a phase-shift frequency multiplier (a). This audio-frequency multiplier (b) quadruples a frequency of 625 Hz to 2,500 Hz. The oscilloscope photograph (c) shows the quality of the output waveform that can be achieved by means of this circuit.

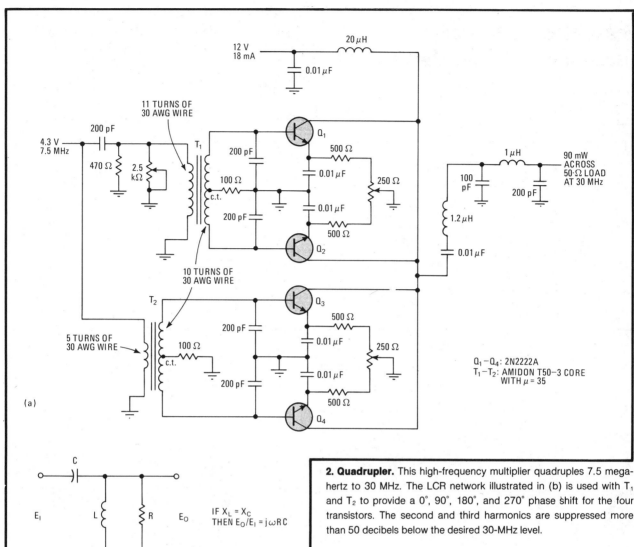

2. Quadrupler. This high-frequency multiplier quadruples 7.5 megahertz to 30 MHz. The LCR network illustrated in (b) is used with T_1 and T_2 to provide a 0°, 90°, 180°, and 270° phase shift for the four transistors. The second and third harmonics are suppressed more than 50 decibels below the desired 30-MHz level.

ly spaced through 360°. These N phases drive N class-C transistors whose outputs are combined to deliver a pulse every 360°/N. The use of N transistors allows the input power to the circuit to be N times as high without saturating the transistors.

This audio-frequency quadrupler (Fig.1b) uses frequency-dependent 90° phase-shift networks $R_1 C_1$, and R_2, C_2. Transistors Q_1 and Q_4 provide pulses that are shifted 0° and 90° in phase at the output. Phase inversion of the pulses is achieved by transistors Q_5 and Q_6, which drive Q_2 and Q_3 to provide pulses that are 180°- and 270°-phase-shifted at the output. The output pulses that are 90° apart are combined to produce the quadrupled frequency. The af multiplier quadruples a frequency of 625 hertz to 2,500 Hz.

The amplitude of the input signal is adjusted for the proper level at the base of Q_4. In addition, level adjustments for Q_1, Q_2, and Q_3 are controlled by R_3, R_4, and R_5. The oscilloscope photograph (Fig.1c) shows the quality of the ×4 output frequency at 2,500 Hz.

Phase-shift frequency multipliers are superior to conventional multipliers at high frequencies in subharmonic suppression. A high-frequency version of this type of multiplier (Fig.2a), also a quadrupler, uses a simple LCR phase-shift network (Fig.2b) to produce a 90° phase shift.

An interesting property of this network is that when the reactances are made equal, the phase shift between the input and output ports will always be 90°, regardless of the value of R. This property allows both amplitude (varying R) and phase (varying L or C) control.

The inductance L is created by the primary winding of T_1; the secondary winding delivers a 90° and 270° phase shift to Q_1 and Q_2, respectively. The 0° and 180° phase shifts are provided by T_2 to Q_3 and Q_4.

In addition, the L-pi network at the output provides an optimum match to the 50-ohm load and a little attenuation of subharmonics. This multiplier, unlike conventional ones, is capable of suppressing subharmonics and therefore does not require output filtering.

A spectrum-analyzer display showed that the second and third harmonics could easily be reduced by more than 50 decibels below the desired fourth harmonic. □

V-MOS oscillator ups converter's switching frequency

by Bill Roehr
Siliconix Inc., Santa Clara, Calif.

The benefits of switching a flyback converter at high frequency to increase its efficiency and minimize its size may be realized by employing a V-groove MOS field-effect transistor as its power oscillator. Unlike bipolar power transistors, where storage-time effects hamper device turn-off, the turn-on and turn-off times for V-MOS units are fast—typically a few nanoseconds. Thus, switching speeds of 250 kilohertz can easily be achieved.

The circuit configuration is very simple, as shown. When the circuit is first energized, a positive voltage is capacitively coupled to the gate, turning on the VN10KM V-MOS device. Enhancement voltage is maintained by the potential across the transformer's primary, which is reflected onto its feedback winding. The FET continues to conduct until the core saturates, whereupon the feedback voltage collapses and turns the device off.

With the FET off, energy stored in the magnetic field surrounding the primary winding is transferred to the secondary winding. Zener diode D_1 clamps the primary winding voltage to the desired potential and limits the voltage across the V-MOS gate to some value below its 60-volt breakdown rating. The energy transferred to the feedback winding has the proper polarity to hold the FET in cutoff. When the transformer comes out of saturation, the operating cycle repeats. Diode D_2 prevents negative spikes from damaging the gate of the FET. Resistor R_1 suppresses any parasitic oscillations caused by switching.

Energy transferred to the secondary winding delivers power to filter capacitor C_2 via rectifying diode D_3. A single 4.7-µF capacitor provides sufficient filtering at the 250-kHz operating frequency. The dc output voltage may be made positive with respect to the main rail by grounding terminal 1 and negative with respect to the main rail by grounding terminal 2.

A dc output of up to 60 V can be developed by simply selecting a zener diode of that same value, although practically any voltage can be obtained by altering the transformer's turn ratio. The supply voltage should be set between 3 and 5 V dc.

Note that the physical size required for this flyback converter will be minimal, since the reactive components will be small and light because of the high operating frequency. □

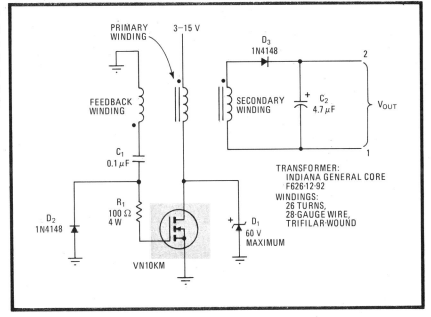

High-flying. V-MOS power-FET converter easily operates at switching frequencies of 250 kHz and can work up to several megacycles, thereby increasing the efficiency of the flyback converter and also minimizing its size. Dc output potentials of up to 60 V may be ordered by appropriate selection of zener diode D_1.

Programmable sine generator is linearly controlled

by S. Awad and B. Guerin
Laboratoire de la Communication Parlée, Grenoble, France

Because the frequency of this sine-wave oscillator is a linear function of a digitally controlled input, it is attractive for microprocessor and speech-synthesis applications. In addition, it is easy to implement—the digital controller, made up of two resistor modules, needs only to be inserted in place of the two passive resistors normally present in a standard sine-wave oscillator.

It can be shown from the basic circuit in the figure that the frequency of oscillation is given by:

$$f_o = (1/2\pi)(R_5/C_1C_2R_1R_2R_6)^{1/2}$$

with the condition for oscillation being given by:

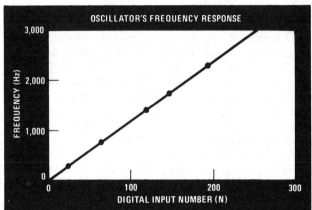

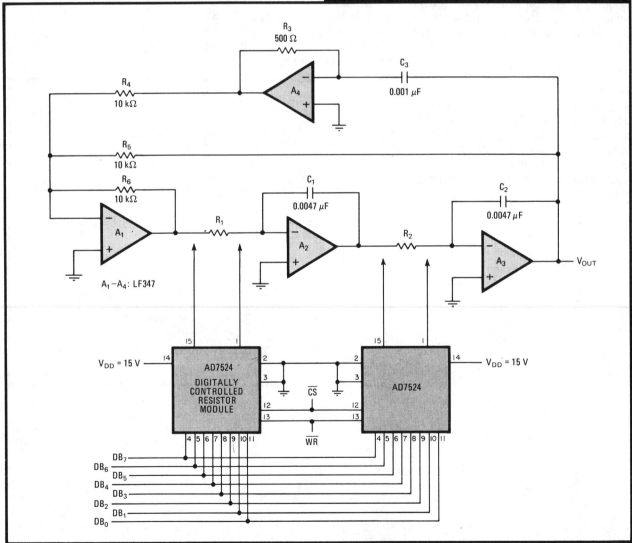

Logically linearized. Oscillator frequency is a linear function of the setting of two digitally controlled resistor modules, making the circuit attractive for remote-control applications of all kinds. An oscillator is easily modified for digital control by the insertion of these modules in place of R_1 and R_2 of a standard sine-wave oscillator's network. Response has straight-line characteristics from zero to 3,000 Hz.

$$C_3R_3R_6/R_4 \geq G$$

where G is a gain constant. As can be seen, the frequency of oscillation is a function of the virtually grounded resistors R_1 and R_2. They can be replaced by two AD7524 digitally controlled resistors, whose output value is given by:

$$R_1 = R_2 = R_0(2^n - 1)/N$$

where n is the number of digital bits available, N is the digital number that is programmed (set), and R_0 is a constant.

Substituting for R_1 and R_2:

$$f_0 = (N/2\pi)[R_0(2^n - 1)(R_5/R_6C_1C_2)^{1/2}] = KN$$

where K is a constant. Thus the relationship between the output frequency and N is perfectly linear.

These theoretical results are confirmed in the laboratory model, as seen by the response (see plot at bottom of figure). These results were obtained with an R_0 of 11 kilohms. □

20. PERIPHERAL INTERFACING

Low-cost interface unites RAM with multiplexed processors

by G. Aravanan and U. K. Kalyanaramudu
Bharat Electronics Ltd., Bangalore, India

This simple interface hooks a standard random-access memory directly, without using latches, to a microprocessor that has a multiplexed bus. The nine-gate, one–flip-flop array is inexpensive, costing only a few dollars.

General Telephone & Electronics' 8114-2, a 1-K-by-8-bit RAM, is joined to an Intel 8085A microprocessor. At the start of the microprocessor's write cycle, the 8085's address-latch–enable (ALE) line moves high and the 74LS74 D-type flip-flop is reset. The RAM's write-enable ($\overline{WE}$) line is then activated via OR gates G_2 and G_3 when the input/output memory control and bus-cycle–status ($IO/\overline{M}$ and S_1) lines are also brought low by the processor.

On the falling edge of the ALE line the RAM is enabled via OR gate G_4, given that the chip-select input for the device is high (G_5 would be low). Then the 8114 latches the address presented on lines AD_0 through AD_7, which are tied to their corresponding address lines on the RAM. Following this, the data is transferred from the processor to the RAM. The 8085's write-control ($\overline{WR}$) line will then go high, causing the flip-flop to latch a high state, and thus the write cycle will terminate when the $\overline{WE}$ and $\overline{CE}$ lines of the RAM go high.

The address should remain on the lines for at least 100 nanoseconds after $\overline{CE}$ line becomes active (low), and the data should be on the lines for at least 100 ns before the $\overline{CE}$ line goes high—for at least 30 ns thereafter. The $\overline{CE}$ line will be active in the cycle for 400 ns.

The read cycle is similar to the write cycle except that the $\overline{WE}$ line remains high during the entire cycle interval. The ALE line of the processor initially goes high, as does its S_1 line, and the RAM's desired address is placed on the AD_0 through AD_7 output bus.

Once the ALE line falls flat, the $\overline{CE}$ line is activated as before, and the address information, which should remain on the bus for at least 100 ns thereafter, is accepted. After 100 ns, the RAM's corresponding data is transferred to the bus and then to the processor. □

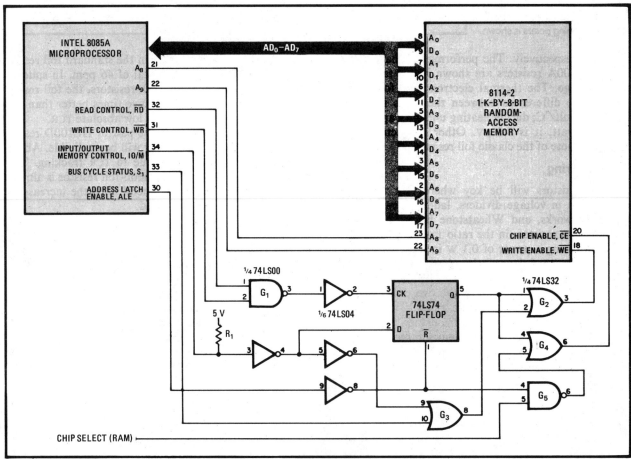

Link. A four-chip interface facilitates an easy exchange of data between processors having multiplexed data-address buses and standard random-access memories. Circuitry is simplified by tying a RAM's data and address lines together. The interface costs under $5.

Enabling a processor to interact with peripherals using DMA

by Trung D. Nguyen
Litton Data Systems Division, Van Nuys, Calif.

Most peripheral devices connected to a microcomputer by the peripheral interface bus can request services from the computer over the interrupt line. This facility, though, is lacking on those input/output devices that use direct memory access to transfer data at high speed between themselves and the processor's main memory. However, a simple interrupt logic circuit will enable such I/O devices to notify the computer when a DMA operation is complete.

As shown in (a), the interrupt-enabling flip-flop U_{1-1} is initially enabled by the enable-service-interrupt ($\overline{ESI}$) command provided by the computer. The computer next sends the external-function-enable ($\overline{EFE}$) command to an addressed I/O device to indicate its readiness for data transfer. If the addressed I/O device chooses to respond to this code, it generates an I/O interrupt request. Interrupt flip-flop U_{2-1} holds this request until the computer

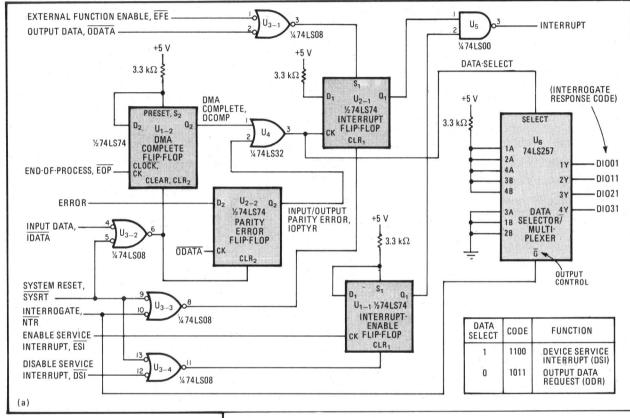

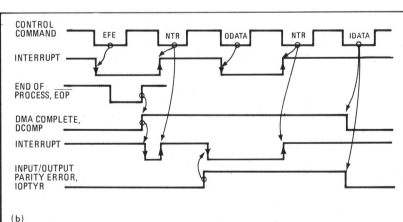

Interrupt. This interrupt logic circuit (a) allows a microcomputer to respond to input/output devices that access its memory directly. The timing diagram (b) of the logic shows that the interrupt line is set when the peripheral device detects an EFE or an output-data command and is reset by the NTR signal. The line is also set by a signal that a direct memory access is complete or that a parity error has been detected.

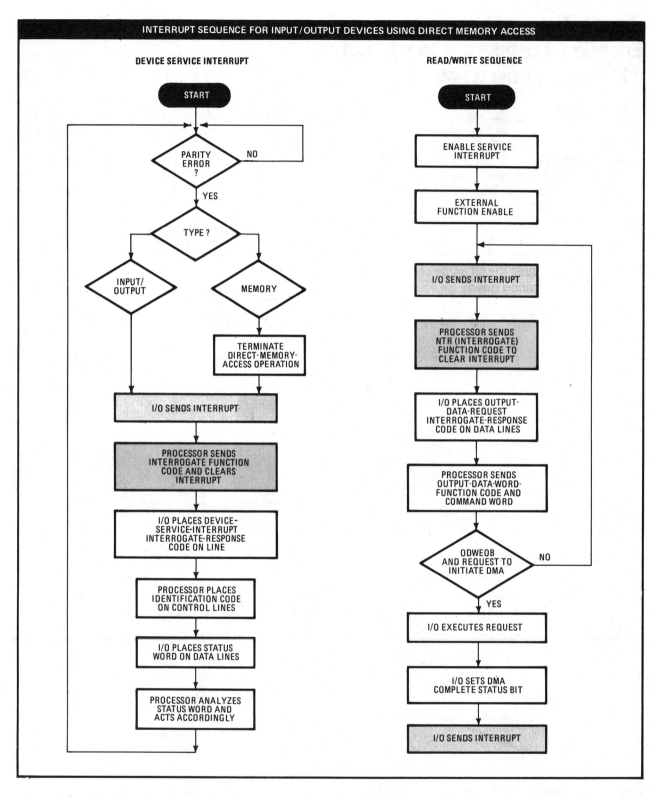

acknowledges it with an interrogate function code.

The I/O device next places the output-data-request interrogate-response code (ODR IRC) on the data lines, whereupon the processor places the output-data-word (ODW) function code and command word on the data lines. Then the I/O device interrupts the processor again for more ODW code, and this procedure continues until all parameters in the command word are sent. The processor indicates completion of the task by sending an output data word with an end-of-block (ODWEOB) signal on the control lines and by requesting the initiation of a DMA operation on the data lines.

Subsequently, the end-of-process (EOP) signal from the DMA indicates that the DMA operation is complete. The I/O device now sets the DMA-complete status bit and issues an interrupt to the processor.

If a memory parity error occurs during the memory-to-memory transfer cycle, the external end-of-process signal (not shown) causes the abrupt completion of the DMA operation and makes the I/O device issue an interrupt. The I/O parity error sets the I/O parity error flip-flop U_{2-2}, which enables the I/O device to generate an interrupt signal (b). The I/O device responds to this parity error interrupt by placing the device-service-interrupt interrogate-response code (DSI IRC) on the data lines. In addition, the processor sends the interrogate command code over the control lines and clears the interrupt. Next, the processor sends the input command code, and the I/O device places the status words on data lines and clears the parity flip-flop. The processor then analyzes the status word and acts accordingly. The table shows the step-by-step interrupt sequence.

The interrupt response codes are generated by data selector–multiplexer U_6. When the processor sends the interrogate ($\overline{NTR}$) command, U_6 selects either device-service-interrupt (DSI) or output-data-request (ODR) on the basis of the state of the data-select line. If a parity error or DMA completion occurs, the data-select line goes high and DSI is selected. If nothing happens, the data-select line remains low and ODR is selected. The codes for each function are placed on the data line bus, and the processor is programmed to recognize them. □

Interface unites Z8000 with other families of peripheral devices

by S. Majundar, K. Kumar, and K. S. Raghunathan
Indian Telephone Industries Ltd., Bangalore, India

Users desiring to link up to the popular and versatile Z8000 microprocessor/microcomputer with peripherals belonging to other families will find the going much easier with this interface, which uses only six low-cost integrated circuits. Whether merging technologies in order to meet system requirements in a hurry or combining them to garner the advantages of both, the user will often find the new system most useful and cost-effective when appropriate software is at hand.

Operation of the circuit shown in the figure may be better visualized with the aid of the legend and the timing diagram shown at the lower right and bottom, respectively. Initialization occurs with the interrupt signal $\overline{HELP}$ for requesting attention from the processor. It is of course assumed that the daisy-chain interrupt line, IEI, is high and also that the vector-interrupt acknowledge line, $\overline{VIACK}$ is inactive, yielding the condition where the 74LS158 multiplexer outputs show $Y_i = B_i$.

With the falling edge of $\overline{HELP}$, the interrupt-pending (IP) flip-flop is set and the vector-interrupt-request line (VI) goes active. After completing execution of the current instruction, the processor will then enter the vector-interrupt-acknowledge cycle and the $\overline{VIACK}$ and data-strobe ($\overline{DS}$) signals will be activated in sequence (see timing diagram).

When the $\overline{VIACK}$ line falls, the 74LS158's Y_i outputs are switched to conform to $Y_i = A_i$, and the following sequence of operations is initialized:

■ $\overline{VI}$ is deactivated and IEO goes low in order to disable other interrupting devices in the chain. The priority resolution process is initiated in the daisy chain, settling down before the data strobe ($\overline{DS}$) line goes active low.

■ Assuming the device under discussion has the highest priority among all interrupting devices, the falling edge

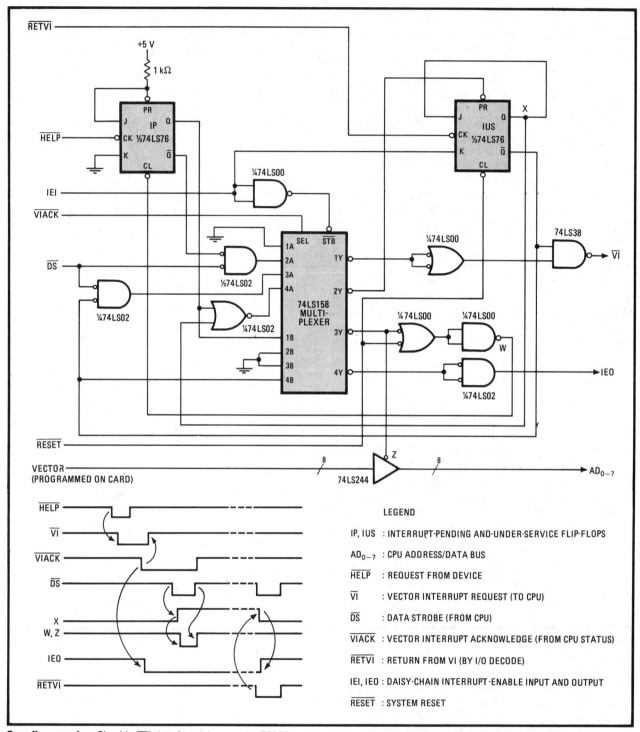

Coupling cousins. Six-chip TTL interface joins popular Z8000 series of microprocessors with peripheral devices from other families, giving users extended facility. System thus becomes more readily adaptable for such tasks as parity checking and high-level diagnostics.

of DS sets the interrupt-under-service (IUS) flip-flop and the IP flip-flop is cleared. The data vector is then gated to the address bus as line Z goes low. Note that IEO is held low by IUS throughout the interrupt service of the device, thereby disabling lower-priority devices.

■ Finally, as VIACK goes inactive (high), the outputs of the 74LS158 go to the inactive state. The same device will be prevented from interrupting its own interrupt because one of the inputs of the 74LS38 is disabled by the IUS flip-flop.

At the end of the interrupt, the return-from-VI (RETVI) signal can be generated by the processor by programming it to generate an input/output write cycle in which the address is decoded and the signal will be run as a bus signal. The falling edge of RETVI releases the IUS flip-flop and IEO goes high. On the other hand, if the generation of a common signal such as RETVI is undesirable, the same function can be generated locally (decoding a unique I/O address for each interrupting device on the card) as the program initiates an I/O operation for resetting the IUS bit of the interrupting device. In the case where the IEI line goes low (as during an interrupt-acknowledge cycle with a higher-priority interrupt), the 74LS158's outputs go inactive, forcing IEO low. □

Adapting a home computer for data acquisition

by Peter Bradshaw
Intersil Inc., Cupertino, Calif.

A personal computer, of the kind made by Apple and Atari, can be easily interfaced with an instrumentation module if a machine interrupt is used to overcome their inherent incompatibility—a condition attributable to the different clock speeds at which module and machine run. The interrupt scheme shown here facilitates the transfer of data by exploiting the data-ready signal available on most modules that generate a multiplexed binary-coded decimal output. Thus, the computer can be freed for other tasks, as in any timeshared system.

As an example, consider the arrangement in the figure whereby the Intersil 7226 multipurpose counter is interfaced with the popular MC6800 or MCS6502 microprocessors, which are at the heart of many personal computing systems. After the counter measures a designated interval, its store output moves low, signaling the MC6820 peripheral interface adapter with an interrupt request. This interrupt should be serviced within 100 milliseconds and control register CRA and the ports PA_0–PA_7 set to the required bits (see table).

Thereafter, an interrupt is generated through the PIA's CA_2 port and the 74LS02 open-collector NAND gates each time the multiplexed display-digit outputs of the counter match that of the bits on the PA_0–PA_7 lines. The BCD output data corresponding to each display digit D_1–D_8 is thus successively applied to the PB_0–PB_3 inputs (at a 4-kilohertz rate) and then to the processor.

The second through ninth interrupts should each be typically serviced in less than 244 microseconds. This task can be easily accomplished if proper priority is assigned to the interrupts. Thus the data will be read in

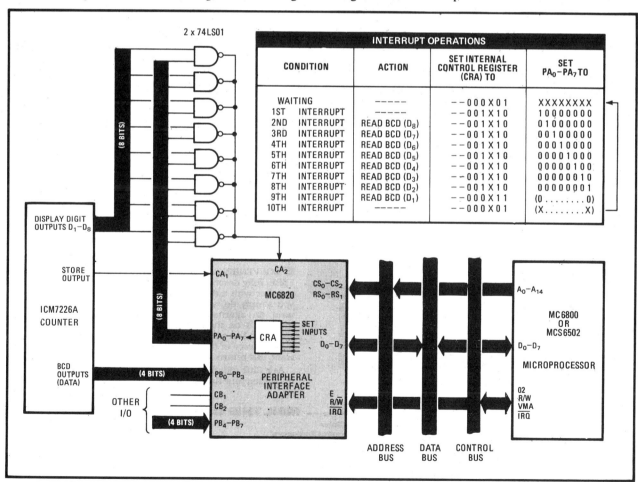

CONDITION	ACTION	SET INTERNAL CONTROL REGISTER (CRA) TO	SET PA_0–PA_7 TO
WAITING	-----	--000X01	XXXXXXXX
1ST INTERRUPT	-----	--001X10	10000000
2ND INTERRUPT	READ BCD (D_8)	--001X10	01000000
3RD INTERRUPT	READ BCD (D_7)	--001X10	00100000
4TH INTERRUPT	READ BCD (D_6)	--001X10	00010000
5TH INTERRUPT	READ BCD (D_5)	--001X10	00001000
6TH INTERRUPT	READ BCD (D_4)	--001X10	00000100
7TH INTERRUPT	READ BCD (D_3)	--001X10	00000010
8TH INTERRUPT	READ BCD (D_2)	--001X10	00000001
9TH INTERRUPT	READ BCD (D_1)	--000X11	(0........0)
10TH INTERRUPT	-----	--000X01	(X........X)

Prompt servicing. Using an interrupt scheme, instrumentation modules and other data-gathering processors with BCD multiplexed outputs can be readily interfaced with home computers such as the Apple and the Atari. Interrupts for reading the data corresponding to each display digit of counter in succession are generated by the microprocessor system itself, which produces a pulse at CA_2 each time the contents of its updated PA_0–PA_7 output register equal that of the scanned display digit outputs D_1–D_8.

less than 4 ms, so there will be no problem with data overruns (200 ms between measurements).

Ideally, the first interrupt should either include a check to ensure that digit D_8 is not high and the input is correct or else be followed by a statement that creates an interrupt when line D_1 is high. In addition, the computer bus will usually require some form of bidirectional buffering to the peripheral interface adapter.

The same system can be used with any processor in the MC6800 series; with other processors, a more complex interrupt-handling scheme is required. □

Serial-communication link controls remote displays

by John Klimek
Pretoria, South Africa

This simple serial-communication circuit provides a remote control for a four-digit display through a two-wire communication link. The transmitter converts parallel input data into serial output data in the form of long and short pulses representing the high and low levels of the shift-register output, respectively, while the receiver decodes the serial input data for the display circuitry. The circuit is useful for applications requiring low-cost remote display units.

The binary-coded decimal data is entered into U_1 and U_2, the two 8-bit shift registers of the transmitter (a). This data is shifted out in accordance with pulse train M (b), which is generated by counter U_3. The counter is driven by a 2-kilohertz square-wave oscillator, which is composed of a CD4093 Schmitt trigger, a 10-nanofarad capacitor, and a 100-kilohm resistor. Depending on the state of output Q_8 of register U_1, waveform M is converted into either 320-microsecond or 100-μs pulses.

On receiving these pulses, dual 4-bit register U_4 and

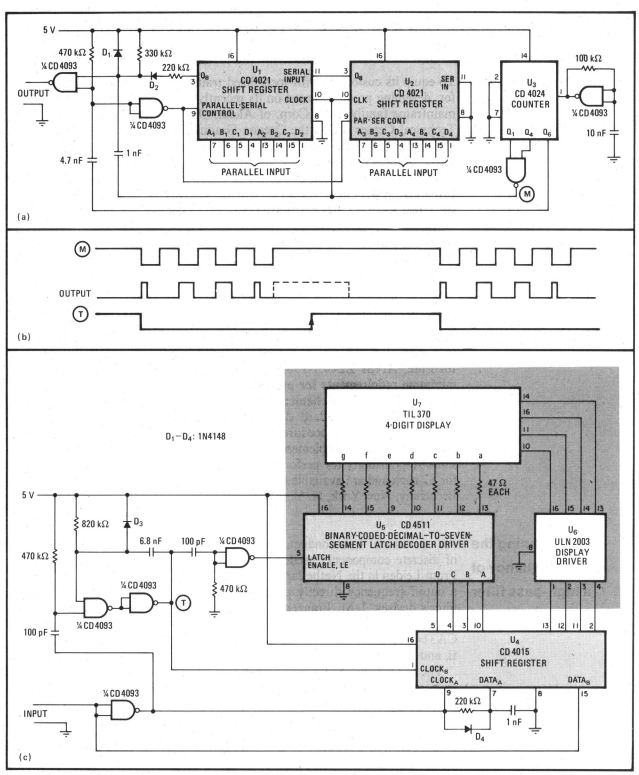

Four-digit display. A four-digit display is controlled remotely by a two-wire communication link comprising a transmitter and a receiver. The transmitter section (a) converts BCD data into high and low bits at the output of shift register U_1. These bits are then transformed into short and long pulses (b) at the transmitter output. The receiver (c) decodes the input pulses for the display circuitry.

the timing network at its input convert the pulses into high- or low-level bits (c). The data is transferred in groups of four pulses and entered into the internal latches of U_5 at the positive edge of pulse-train T. This train also clocks the other half of U_4. The B section of U_4 generates a high-level pulse in sequence, which enables the desired character via driver U_6 to be in phase with the segment-enabling pulses generated by the A section of U_4. After every 16 data pulses, converted into 4 output pulses, a fifth pulse is generated by Q_6 of U_3, as is shown on the timing diagram. The high-level bit is then recorded from this pulse. □

Transferring data reliably between core and CPU RAM

by Trung D. Nguyen
Litton Data Systems Division, Van Nuys, Calif.

Magnetic-core memories are still one of the most dependable ways to store large amounts of nonvolatile digital data (up to 1 megabyte). This circuit is a conduit for data transfer between the core memory and the random-access memory situated in the central processing unit. A word in the core is composed of 3 bits and is matched to the RAM's 18-bit words. Because of the low bits-per-word count in core, circuit reliability is high and costs are low.

The data that will be written into or read from core is temporarily stored in the data-register circuit. During the write cycle, input data D_{I/O_1} through $D_{I/O_{18}}$ is loaded into the core write-data register, which is composed of flip-flops U_1 through U_3 (Fig. 1). The two most significant bits—17 and 18—are used as parity bits for low-

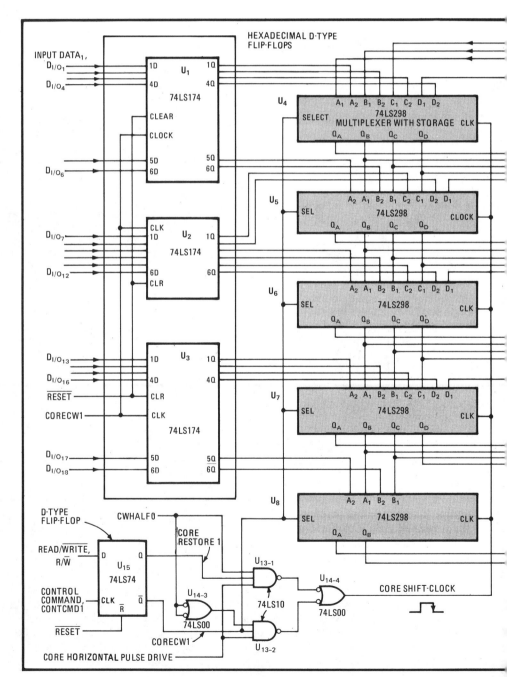

1. Shifting data. This circuit allows two-way transfer of digital data between magnetic-core memory (3 bits/word) and the random-access memory (18 bits/word) located in the central processing unit. Data shifting is controlled by the horizontal pulse-drive signal supplied to the core, and the direction of data transfer is controlled by the read/write pulse from the CPU.

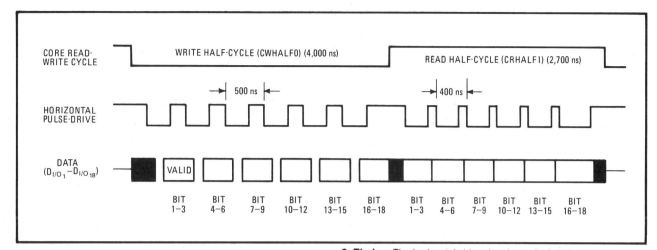

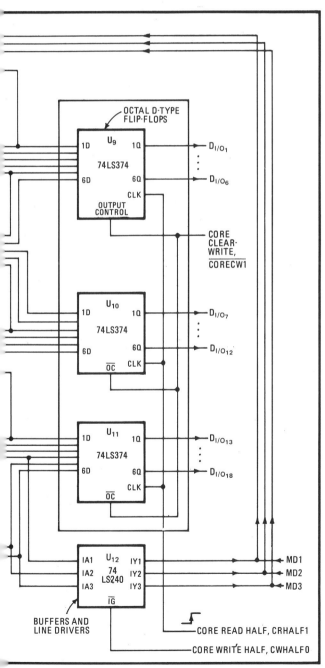

2. Timing. The horizontal drive signal supplied to the core is pulsed six times during each read or write half-cycle to shift data from the core to the RAM and *vice versa*. A typical R/W core-cycle lasts 6.7 microseconds, thereby permitting a data-transfer rate of 150,000 words per second with each word comprising 18 bits.

and high-order data bytes. In addition, on the leading edge of the core clear-write pulse (CORECW1), the data is transferred to the storage-multiplexer circuit, which comprises multiplexers U_4 through U_8. This pulse clears the location of the addressed word in the core memory.

Serial data is sent to the core read-data register (flip-flops U_9 through U_{11}) during the read cycle. This data travels through the storage-multiplexer circuit, which now functions as a data-shift register.

The word-select input is low during the read half-cycle. As a result, a 3-bit word from the core is shifted through the data-shift register by the core's horizontal pulse-drive signal and stored in the core's read-data registers. When the core clear-write pulse is low, this data reaches the RAM. The shift registers are clocked by the shift clock's (CSHIFTCH) negative-going edge, which occurs for every positive-going core horizontal pulse. The horizontal drive signal pulses six times for each memory half-cycle (Fig. 2).

In contrast, during the write half-cycle, the multiplexer word-select input is high and the RAM's 18-bit word is shifted through registers U_4 to U_8 by the horizontal drive pulse and stored in the core's write data register. When the core clear-write pulse is high, the data reaches the core memory.

For proper core-memory operation, the core read/write and horizontal pulse-drive signals are generated by the programmable read-only memory. Data shifting occurs on the leading edge of the core's horizontal drive pulse through its R/W cycle. In addition, the transfer of data during R/W operation is also controlled by NAND gates U_{13-1} and U_{13-2}. The average data-transfer rate per second is 150,000 words, at 18 bits/word. However, if the number of bits per word differs from those mentioned, the horizontal drive pulse, which serves the core module as a timing signal, must be modified. □

Interfacing C-MOS directly with 6800 and 6500 buses

by Ralph Tenny
George Goode & Associates Inc., Dallas, Texas

Most complementary-MOS devices used with the early members of the 6800 and 6500 microprocessor family are interfaced with the processor bus through programmable interface adapters because bus output data is available for only a short period of time. However, C-MOS devices can be added directly to 6800 and 6500 systems with this circuit. It is assumed some memory-address space is available.

Bus input timing is no problem. It is such that C-MOS latches with output-disable capability can be enabled with an AND/NAND gate of an address-decoding strobe and READ/WRITE signals.

When unassigned blocks of memory-address space exist, C-MOS latches and registers can serve as output registers residing on the processor bus. This technique is demonstrated through a 256-byte block of unused memory that interfaces bus data with a two-character hexadecimal display that is resident on the microprocessor bus (Fig. 1a).

With this design, the high-order address lines are decoded for port selection and the low-order address lines are used to transmit data to that port. Thus the 4-bit binary word is displayed as the equivalent hexadecimal word on the processor bus. The timing information (Fig. 1b) uses worst-case delays associated with C-MOS parts.

Address lines A_4 to A_7 can be decoded through the use of a dual four-input OR gate for U_2 so that a 16-bit output port occupies 256 bytes of memory-address space. The four C-MOS D-type registers U_5 through U_8 use the basic strobe to clock data into the latch (Fig. 2). The four address lines are now used as device enables, so that 4-bit data is entered into one of the four C-MOS parts, resulting in a 16-bit output port. □

1. C-MOS interface. The logic in (a) allows a C-MOS latch to reside directly on the 6800 and 6500 microprocessor buses. It uses unassigned memory-address space to mix address and data information for a port residing on the processor bus. The bus timing (b) shows that microprocessor data is available for a very short time.

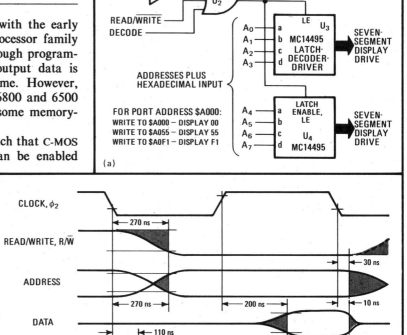

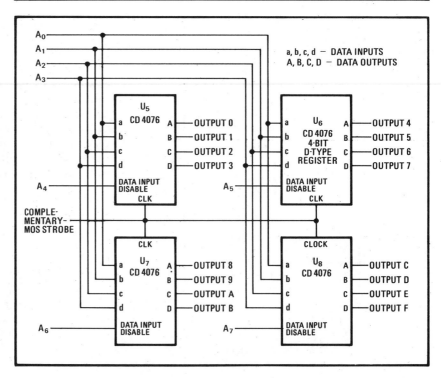

2. Expansion. A 16-bit C-MOS parallel output port uses low-order address lines for both data and address. The four C-MOS devices U_5 through U_8 use the strobe to clock the data into the latch. A 16-bit output occupies 256 bytes of address space.

Simple patch reconciles parity flags in Z80, 8080

by Zvi Herman
Elbit Computers Ltd., Haifa, Israel

Since the Z80's instruction set is a superset of the 8080's, a program written for the latter can be run on the former. However, there are subtle incompatibilities that may cause unexpected behavior when such a program is run on the Z80—one of them being the definition of the parity flag.

In the 8080, this flag indicates the parity of the result, whether the operation is arithmetic or logical. But the meaning of the flag has been modified in the Z80 so that it indicates the parity of the result after a logical operation only.

When an arithmetic operation is executed, the parity flag (called P/V in the Z80) instead indicates an overflow condition. Hence, when an 8080 program relies on the parity resulting from an arithmetic operation, that piece of code could produce erroneous results when run on the Z80.

For instance, suppose the content of the accumulator is 55_{16} and the following instructions are executed:

ADI 11H (add 11_{16} to the contents of the accumulator)
JPE NEXT (if parity is even, jump to NEXT)

The 8080 adds 11_{16} to 55_{16}, the result is even, and the condition for the jump is true. But when the Z80 adds 11_{16} to 55_{16}, the result causes no overflow and the P/V flag stays off. Now the condition of the jump is false, and the Z80 continues right through the next instruction, rather than jumping to NEXT.

It is obvious that a patch is needed here. This patch (between the ADD and the JUMP) should leave the accumulator and the flags intact. The only change required is modification of the P/V flag to reflect the parity result from the arithmetic operation.

There is no single instruction in the Z80, however, that can do all that. All instructions that modify the P/V flag, in the sense of parity, change either the accumulator or other flags; a series of instructions, however, will produce the desired result.

The following sequence provides the solution:

```
ADD  A,11H
RLD       (rotate left one digit [nibble] to the left)
RLD
RLD
JP   PE.NEXT
```

The RLD instruction rotates the least significant (4-bit) digit of the accumulator through the memory location pointed to by register HL (see figure). It modifies the P/V flag to reflect the parity of the accumulator and does not affect the carry (CY) flag.

If the instruction is used three times, the accumulator is returned to its initial value and the CY, sign (S), and zero (Z) flags are restored. Note that RRD may be used to rotate the digit right with the same success.

This patch, however, does contain a few pitfalls. For one, the half-carry flag is zeroed by the RRD/RLD instruction. When the parity of the result of an arithmetic instruction is desired, the chance that the half carry is needed somewhat later is small, but it does exist.

When HL points to locations in read-only memory, this patch is not suitable, since ROM locations cannot be written into and thus cannot be used for temporary storage. Finally, the patch adds execution time to the program, and that factor should be considered when software timing is important. □

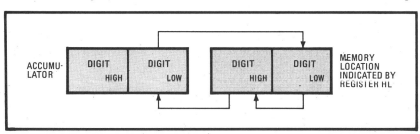

Bit brigade. Parity is flagged for arithmetic operations in the Z80 by rotating the least significant nibble in the accumulator through memory and back. Three executions of the RLD or RRD instructions perform this task.

Current loop supports remote distributed processing

by Akavia Kaniel
Measurex Inc., Cupertino, Calif.

Long-distance serial communications between distributed processors, such as Intel's 8051 single-chip microcomputer, is best achieved by using one of the data-transmission standards—a 20-milliampere current loop, for instance. The circuit shown in the figure uses the universal asynchronous receiver/transmitter and the bit-rate generator built into the 8051 to implement a full-duplex, 20-mA current loop that operates at speeds of

253

COMMUNICATION SUBROUTINES FOR 8051 MICROCOMPUTER	
Program	Comments
INIT: MOV TL1,#0FDH	PRESET COUNTER 1 TO 253
MOV TH1,#0FDH	FOR 9,600-b/s RATE.
MOV SCON,#0C0H	UART SET TO ASYNCHRONOUS, 1 STOP, 1 START, 8 DATA BITS, 1 EVEN PARITY BIT; RECEIVER NOT ENABLED YET.
MOV TMOD,#25H	TIMER 1 IS BAUD-RATE GENERATOR, TIMER 0 IS A 16-BIT COUNTER.
SETB TR1	ENABLE BAUD-RATE GENERATOR.
SETB REN	ENABLE RECEIVER.
RET	
SEND: JNB TI,SEND	WAIT HERE TILL THE TRANSMIT BUFFER IS EMPTY.
MOV A,CHAR	STORE CHARACTER TO BE SENT IN THE ACCUMULATOR.
MOV C,P	MOVE THE EVEN-PARITY
MOV TB8,C	BIT FROM THE PROCESSOR STATUS WORD TO TB8 = PARITY BIT IN TRANSMIT BUFFER.
CLR TI	CLEAR THE TRANSMIT-BUFFER-EMPTY FLAG.
MOV SBUF,A	SEND THE DATA TO THE TRANSMIT-BUFFER.
RET	
RCV: JNB RI,RCV	WAIT HERE TILL THE BYTE-RECEIVED-FLAG IS HIGH.
CLR RI	RESET THE FLAG.
MOV A,SBUF	STORE THE RECEIVED BYTE IN THE ACCUMULATOR.
MOV C,RB8	STORE RECEIVED PARITY BIT.
JNB P,N_PAR	CHECK IF THE PARITY
JB CY,PROC	OF THE RECEIVED BYTE
CALL ERROR	EQUALS THE
JMP PROC	PARITY BIT IN THE
N_PAR: JNB CY,PROC	RECEIVED BYTE; IF
CALL ERROR	NOT CALL THE ERROR-
PROC: RET	ROUTINE.

110 to 9,600 bits per second.

In addition, there are three canned software routines that perform initialization, data transmission, and data reception and are meant to be called by the user software when needed. These routines are given in the accompanying table.

The figure also shows how a string of 8051-based nodes situated along a four-wire transmission line is configured. Each node represents a point where some processing is accomplished, as, for example, in an industrial plant. At least 10 nodes can be driven with the system shown, this limit being set by the current-sinking capability of transistor Q_1 in the multidrop master.

Besides data, the transmission line also carries power to each node, 12 volts dc in this case. If necessary, this 12 v dc can be converted to other voltage levels that are needed locally by the electronics at each of the nodes. As illustrated in the figure, each node is also coupled to the transmission line through an optical-coupler circuit using a 1N5308 constant-current diode. □

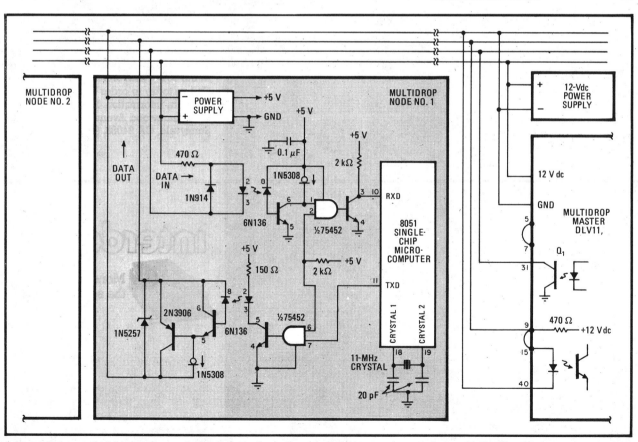

1. Long distance. Individual nodes are linked along the four-wire transmission system in accordance with the 20-mA–current-loop standard. At least 10 nodes can be driven by this system. In addition, the transmission line supplies 12 volts dc to each node for local electronics.

Serial-data interface eases remote use of terminal keyboard

by Robert Nixon
Nixon Engineering Co., San Jose, Calif.

A keyboard that can be detached and operated at a distance from its terminal is often desirable in a computer-based system, but the job of developing the proper cable and connector can be hard and time-consuming. The interconnecting wire count can be reduced to just two leads, however. The trick is to choose a keyboard that generates a serial output for each keystroke and use it with the current-modulating interface shown. A minor modification will adapt the circuit to the RS-232 bus.

Here, the microprocessor-based keyboard is the Micro Switch 103SD24-2, which delivers the serial ASCII data at 110 bits per second. Power to the keyboard, which requires only 5 volts, is provided by the µA7812 voltage regulator. This device also serves as a current limiter, protecting the interface against damage caused by shorts in the connecting cable. A second regulator, the µA7805, smoothes out the small variations in voltage generated by the current modulator.

During the mark states of the data stream, transistor Q_1 is held off because the keyboard output is active—the total current through R_1 and the keyboard is 360 milliamperes, which represents the difference between the supply voltage and the keyboard's optocoupler output stage. During the space interval of the serial data stream, the keyboard output is inactive, and Q_1 is turned on. The total supply current then becomes 610 mA. Diodes D_1 and D_2 allow for the small voltage drop between the Darlington-coupled output of the keyboard, ensuring that Q_1 is switched properly. Capacitors C_1 and C_2 aid in stabilizing the µA7805 regulator during modulation.

The two-state current variation is detected on the receiver side by Q_2, and associated resistors R_3–R_5. During the mark state, the drop across R_3 is about 550 millivolts, which is not enough to turn on the transistor. During the space state, however, the additional current produced will switch on Q_2 and generate a positive voltage level at the output. Note that R_4 and R_5 prevent the transistor from being damaged by a short circuit.

For TTL compatibility, the output signal should be introduced to a Schmitt trigger to eliminate hash and ill-defined switching levels. To interface with the RS-232 bus, it is only necessary to increase R_7's value to 3 kilohms and connect its lower end to −12 V instead of ground. □

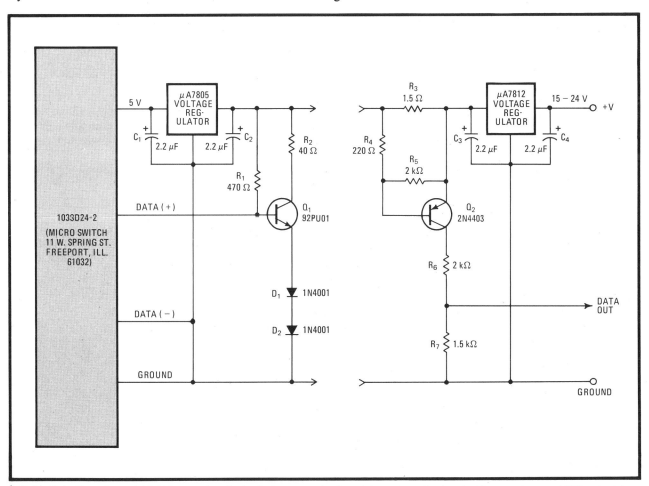

Interim link. Two-wire serial interface makes it simple to remove keyboard from computer terminal for remote use. ASCII output from keyboard is current-modulated by Q_1. Two-state output is recovered by Q_2. Minor modification makes board compatible with RS-232 bus.

8-bit DMA controller handles 16-bit data transfers

by Trung D. Nguyen
Litton Data Systems Division, Van Nuys, Calif.

Many microcomputer systems today are 16 bits wide. But only 8-bit controllers are available to handle direct memory access for them should they need it. This circuit interfaces an 8-bit DMA controller (like a AM9517) with a 16-bit system bus.

As the figure shows in (a), a data strobe loads the contents of an 8-bit data bus into latch 74LS373, and the address enable moves that data onto the address bus. The four least significant bidirectional address lines are inputs from the system data bus (D_8 through D_{11}) when the direct-memory-access controller is idle and are outputs when it is active. The controller's data bus is enabled to input data from D_0 through D_7 during input/output direct-memory-access write, allowing the processor to program the DMA control register.

The circuit (b) allows the 8-bit controller to transfer data between several 1-megaword core-memory storage modules and a 16-bit system bus. Initially, the outputs of counters A_5 and A_6 are reset to all 0s by the reset line.

Data transfer. The circuit (a) interfaces the 8-bit DMA controller Am9517 with a 16-bit bus. The register 74LS373 is used for latching an 8-bit controller data bus to complete the 16 bits of the address bus. The four LSBs of the high-order system data bus are used to program the DMA control register. The logic (b) allows the 8-bit controller to handle data transfer of several 1-megaword modules.

The A inputs selected by multiplexer A_7 are decoded by A_8 to select a desired module when the module select signal (MSMSELO) is active. Data (not shown) is stored at the addressed location selected by controller's address lines (AD_0 through AD_{15}). The busy signal (MSMBUSY) switches to the busy state when the controller initiates either the reading of data (MEMRO) from the storage module during a read-restore (MSMRRO) or writing of data (MEMWO) into the storage module during a clear-write (MSMCWO).

If an address location larger than 64-K is needed to access the storage module, the controller's addresses AD_0–AD_{15} are all high, causing the gate A_3 output to go low. The leading edge of the A_4–A_3 output advances counter A_5 whose output (AD_{16} through AD_{19}) together with AD_0 through AD_{15} is used to access the entire 1-megabit address of the module.

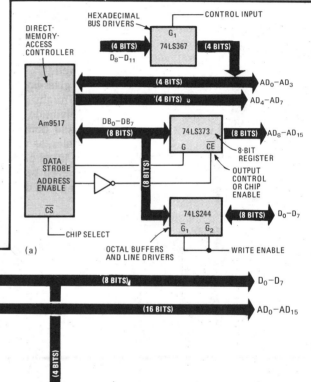

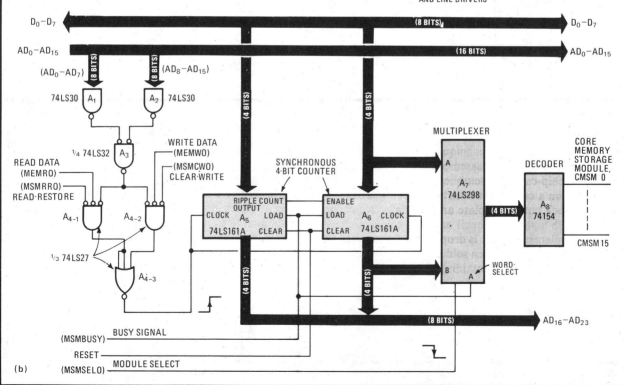

If more than a 1-Mb address location is required, the ripple count output of A_5 goes high, enabling counter A_6 to select the next–higher-order module. The outputs AD_{20}–AD_{23} of A_6 are used for higher address locations. Decoder A_8 selects up to 16 modules. Extra counters and decoders may be added for more modules. □

Driver simplifies design of display interface

by Wes Freeman
Teledyne Semiconductor, Mountain View, Calif.

This light-emitting-diode display interface uses a monolithic-LED driver chip to demultiplex the 3½-digit analog-to-digital converter data lines and provides a display with just a few parts. Because the driver powers the LED statically, the high transient currents associated with multiplexed LEDs are reduced. As a result, the design eases power supply and component layout requirements and minimizes the noise at the input.

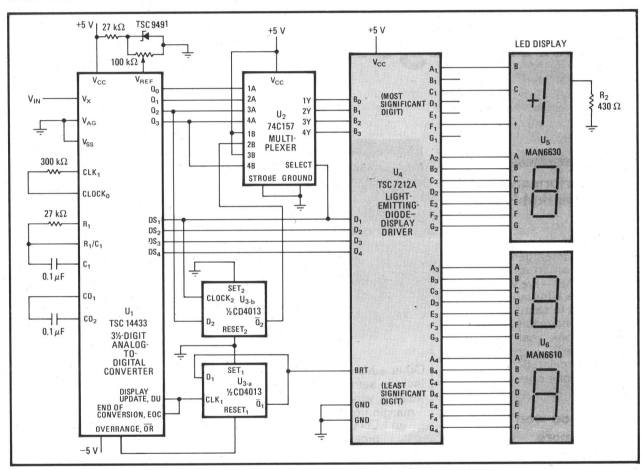

Display. With Teledyne Semiconductor's LED driver TSC7212A, this display interface has a low component count and high performance. Outputs A_1 and C_1 of U_4 drive the half digit, while the polarity indicator is driven by F_1. In addition, external resistor R_2 turns on the negative segment of the polarity sign.

TABLE 1: DATA OUTPUT OF U_1					
Coded condition of most significant digit (MSD)	Q_3	Q_2	Q_1	Q_0	Binary-coded-decimal-to-seven-segment decoding
+0	1	1	1	0	blank
−0	1	0	1	0	blank
+0 UR	1	1	1	1	blank
−0 UR	1	0	1	1	blank
+1	0	1	0	0	4 → 1
−1	0	0	0	0	0 → 1 hook up only segments b and c to MSD
+1 OR	0	1	1	1	7 → 1
−1 OR	0	0	1	1	3 → 1

Notes:
Q_3: ½ digit, low for 1, high for 0
Q_2: Polarity: 1 = positive, 0 = negative
Q_0: Out-of-range condition exists if Q_0 = 1; when used in conjunction with Q_3 the type of out-of-range condition is indicated; Q_3 = 0 → OR or Q_3 = 1 → UR

For the three least significant digits, circuit operation is straightforward. Data is produced at pins Q_0 through Q_3 with a-d converter U_1 when the appropriate digit strobe goes high. At this point, data latches into LED driver U_4. However, obtaining proper data for the half-digit and polarity sign is more complex. Table 1 shows the data output of U_1 when DS_1 goes high. Because Q_3 and Q_2 directly influence the display, only a combination of these signals lights the proper segments of U_4.

Table 2 shows the display that results when the output of Q_2 is inverted and shifted to Q_1 while Q_0 and Q_2 are

TABLE 2: DECODED OUTPUT OF THE DISPLAY DRIVER							
Coded condition of most significant digit	Output of U_1		Input to U_4 with $\overline{Q}_2$ shifted to Q_1 and $Q_0, Q_2 = 1$				Segments displayed
	Q_3	Q_2	B_3	B_2	B_1	B_0	
−1	0	0	0	1	1	1	5
+1	0	1	0	1	0	1	7
−0	1	0	1	1	1	1	blank
+0	1	1	1	1	0	1	L

high. Because outputs A_1 and C_1 of U_4 drive the half digit and F_1 drives the polarity indicator, no external LED drivers are required. In addition, Q_2's output is inverted by D-type flip-flop $U_{3\text{-}b}$ and is shifted to Q_1 by two-input multiplexer U_2. When data strobe DS_1 is low, data passes through the multiplexer unchanged. However, when DS_1 is high, multiplexer U_2 inverts Q_2 on pin 2Y and sets 1Y and 3Y high. U_4 decodes these inputs and the correct segments turn on. External resistor R_2 turns on the negative segment of the polarity sign.

A flashing display indicates the overrange and is implemented with flip-flop $U_{3\text{-}a}$. An analog input exceeding full-scale causes $\overline{OR}$ output of U_1 to go low, thereby disabling the reset. As a result, the flip-flop toggles at the end of each conversion and causes the display to flash. □

21. POWER SUPPLIES

Stabilizer boosts current of ±dc-dc converter

by Gerald Girolami, *Université Pierre et Marie Curie, Physique et Dynamique de l'Atmosphère, Paris, France*

The power-handling capacity of Intersil's popular positive-to-negative voltage converter can be increased by a factor of 10 with this circuit. Using an efficient, low-cost stabilizer to hold output voltage constant for an operating current generally above that which the ICL7660 is normally used, the circuit is a viable alternative to using an expensive modular dc-dc converter when a 0-to-10-volt, 0-to-100-milliampere source is required.

Although the ICL7660 is used almost exclusively in low-power applications, its maximum load current is in excess of 50 mA for a 5-v, 100-mA input. At this level, however, the output voltage drops from the expected −5-v value to near zero. By using the LM301 operational amplifiers (A_1 and A_2) in a feedback loop, however, the output voltage, V_{out}, can be made to equal the negative of the given reference voltage, $-V_{ref}$, at the desired output current.

As seen, the output voltage, V_{out}, is compared with V_{ref} at op amp A_1. As a consequence of the circuit configuration:

$$V_1 = V_{ref}\left[1 + \frac{R_4}{R_3 \| R_5}\right]\left[\frac{R_2}{R_1 + R_2}\right] - \frac{V_{ref} R_4}{R_3} - \frac{V_{out} R_4}{R_5}$$

where $R_3 \| R_5 = R_4$, $R_1 = R_2$, and $R_3 = 2R_4$. Thus $V_1 = \frac{1}{2}(V_{ref} - V_{out})$. Similarly, op amp A_2 generates a voltage:

$$V_2 = V_{ref}[1 + (R_7/R_6)] - V_1 R_7/R_6$$

where $R_6 = R_7$, and so $V_2 = 2 V_{ref} - V_1$. Combining the final expressions of V_1 and V_2 yields $V_2 = 3/2\, V_{ref} + \frac{1}{2} V_{out}$.

Thus, an increase in V_{out} will cause V_2 to increase. Because $V_2 = -V_{out}$, as specified under normal conditions for the ICL7660, an increase in V_2 will cause V_{out} ultimately to decrease, and vice versa, and negative feedback is achieved. Thus, under conditions for stabilization, $V_2 = -V_{out} = 3/2\, V_{ref} + \frac{1}{2} V_{out}$, or $V_{out} = -V_{ref}$, and only V_{ref} will have any effect on V_{out}, with the required driving and load currents provided by A_1 and A_2 and power transistor Q_1.

In practical applications, the output voltage is adjustable from −2 v to −10 v for a positive V_{ref} of similar range. The maximum output current is 100 mA at $V_{out} = -5$ v and drops thereafter. But it is possible to use two such circuits and two diodes to double the output current. That may not seem too interesting because many good dc-dc converters that deliver high current are available, but note that this circuit allows selection of the output voltage, a feature not usually found in commercial units. The regulation is good and the mean output voltage variation will be only 70 millivolts at $I_{load} = 100$ mA. This specification is adequate for digital circuits that can withstand a ±5% variation (that is, 250 mV) at the 5-v level.

Note that the MC1776CG is suitable for use as A_1 and A_2. It is necessary to add an 8.2-megohm resistor between pins 4 and 8 of each device, however, to set the operating current in the amplifier. □

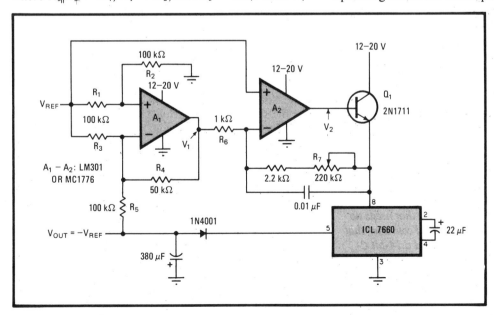

Steady power: Negative feedback provided by A_1 and A_2 stabilize output voltage of ICL7106 converter for medium-range currents giving user a 2-to-10-V negative voltage source capable of delivering 100 mA. Reference voltage controls output such that $V_{ref} = -V_{out}$. Output voltage regulation is only 70 mV at 5 V.

Milliampere current source is voltage-controlled

by William J. Mundl
Concordia University, Department of Psychology, Montreal, Canada

The constant small positive or negative current produced by this voltage-controlled source is useful for a variety of low-level measurements. As the current also has a virtually linear relationship with the input voltage, it may be modulated as desired by a given input waveform or, still more usefully, be put under microprocessor control to create an automated test system.

As seen in (a), incoming signals in the range of 0 to 10 volts are buffered by the LM358 micropower operational amplifier A_1 and then introduced to A_2, which with transistor Q_1 makes the current-monitoring feedback circuit. For a given input voltage, A_2 amplifies and inverts Q_1's emitter-to-base voltage variations, so that any increase or decrease in current due to temperature or load variations is counteracted. As a result, the current will rise linearly from zero to I_{max}, where I_{max} is determined by resistor R, with the variation of I for a given V_{in} being about 2%. In this circuit, the maximum attainable value of I_{max} is approximately 4 milliamperes, obtained with a 1.5-kilohm load impedance.

Calibration of the circuit is simple. Potentiometer R_1 need only be adjusted to null the output current for V_{in} = 0. For convenience in setting the output current, an oscilloscope can be placed across a 1-kΩ resistor.

The layout for this circuit's counterpart, a negative current sink, is similar, as shown in (b). Q_1 becomes an npn transistor, the supply potential on the circuit is reversed, and an inverting stage is added at the input.

Although the circuit is relatively insensitive to variations in the supply voltage, use of a regulated power supply of the simple series type is recommended. □

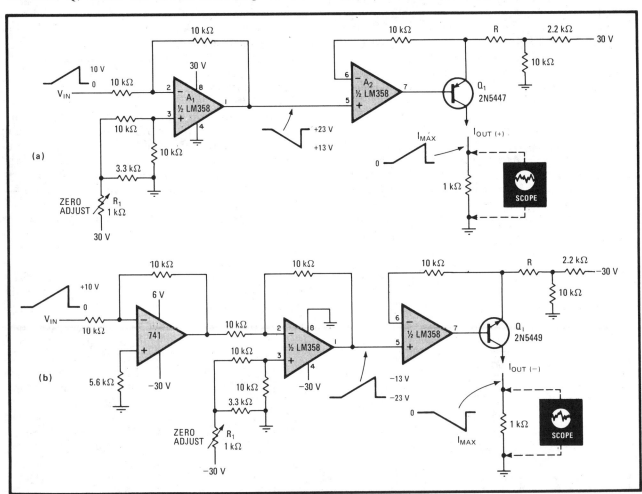

Milliampere magnitudes. The circuit's input voltage sets its constant output current to within 5% of the desired value, adjustable from 0 to 4 mA if R = 1.5 kΩ. Its linear response makes it attractive for microprocessor-based (automated) tests and measurements. The positive source in (a) or negative sink in (b) is simply calibrated by means of resistor R_1 and an oscilloscope monitor.

Low-cost timers govern switched-mode regulator

by Luces M. Faulkenberry
Texas State Technical Institute, Waco, Texas

This step-down switching power supply, which uses 555 timers for pulse-width modulation, combines good performance with very reasonable cost. Providing an output of 12 volts at 1 ampere for an 18-v input, the unit offers input-current limiting, 0.1%/v line regulation, 0.5% load regulation, and an output ripple of only 20 mV. However, the design equations given here enable the user to specify his own requirements. The supply can be built for less than $15.

Operating as an astable multivibrator at 20 kilohertz, timer A_1 generates the trigger pulses needed to switch the output of monostable multivibrator A_2 to logic 1 during each cycle. Modulating the control pin of one-shot A_2 with the output of the 741 operational amplifier controls the width.

The op amp compares a preset fraction of the supply voltage, V_{out}', with the 6.8-v reference, V_{ref}. When $V_{ref} > V_{out}'$, the control pin of the one-shot moves high and each pulse from the output of A_2 is lengthened accordingly until the reference and supply voltages are virtually equal. Similarly, if $V_{ref} < V_{out}'$, the output pulses are shortened.

As seen, transistors Q_1 and Q_2 in the simple feedback loop perform the switching function. Monitoring transistors Q_3 and Q_4 limit the current by bringing A_2's reset pin low when the design-maximum peak current through the inductor is reached, thereby shortening the width of the output pulses until the cause of the trouble is removed. Q_3 can also serve in a dual capacity as a switch to turn off the supply during overload conditions. For example, should automatic shutdown of the supply be necessary, Q_3 could be used to fire a silicon controlled rectifier in order to hold the reset pin of oscillator A_1 low permanently. In these cases, a simple circuit would also be needed to reset the supply manually. □

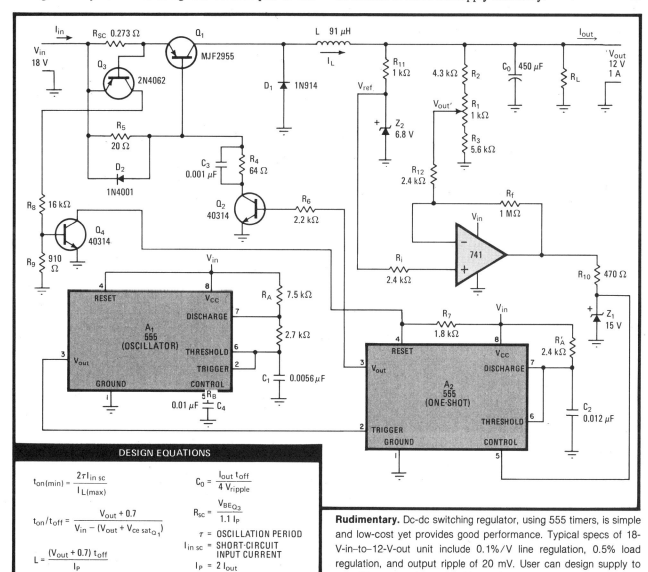

DESIGN EQUATIONS

$$t_{on(min)} = \frac{2\tau I_{in\,sc}}{I_{L(max)}}$$

$$t_{on}/t_{off} = \frac{V_{out} + 0.7}{V_{in} - (V_{out} + V_{ce\,sat_{Q_1}})}$$

$$L = \frac{(V_{out} + 0.7)\,t_{off}}{I_P}$$

$$C_0 = \frac{I_{out}\,t_{off}}{4\,V_{ripple}}$$

$$R_{sc} = \frac{V_{BE_{Q_3}}}{1.1\,I_P}$$

τ = OSCILLATION PERIOD
$I_{in\,sc}$ = SHORT-CIRCUIT INPUT CURRENT
$I_P = 2\,I_{out}$

Rudimentary. Dc-dc switching regulator, using 555 timers, is simple and low-cost yet provides good performance. Typical specs of 18-V-in–to–12-V-out unit include 0.1%/V line regulation, 0.5% load regulation, and output ripple of 20 mV. User can design supply to meet his own requirements with aid of given equation set.

Ni-Cd–battery charger has wide range of features

by Huynh Trung Hung
Paris, France

Though there are many ways of charging nickel-cadmium batteries efficiently, this circuit is unique in uniting just about all their diverse advantages. It provides a constant charging current that can range from 400 milliamperes on up to 1 ampere. It can be operated from either a 220-volt ac line or a 12-v battery. It protects the battery against overcharge by turning it off automatically when a predetermined voltage is reached. Moreover, this preset voltage is adjustable. Finally, the circuit is inexpensive and protected against shorting out.

When the battery is discharged, the voltage at the inverting input of operational amplifier U_1 is lower than noninverting voltage V_{in}, which is set by potentiometer R_1 (see figure). As a result, the output of U_1 swings to the positive rail and turns transistor Q_1 on. Such a swing causes Q_2 to switch on and provides a constant charging current given by $(V_d - V_{be})/R_6$, where V_d is the voltage at the base of Q_2 and V_{be} is the base-to-emitter voltage of Q_2. This current flows through diode D_8 to charge the Ni-Cd battery. Under this condition, light-emitting diode D_7 glows to indicate the charging process, thus serving as a pilot indicator.

While the battery is being charged, the voltage across it rises and causes the voltage at the inverting pin of U_1 to increase until it is equal to V_{in}. At this point, the output of U_1 is grounded and Q_1 and Q_2 turn off, preventing the battery from being overcharged. The preset output voltage can be calculated from the relationship $V_{out} = V_{in}(R_7 + R_8)/R_8$.

For the component values shown, the circuit provides a charging current of 400 mA, which can be modified by varying R_6 to furnish a maximum of 1 A. The predetermined charge voltage must be set with the battery disconnected. Diode D_8 prevents back discharging when the line or the 12-v supply is disconnected. For a 7.2-v Ni-Cd battery, the predetermined charge voltage is set between 7.9 and 8.0 v. Power transistor Q_2 is mounted on a large heat sink. □

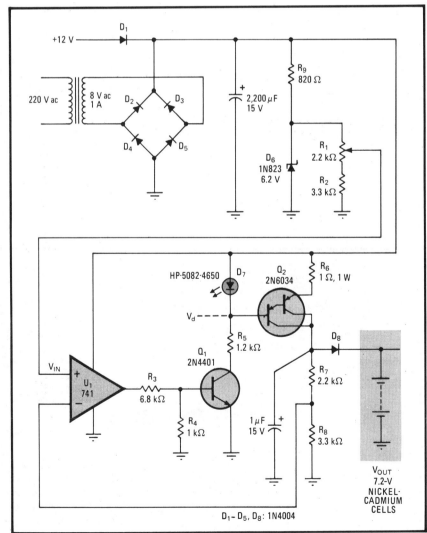

Auto charge. Designed to charge a 7.2-volt, 4-ampere-hour nickel-cadmium battery, this circuit provides a constant charge current with automatic shut-off when the battery reaches a preset charge voltage. The preset voltage is adjustable. For a 7.2-V battery, the end-of-charge voltage is set between 7.9 and 8.0 V. The charging current can be raised to 1 A by varying R_6.

External pass FET boosts regulated output voltage

by H. F. Nissink
Australian Maritime College, Tasmania, Australia

With a floating regulator and a power field-effect transistor acting as the external pass transistor, this power supply can deliver limited current and regulated high voltages. The supply's output voltage and current are adjustable, and the low drive requirement of the FET eliminates the need for Darlington-type stages. Regulators with low drop-out voltages can be designed easily because the power FET has a low on-resistance.

Floating regulator U_1's internal constant-current source of 1 milliampere develops a maximum voltage of 300 volts, serving as the regulator's reference voltage, across resistance chain R. The output voltage is compared with this, and an error voltage is applied to the gate of the power FET Q_1 to keep the load's current and voltage constant. Current-sense resistor R_s determines the maximum constant-current output, while potentiometer R_1 is used to adjust the output current between zero and the maximum value.

The regulated output voltage is adjustable in steps of 10 V from 200 to 300 V, and the maximum constant current is 50 mA. Switch S_1 selects the desired output voltage. Zener diode D_1 connected across the gate-source junctions of Q_1 provides extra gate protection. For a continuous output voltage control, a 100-kilohm potentiometer may replace R. □

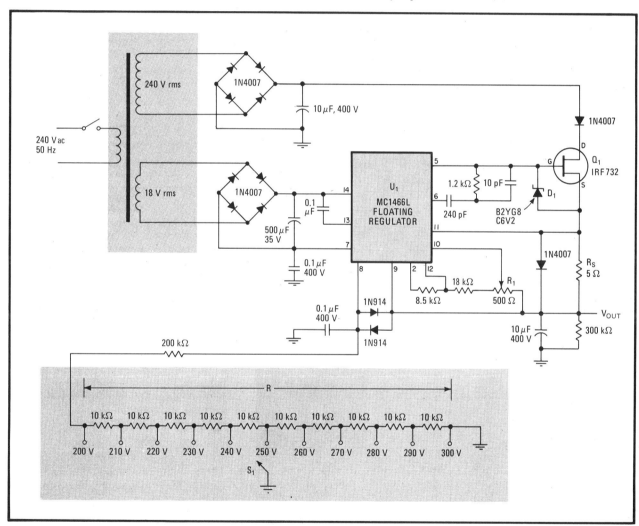

High-voltage regulator. Motorola's floating regulator MC1466 and power FET IRF732, serving as an external pass transistor, provide current-limited high voltage that is adjustable in steps of 10 V from 200 to 300 V. The maximum constant output current is 50 mA, and the FET's low drive requirement enables the regulator to boost the output regulated voltage.

Comparator circuit regulates battery's charging current

by Ajit Pal
Indian Statistical Institute, Calcutta, India

As charge builds up in a battery, its effective plate-charging area gradually decreases. To prevent damage, a good battery charger should continuously limit the charging current from the power line as a function of time. This completely solid-state charger performs the required regulation for a 12-volt automobile battery using a simple circuit built around the μA710 comparator. Although designed for 220-v operation, the charger is easily adapted for 110-v service, making it suitable for application in the U.S.

The comparator automatically adjusts the charging current by sensing the battery voltage, which increases as charge accumulates. The 710 also regulates the current by controlling the on-off switching times of a thyristor that is placed in series with the battery.

As shown in the figure, a dc voltage proportional to the battery voltage is applied to pin 3 of the comparator, with potentiometer R_1 determining the actual value. Simultaneously, a ramp signal that is derived from the power line is fed to pin 2 of the 710, with R_2 setting the slope of the ramp.

When the battery is being discharged, the voltage at pin 3 of the 710 is nearly equal to the lowest instantaneous ramp voltage, and so the output of the 710 is virtually always high. Thus, the thyristor is on for almost the entire 180° switching cycle.

At the other extreme, when the battery is almost fully charged, the voltage at pin 3 is practically equal to the highest instantaneous ramp voltage, and so the thyristor is on for only a small portion of the cycle. For intermediate conditions, the thyristor will be on from between 0° and 180° of the cycle. The maximum charging current is limited by the resistor R_3. □

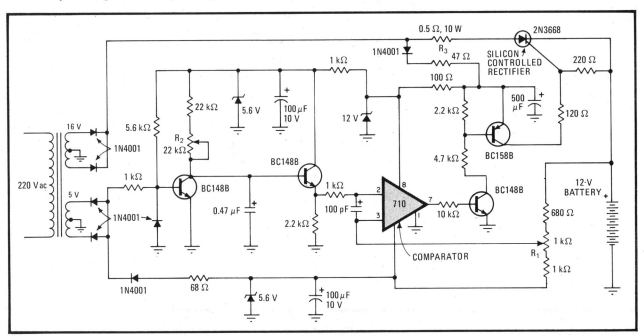

Cutting down. This circuit progressively limits the amount of charging current through a standard 12-V automobile battery as it attains its nominal terminal voltage from its discharged condition, thus avoiding cell damage. The single 710 comparator performs comparison regulation functions. Other circuitry sets conditions where the thyristor can be fired over a 0° to 180° cycle.

Current mirror stabilizes zener-diode voltage

by Alan Rich
Analog Devices Inc., Norwood, Mass.

All too often, designing a precision voltage regulator using a zener diode becomes a tedious task when the diode is operated at low current levels because the device's dynamic resistance generates an error voltage that is directly proportional to the zener current (a). This problem is usually corrected by applying a current that is regulated with a transistor or operational-amplifier circuit or a field-effect-transistor current-regulator diode. However, each of these method has disadvantages with respect to temperature stability, dc voltage availability, cost, and parts availability.

This circuit alleviates these problems and presents a well-regulated current to the zener diode. In addition, the current regulator is simple and costs a mere 75¢, requires a supply voltage that is only 1.5 volts above the zener voltage, and employs readily available parts.

To start up when power is applied, the circuit (b) uses resistors R_1 and R_2 and diode D_1. For proper regulation, start-up voltage V_s must be less than the zener voltage. Transistor Q_1 and resistor R_3 generate current $I_e = (V_z - V_{be})/R_3$, which is stabilized by zener voltage V_z. Monolithic current-mirror U_1 multiplies current I_e by four and supplies zener diode D_2 with a regulated current. The load current must be kept constant by either using a buffer so that $I_{load} = 0$ or employing a fixed load resistance. As a result, output voltage $V_o = V_z + [V_{cc} - V_z][R_z/(R + R_z)]$. The last term of the equation represents the error voltage.

The circuit may be adapted to different zener voltages by selecting R_1 and R_2 for proper start-up voltage and calculating R_3 for desired zener current. For the components shown, the circuit generates a negative voltage. If a positive regulated voltage is required, either U_1 should be replaced with a pnp current mirror or an inverting amplifier may be used. The error voltage for the compensated circuit is compared with the fundamental regulator (c). The error curve shows that the current mirror substantially reduces the error voltage in the compensated regulator and stabilizes the zener voltage. □

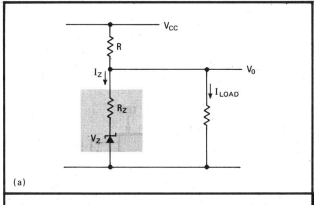

Stable. The fundamental voltage regulator circuit (a) is subject to error when operated at low current levels. The circuit (b) offers a precision voltage regulator by employing current mirror U_1 to provide a well-regulated current to zener diode D_2. Comparison of the above two circuits (c) shows that the error in the compensated voltage regulator is substantially reduced.

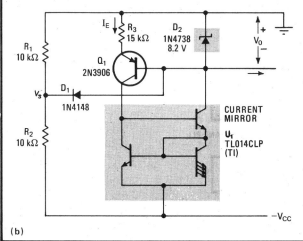

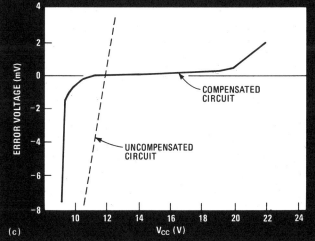

Reed-coil relay is behind flexible fault detection

by Daniel Appiolaza
Mendoza, Argentina

Mechanically providing such functions as undercurrent and overcurrent protection for power supplies and fault indication for an automobile's turn signals or stoplights is easier to achieve inexpensively with relays having a separate reed and coil. Using the coil as a remote current-sensing device also makes the relay flexible enough to do a myriad of other jobs not possible with self-contained units.

Consider the example of current-overload monitoring (a). Here, the normally open reed switch serves to activate the shunt formed by the light-emitting diode, resistor R_1, and the zener diode when excessive supply current flows.

The coil, made from four turns of No. 12 gauge wire, is tightly wound over the reed so that an instantaneous line current equal to or larger than approximately 5 amperes dc will close the reed relay and trigger zener diode D_1. Thus the reference voltage will drop to zero until the line current is reduced and the reset switch is depressed.

A second example is the fault detector of an automobile's brake signals (b), where it is important to know when a stop or turn lamp has failed (a feature not supplied by auto makers). Here, two reed/coil assemblies are required, with the reed contacts being normally closed.

If for any reason either of the stoplights does not turn on when the brake is applied, no current can flow through either coil. Consequently, the reed switches will not open and the panel LED will indicate trouble with the signalling system.

With normally closed relay contacts, however, the stoplights may still turn on despite a failure in the reed circuit or even the LED/resistor itself. Alternatively, it might be better if the reeds are of the normally open type. Then the circuit can be wired to switch on the panel LED only when the car's stoplights become active.

This latter arrangement will positively indicate a failure in the system. If the LED does not turn on when the brake is applied in the normally open reed system, however, it does indicate difficulty with either the lamp or the monitor circuit. □

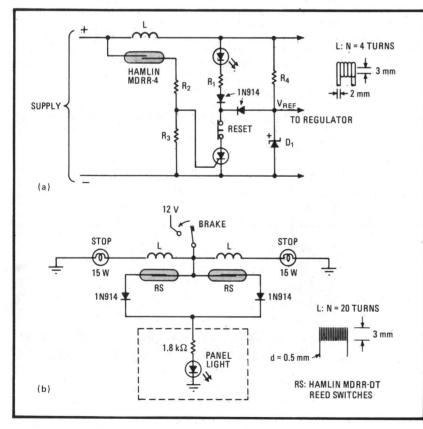

Switching separates. Two-element relay having remote four-turn current-sensing coil and normally open reed switch (a) provides inexpensive overcurrent protection for power supply. When implemented in car's brake-signal system, relay detects faulty turn or stoplights. A normally open relay circuit might be preferable to the normally closed configuration shown.

Voltage translator switches auxiliary voltages when needed

by Ralph Tenny
George Goode & Associates Inc., Dallas, Texas

Low-drain auxiliary voltages, which are needed for short periods of time in many devices, can be supplied by dc-to-dc converters provided they are used continuously. If not, their presence constitutes an unnecessary and excessive power drain. However, this circuit allows these voltages to be switched on only when they are needed, thereby reducing power consumption and optimizing conversion efficiencies. In addition, the circuit's input drive is programmable, and the translator's operation may be controlled remotely.

Only the converter stage that is preceded by voltage translator Q_1 is shown (see figure) because the circuit's input can be driven from many sources—the bit-per-second rate generator of a microcomputer board, a peripheral-interface adapter port, TTL, and the output of a free-running timer, to mention only a few.

However, the main power input to the converter is a separate voltage supply. For example, it could be that only a 5-volt bus is available, and then the doubler shown may have to be replaced with tripler or quadrupler circuitry. Assuming a 9-v source is available, +15v is possible using this circuit; +18v would require a tripler.

This circuit produces approximately +8 v and −8 v at 30 milliamperes each when powered with 5 v. The conversion efficiency is about 75% and depends upon the saturation voltages of transistors Q_2 and Q_3 and the voltage drop across the diode, as does the output voltage. To reduce the voltage drop across the rectifier, the silicon diodes should be replaced with Schottky rectifiers. This change improves both conversion efficiency and output voltage by about 8%. □

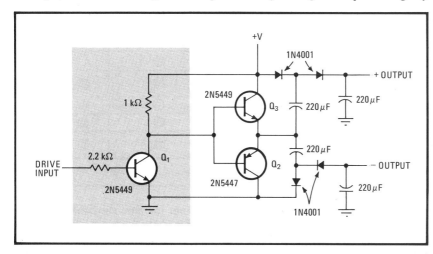

Part-time. When voltage translator Q_1 precedes the dc converter stage, the drive input can be generated by the logic signals regardless of the converter input voltage. The circuit allows the auxiliary voltages to be switched on when needed. Also, the circuit's operation can be controlled remotely.

Three-level inverter conserves battery power

by Geert J. Naaijer
Laboratoires d'Electronique et de Physique Appliquée, Limeil-Brévannes, France

Converting 12 volts dc into a well-regulated 220 v ac at high efficiency is especially desirable when the input power source is derived from solar cells or a car battery. Often, too, output harmonics generated by the conventional square-wave inverter must be reduced to a point where it does not create other practical drawbacks. These design requirements are met with this inverter, which maintains efficiency and reduces harmonics at all power levels by ensuring that the driving current is almost linearly proportional to the output power and by producing a three-level output waveform that more closely approximates a sine wave.

As shown, the CD4013 dual flip-flop generates a 100-hertz pulse train (duty cycle is ⅙) and a 50-Hz square wave (the oscillator is easily modified for 60 Hz), both of which are independent of input voltage and load variations. The AND gates formed by the diode-resistor networks at the output of the flip-flops apply these signals to the power transistor drive circuitry (BC548, MJE801, etc.), which, configured in a two-phase switched-mode arrangement that sends a pulse through the primary of the line transformer every quarter cycle at 50 Hz, generates a three-level waveform.

Because the positive and negative output swings each last one third of a cycle, separated by sixth-period intervals at the zero crossings, third-harmonic attenuation is theoretically reduced to zero. By appropriate modulation

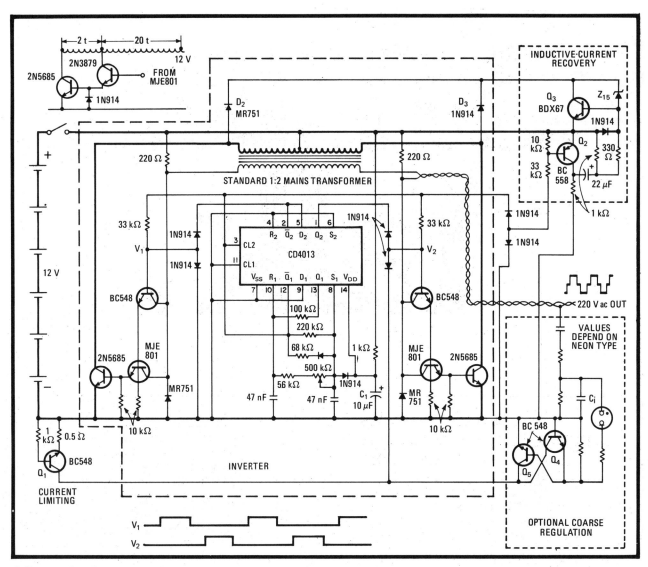

Cell saver. Switched-mode inverter attains high efficiency and low harmonic distortion over a wide range of power levels, thus conserving solar cells and batteries. Output frequency is independent of input fluctuations. Current limiting and voltage regulation are provided.

of these zero-voltage intervals, good regulation is achieved at low loss without reintroducing excessive third harmonic energy. Although this method is not as elegant as the transformerless method of summing phase-shifted square waves or stepped sine waves,[1] it is much simpler.

D_1 and C_1 ensure reliable startup. Current limiting is provided by Q_1, which diverts drive current if output current soars. Output impedance is kept virtually constant during all parts of the cycle, including zero-voltage periods, where Q_2 saturates Q_3, thus feeding an inductive current back into the battery via diodes D_2 and D_3. The zener diode provides spike suppression. A neon lamp or similar symmetrical-breakdown device allows simple voltage regulation. When V_{out} soars, C_i attains breakdown before the normal zero-voltage period, and Q_4 or Q_5 diverts current.

Because this is a switched-mode inverter, efficiency is excellent at nominal output power. But, in contrast to other inverters, its high-efficiency characteristic extends down to very low output-power levels. This is achieved by forcing output current to flow not only through the transformer's secondary, but also through the base-emitter junctions of the 2N5685 power transistors. Thus, base drive closely tracks the output power requirements. As long as the transformer winding ratio is compatible with the gain needed to saturate the power transistors, significant power can be saved at the lower power levels. Efficiency exceeds 88% at 120 volt-amperes, 75% at 175 VA, and 50% at 15 VA. No-load loss is only 12 watts.

Still better performance can be obtained by adopting the power Darlington configuration, as shown at the upper left. Using the Darlington, the inverter's efficiency will be 89% at 150 VA, 70% at 225 VA, and 50% at 15 VA. Lower transformer step-up ratios further improve performance, because the output transistors switch more totally into saturation. Thus, 12-v dc-to-110-v ac and 24-v-dc-to-220-v-ac designs are more efficient than this 12-v-dc-to-220-v-ac circuit. □

References
1. Geert J. Naaijer, "Transformerless inverter cuts photovoltaic system losses," *Electronics*, Aug. 14, 1980, p. 121

Capacitive voltage doubler forms ±12-to-±15-V converter

by Tom Durgavich
National Semiconductor Corp., Santa Clara, Calif.

Pairing a capacitive voltage doubler with a regulator provides a simple solution to the problem of converting ±12 volts dc into ±15 v dc. Such a conversion is often required in systems using Intel's Multibus, for example, which puts out only ±12 v dc. Such a conversion is often tional amplifiers and data converters to a guaranteed voltage swing of ±10 v. Unlike conventional dc-to-dc converters, this approach is inexpensive and occupies little board space.

The circuit (a) uses a 5-watt audio power amplifier (LM384) to drive capacitive voltage doublers that generate ±18 v. One doubler, consisting of capacitors C_1 and C_3 and diodes D_1 and D_3, generates +18 v and the other, consisting of C_2 and C_4 and D_2 and D_4, −18 v. The saturation voltage of the 384 op amp along with the voltage drop across the diodes prevents these voltages from ever reaching ±24 v.

The power-amplifier clock input derived from the system clock keeps the switching waveform synchronous and random noise to a minimum. The clock input voltage and frequency can vary from 2 to 12 v peak to peak and 3 to 20 kilohertz, respectively. The 7-kHz square-wave oscillator (b) is used when the system clock is not available or synchronous operation is not desired.

The output current of this converter is limited to 100 milliamperes but can be slightly increased by providing the op amp with a good heat sink. The ±18 v unregulated voltages may be increased to greater then ±30 v by connecting diodes D_1 and D_2 to +12-v and −12-v sources instead of to a ground connection. ☐

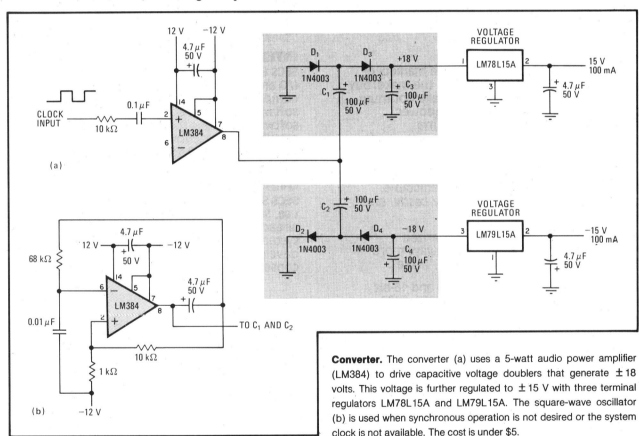

Converter. The converter (a) uses a 5-watt audio power amplifier (LM384) to drive capacitive voltage doublers that generate ±18 volts. This voltage is further regulated to ±15 V with three terminal regulators LM78L15A and LM79L15A. The square-wave oscillator (b) is used when synchronous operation is not desired or the system clock is not available. The cost is under $5.

External transistor boosts load current of voltage regulator

by Dan Watson
Intersil Inc., Cupertino, Calif.

The current capability of Intersil's new low-power programmable voltage regulator may be increased from 40 milliamperes to 1 ampere through the use of an external npn pass transistor (a). The device is connected in parallel with the ICL7663's internal transistor.

The total current supplied by the regulator (I_r) is equal to the base current of the external pass transistor plus the load current of the internal pass transistor. The latter's emitter is situated at pin 2. A 100-ohm resistor is placed between the emitters of the two transistors, so that most of the load current will flow through the external device.

In addition, the circuit does not alter the programming ability of the regulator whose output (V_{output}) equals $(R_2/R_1)V_{set}$, where V_{set} = 1.3 volts. The device can regulate any voltage from 1.3 to 15.5 V for a load current up to 1 A. The load-current versus regulator-current characteristic (b) shows that for a 1.0-A load current, the regulator supplies only 16 mA, which is well within its operating range. A logic 0 or 1 at pin 5 turns the circuit on or off. □

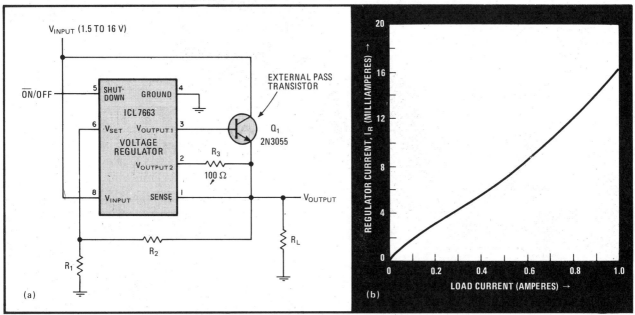

Booster. The circuit (a) uses an external npn pass transistor to boost the current capability of the voltage regulator ICL 7663. This transistor carries the bulk of the load current. The graph (b) shows that for a load current of 1 ampere, the regulator supplies only 16 mA.

Linear controller attenuates switching-supply ripple

by Christopher S. Tocci
Clarkson College of Technology, Potsdam, N. Y.

One of the big disadvantages of a switching power supply is ripple. However, at the cost of a slight efficiency loss and a few parts, this linear controller (a) rejects any ac variations in its dc output when placed at the output of the switcher.

A Laplace voltage-transfer function for the general scheme can be written as:

$$\frac{V_{out}(s)}{V_{in}(s)} = \frac{1}{s([L_1 + k_1k_2(L_1L_2)^{1/2}R_L]/R_L) + 1}$$

where k_1 = the coefficient of coupling, k_2 = the transconductance of the voltage to the current converter, R_L = the load resistance, and L_2 and L_1 are the respective primary and secondary inductances of transformer T_1. When k_1 = 1 and $L_2 = n^2L_1$, where n is T_1's turns ratio, $V_{out}(s)/V_{in}(s)$ may be expressed as:

$$\frac{1}{(sL_1/R_L)(k_2nR_L + 1) + 1}$$

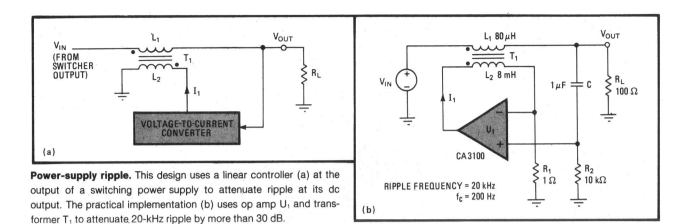

Power-supply ripple. This design uses a linear controller (a) at the output of a switching power supply to attenuate ripple at its dc output. The practical implementation (b) uses op amp U_1 and transformer T_1 to attenuate 20-kHz ripple by more than 30 dB.

This equation is for a first-order, low-pass filter having cutoff frequency $f_c = R_L/2\pi L_1(k_2 nR_L + 1)$ hertz. To reduce ripple by more than 20 decibels, f_c must be at least a decade below the switching frequency.

A possible implementation of the idea is shown in (b). As an example, the design uses a ripple frequency of 20 kilohertz and a load impedance of 100 ohms. With n = 10, k_2 = 1, and f_c = 200 Hz, L_1 = 80 microhenrys and L_2 = 8 millihenrys. Also, when $1/2\pi R_2 C \ll 200$ Hz, R_2 = 10 kilohms and C = 1 microfarad, where R_2 and C may be arbitrarily chosen. The ripple contents measured at the output were found to be below 30 dB. If the ripple frequency is much higher, other parts should be substituted as needed. □

Stacked voltage references improve supply's regulation

by Wes Freeman and George Erdi
Precision Monolithics Inc., Santa Clara, Calif.

By combining low-cost precision voltage references, inexpensive yet accurate power supplies that work over a wide range of voltages may be built. When suitably stacked, these voltage references even improve the regulating performance of the supply.

Consider the circuit in Fig. 1, which can be built for approximately $10. It uses two 10-volt references so combined that the supply will work over a range of 0 to 20 v, with switch S_1 selecting the 0-to-10- and 10-to-20-v ranges.

An operational amplifier isolates potentiometer R_1, which sets the output voltage to within 300 microvolts of the desired value. The op amp's short-circuit current,

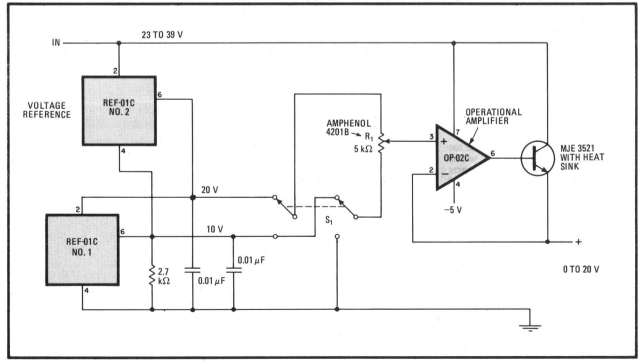

1. Piggyback. Two series-connected voltage references may be united to yield an extended supply output range with significantly improved line regulation at the lower range. The circuit's output can be set to within 300 μV of the desired value. Maximum output current is 1 A.

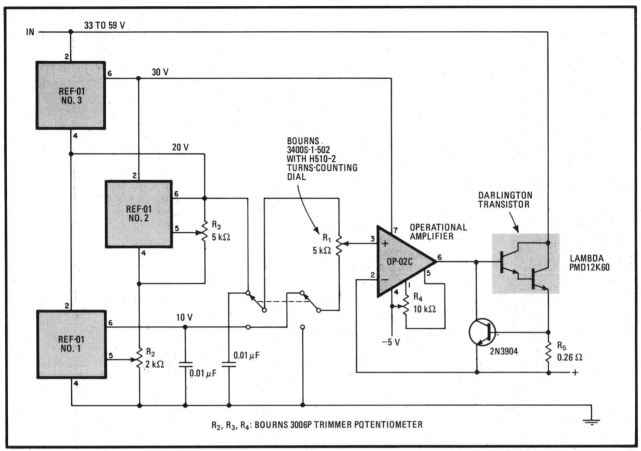

2. Extension. When another reference is added, both ranges become extremely well regulated. Load-handling capability and supply precision are improved. Substituting a linear precision pot and turns-counting dial for R_1 permits setting output to within ±2mV over 0 to 20 V.

approximately 22 milliamperes, limits the maximum base current available to the power transistor. As a result, the maximum available output current is nearly 1 ampere.

The supply's line regulation is within 0.005% of scale reading per volt in the 10-to-20-V range. In the 0-to-10-V range, line regulation is significantly improved to within 0.0001%/V and is mainly limited by the op amp's supply rejection ratio because the output of the second reference regulates the line voltage of the first.

Load regulation is determined by the change in the op amp's open-loop gain versus load current. In this circuit, measured values were ±0.001%/A in the 0-to-800-mA range. Output voltage drift due to temperature is ±0.002% of scale reading per °C.

At an increase in component count and hence also in cost, the performance of the supply may be improved appreciably, as seen in Fig. 2. The addition of a third reference regulates both the 0-to-10- and 10-to-20-V ranges. A Darlington power-output transistor permits a 4-A load current.

As a result, the total change in output voltage is less than ±0.001% for a change in load current of 0 to 2 A and a change in line voltage ranging from 33 to 59 V. Potentiometers R_2 and R_3 adjust the output voltage for the 10-V and 20-V ranges, respectively, while R_4 nulls the op amp's offset voltage.

The substitution of a highly linear precision potentiometer and turn-counting dial for R_1 permits a dial accuracy of ±2 mV from 0 to 20 V, with a resolution of 200 μV. Moreover, if a better grade of reference (REF-01) is employed, the temperature coefficient is ±0.001% of scale reading per °C. □

High frequencies, winding setup improve voltage conversion

by David W. Conway
Universal Engineering Corp., Cedar Rapids, Iowa

Designed to meet high-voltage requirements, this efficient and inexpensive dc-to-dc converter finds use in devices like photoflash equipment and capacitive-discharge ignition systems. The design produces 450 volts at 250 milliamperes from a 12-to-16-v dc supply. The high-power output and high efficiency of the circuit is attributed to high-frequency operation and the use of insulating tape between the layers of the transformer's secondary winding.

A push-pull–amplifier configuration comprising power field-effect transistors Q_3 and Q_4 is driven by integrated circuits U_1 through U_3. Oscillator U_1 generates 150 kilohertz, which is halved to 75 kHz by flip-flop U_2. The high-voltage output is adjusted with potentiometer R_1. When R_1 is at its minimum resistance, the output voltage is at its highest.

This converter is much smaller than competing designs because high frequencies are used. The output transformer's secondary is first wound on the bobbin. In addition, 400 turns of the secondary winding are composed of eight layers of about 50 turns a layer, with each layer insulated from the other. 3M's Permacel 0.0025-inch–thick insulating tape separates the layers.

The two layers of wire that make up the primary winding are referred to as sections A and B. A 48-in. piece of AWG No. 18 wire is folded in half to obtain a bifilar strand, 24 in. long, that is wound as section A on the bobbin. Six bifilar turns fill the cross section of the bobbin to form this layer. Similarly, section B is wound with six bifilar turns of AWG No. 18 wire on top of Section A. The windings are interconnected in series and properly phased to form a 24-turn primary winding. Scaling the turns ratio can provide additional output voltages as necessary. □

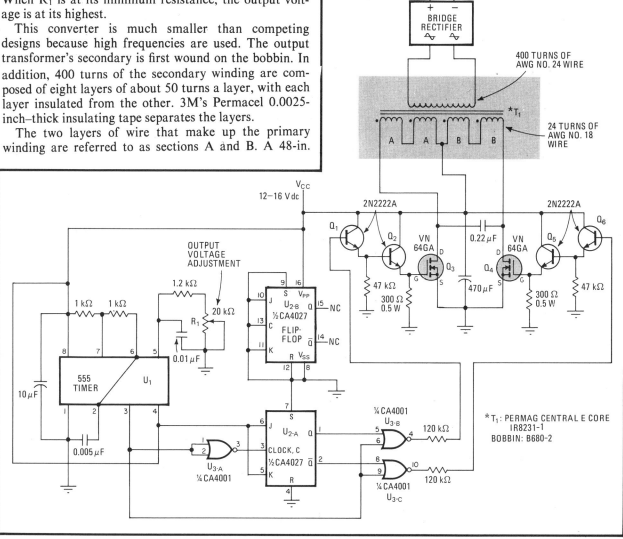

High-voltage converter. Oscillator U_1, frequency divider U_2, and buffer U_3 drive power FETs Q_3 and Q_4 in a push-pull configuration to generate 450 V at 250 mA. A power efficiency of about 90% is attributed to transformer's winding techniques and the use of high frequencies.

22. PROTECTION CIRCUITS

One-chip alarm scares auto thieves

by Andrei D. Stoenescu
Bucharest, Rumania

Most of the burglar alarms that have been designed in recent years to discourage automobile thieves contain too many parts (thus raising doubts about their reliability), need special parts, draw too much current, or cost too much. Actually, a circuit that provides the very same features as most alarms now available can be built around only one integrated circuit and a power transistor. Such a circuit is shown here. Its power drain is only

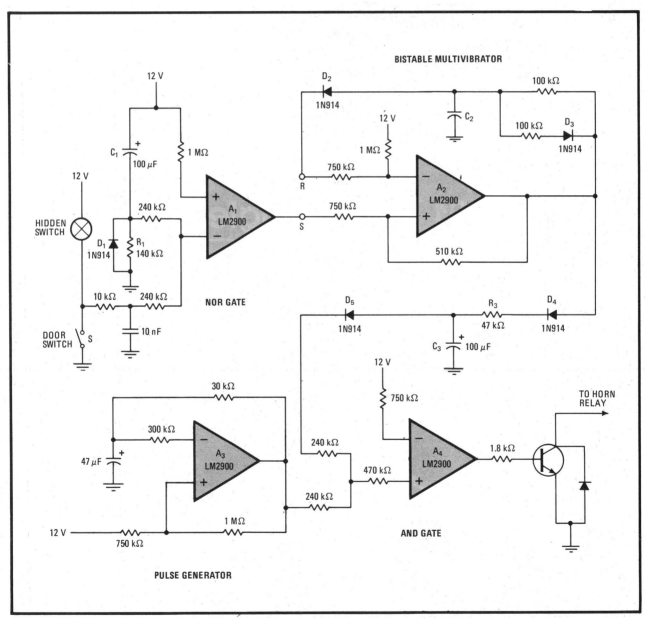

Sedan sentry. This simple alarm uses only two active devices—an LM2900 quad amp and a power transistor—yet performs as efficiently as some of the more complex and expensive circuits. The circuit is activated only after the owner leaves the car and can be disarmed before sounding after he returns. For an intruder, the horn sounds after 10 s and remains on for 400 s. The device can then retrigger.

6.2 milliamperes in the idle state, and its cost can be held to under $10.

Once activated by a hidden switch inside the car, the alarm, which uses the LM2900 quad amplifier:
- Will be inhibited for a few seconds to enable the driver and passengers to exit from the vehicle.
- Will sound 10 seconds after the opening of any door and remain on independently of the position of any door.
- Will time out after about 400 seconds unless a door is open, in which case the alarm will continue to sound.

When the circuit is initialized, capacitor C_1 begins to charge and turns the NOR gate A_1 on after a period equal to R_1C_1. After this time, during which the driver leaves the car, a door switch closure to ground is required to set bistable multivibrator A_2 low. The circuit is thus armed.

Opening any door sets the output of the bistable device high, thus allowing capacitor C_3 to charge through resistor R_3. Unless the circuit is disarmed through the hidden switch, the alarm will sound after 10 seconds because the voltage across C_3 will be sufficient to turn on the AND gate A_3, passing the control output of pulse generator A_4 through to the horn relay.

The R_2C_2 combination determines the bistable multivibrator's reset time, which in this case is approximately 400 seconds. Thus the bistable will again be set high if any door is open, and the horn will continue to sound.

Of course, the same logic functions as are needed for this alarm could also be designed using a quad comparator, such as the LM239. The circuit would then require an idling current of only 0.8 mA. Several additional resistors would have to be added, however. □

Noise-immunized annunciator sounds change-of-state alarm

by K. Soma
Singapore Electronic & Engineering Ltd., Sembawang, Singapore

To determine the status of a multichanneled telemetry receiver, relays are usually employed to activate lamps or light-emitting diodes set in the face of a remotely sited operator panel. However, if the call annunciator must sound an alarm each time a change occurs in the receiver's state, it often becomes vulnerable to noise pickup from external sources and produces false alarms. This circuit immunizes the receiver panel against the effects of that noise.

The simplest way to sense a change in status would be to link 39 capacitors at points P_1 through P_{39} in the figure and to connect their common output to the gate of a thyristor at point J. Indeed, this method is often used. But although the circuit will work, it is extremely susceptible to noise pickup from such sources as an electrical drill because all 39 capacitors form a series-parallel network that filters out most of the 12-volt pulse switchover signal when a lamp changes from green to red or *vice versa*. Thus, only about 100 millivolts of the original pulse is picked up at point J to trigger the thyristor.

A better method is to use transistor switches, as shown, for capacitor-to-thyristor storage and isolation to ensure that a sizable spike—not just a minor glitch from an external noise source—will trigger the alarm. A standard versatile transistor such as the 2N3904 may be used as a switch.

The standby current contributed by each transistor is less than 3.5 microamperes, and the peak current during switch-on is as much as 1 milliampere, which is sufficient to drive most current-gated thyristors—for example, the C103YY.

Note that each lamp in the display panel is effectively monitored by a transistor. Each capacitor is wired so that at least one transistor is momentarily energized only if there is a change in the lamp display input. The capacitors should have a fairly high value, which in this

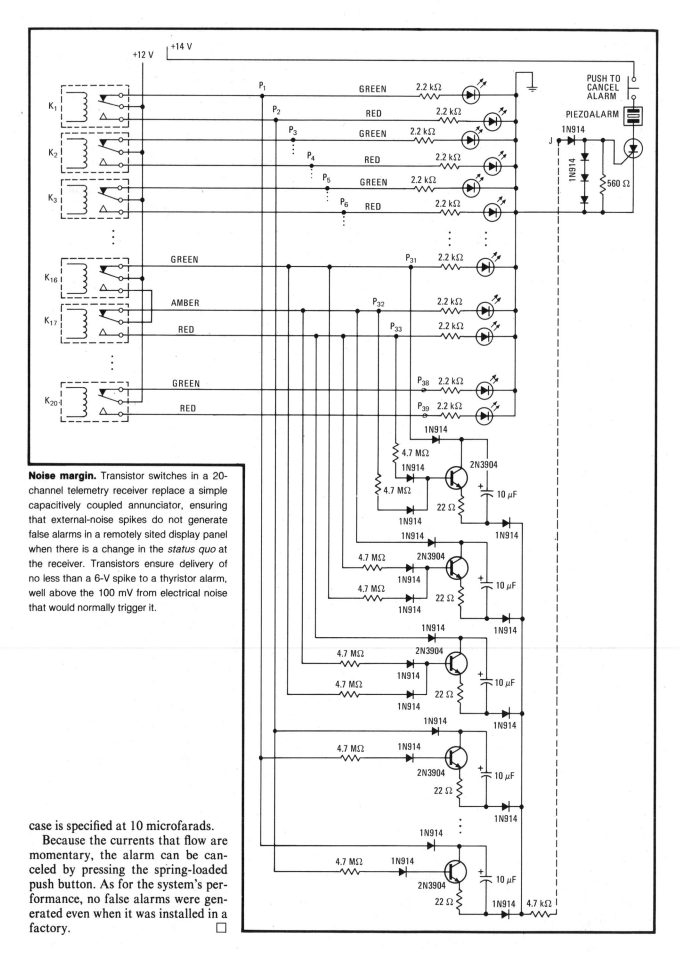

Noise margin. Transistor switches in a 20-channel telemetry receiver replace a simple capacitively coupled annunciator, ensuring that external-noise spikes do not generate false alarms in a remotely sited display panel when there is a change in the *status quo* at the receiver. Transistors ensure delivery of no less than a 6-V spike to a thyristor alarm, well above the 100 mV from electrical noise that would normally trigger it.

case is specified at 10 microfarads.

Because the currents that flow are momentary, the alarm can be canceled by pressing the spring-loaded push button. As for the system's performance, no false alarms were generated even when it was installed in a factory. □

Phase-reversal protector trips main contactor

by Leuridan Carty
Lima, Peru

When the direction of phase rotation is critical, as in a three-phase motor, the reversal of two phases can cause disastrous damage. The advantage of this low-cost phase-reversal protector is that it is independent of line frequency and requires no transformers. Only one chip and few discrete components are needed for the simple circuit, which can also be easily modified to obtain a phase-sequence detector.

The line voltage is directly sensed and applied through transistors Q_1 and Q_2 to the data and clock inputs of flip-flop U_1 (see figure). The reversal of one or more input phases produces a reset pulse at output $\bar{Q}_1$ of U_1. This is delayed by 1 second by means of R_1 and C_1 and then is fed into flip-flop U_2. U_2 is now set to produce an output at Q_2, which is used to drive triac Q_3 into an on-state, so that the main contactor is disabled and the equipment protected.

The phase-sequence detector is obtained by connecting U_1 and U_2 in parallel and by feeding the Q and $\bar{Q}$ outputs to two light-emitting diodes. A normally open pushbutton switch is connected in series with input phase T. If this switch is pressed, one of the LEDs glows, indicating the direction of input phases. The Q output corresponds to clockwise rotation and $\bar{Q}$ to anticlockwise. □

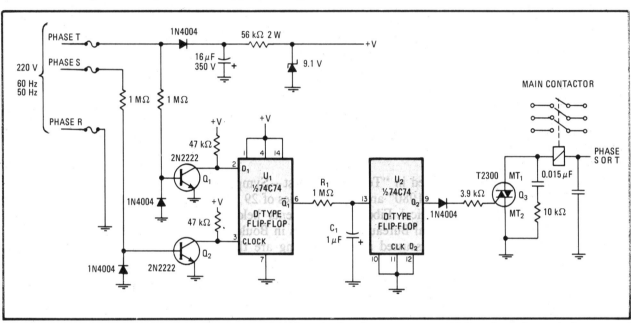

Protection. The circuit links D-type flip-flop 74C74, triac T2300, and a few discrete components into a simple, low-cost phase-reversal protector. Reversal of one or more input phases enables triac Q_3, which in turn disables the main contactor and protects the equipment.

23. PULSE GENERATORS

Deglitcher–delay circuit serves also as pulse generator

by B. Seastrom and G. Goodwin
Sylvania Systems Group, GTE Products Corp., Needham Heights, Mass.

Sustaining its input pulse for a number of clock cycles before translating it into an output pulse, this circuit provides an effective means of discriminating between valid data and spurious pulses or glitches. The designer who uses the circuit has numerous options for adjusting the delay between the input and output transitions, as well as controlling the duration of the output pulse. Furthermore, it triggers on either a rising or a falling edge and generates complementary outputs.

Data entering serial shift register A_1 is sampled at the clock rate and shifted along from output Q_A to Q_H. Meanwhile, for the complement of the input at the output of A_2, the same process occurs at the shift register A_3. Since all 1s are required at gates A_4 or A_5 to toggle cross-coupled NAND gates A_6 and A_7, there is a delay in the leading edge of the output, as well as in the pulse duration. The delay and the pulse duration depend on how many and which taps are connected from the shift registers to the eight-input NAND gates.

It is apparent that noise—in fact, any changes in input level—will be subject to successive samples, whose number is equal to the tap count, before it results in a change in output. By employing different numbers of taps on registers A_1 and A_3, the criterion will be different for different polarity edges—A_1 controls the positive edges and A_3 controls the negative ones—and therefore the circuit is highly noise-resistant. Further, by starting with a tap other than Q_A, initial edge delays can be built in, again selectively for either positive or negative edges.

A version of the circuit allows it to modify the input pulse width. Feeding selected output taps of A_1 into A_5 (eliminating inverter A_2 and serial register A_3) controls the time at which the trailing edge of the output pulse occurs. By judiciously choosing which taps go to A_4 and which go to A_5, the designer can exercise control over the width of the output pulse.

The circuits' applications are enhanced by expanding on the basic concepts. Thus, smaller NAND gates may be used with fewer samples, and expanded gates may be used with more shift registers in tandem. Finally, additional timing signals may be generated by connecting additional sets of gates to the shift registers. □

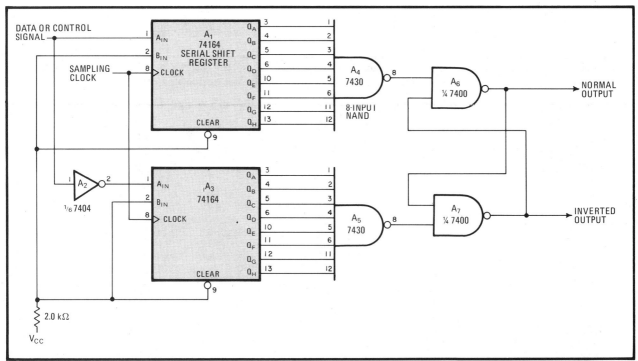

Tap dance. By changing the quantity and position of the shift register outputs into which the multiple-input NAND gates tap, a designer can mask unwanted spikes, as well as exercise a wide range of control over the output pulse width and rising and falling pulse edges.

Two-chip pulse generator operates at 75 MHz

by M. U. Khan
Systronics, Naroda, Ahmedabad, India

Built from integrated circuits in the emitter-coupled-logic family, this pulse generator can provide independent control of delay and width (variable from 5 nanoseconds to 0.1 second) over the frequency range of 10 hertz to 75 megahertz. Only two chips are required—a quad line receiver and a dual D-type flip-flop.

The MC1692 line receiver, A_{1a}, configured as an astable multivibrator, provides a steady stream of pulses, at a frequency determined by R_1C_1, to the delay portion of the circuit. This section, which uses a second line receiver and one half of the MC10231 dual flip-flop, generates a corresponding pulse at the output of A_{2a} whose duration is proportional to R_2C_2. Its maximum duty cycle is greater than 80% at 10 MHz and decreases progressively to about 50% at 75 MHz. After inversion by A_{1c}, the signal is introduced to flip-flop A_{2b}.

A_{2b} is triggered on the positive-going edge of the signal, and so pin 15 of the flip-flop moves high after a time proportional to R_2C_2, thus effecting the delay time. The duration of the pulse emanating from A_{2b} (that is, its width) is set by the A_{2b}–A_{1d} combination, which is identical to the A_{2a}–A_{1b} configuration. Note that the polarity of the output appearing at Q_2 of A_{2b} matches that of the input signal, because the width-determining one-shot works on an inverted version of that signal.

If the flip-flops are replaced by two MC1670 types, the circuit will work beyond 100 MHz. In either case, the circuits used should be mounted on suitable heat sinks. □

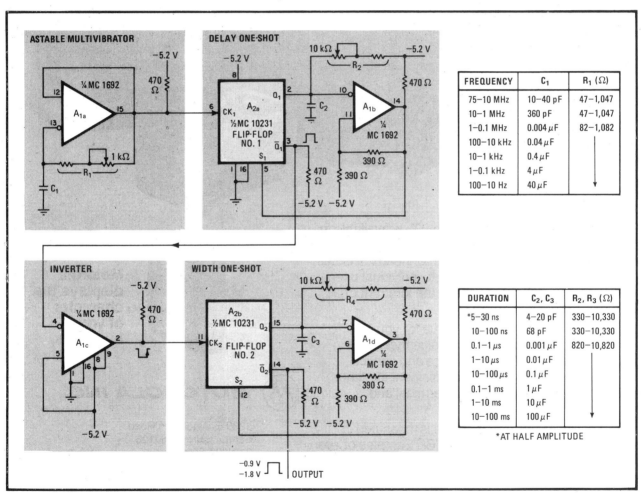

Fast and flexible. A simple ECL pulse generator provides independent control of pulse width and delay and works to 75 MHz. The tables outline the component values. Operation can be extended to 100 MHz by substituting an MC1670 flip-flop for A_2.

FREQUENCY	C_1	R_1 (Ω)
75–10 MHz	10–40 pF	47–1,047
10–1 MHz	360 pF	47–1,047
1–0.1 MHz	0.004 μF	82–1,082
100–10 kHz	0.04 μF	↓
10–1 kHz	0.4 μF	
1–0.1 kHz	4 μF	
100–10 Hz	40 μF	

DURATION	C_2, C_3	R_2, R_3 (Ω)
*5–30 ns	4–20 pF	330–10,330
10–100 ns	68 pF	330–10,330
0.1–1 μs	0.001 μF	820–10,820
1–10 μs	0.01 μF	↓
10–100 μs	0.1 μF	
0.1–1 ms	1 μF	
1–10 ms	10 μF	
10–100 ms	100 μF	

*AT HALF AMPLITUDE

Three-chip circuit produces longer one-shot delays

by Samuel C. Creason
Beckman Instruments Inc., Fullerton, Calif.

One-shot delays of more than a few minutes are difficult to obtain without high RC timing values. However, this three-chip circuit multiplies a one-shot's delay to produce lapses of up to 160 minutes with just a 715-kilohm resistor and a 1-microfarad capacitor. In addition, the circuit can produce even longer delays without changing the values of R and C.

Once the circuit is turned on, counter U_1 is reset and remains in that state until a low level is applied to the input terminal. When the input is low, U_1 begins to count the negative edges of a 1-hertz pulse train that is generated by astable multivibrator U_2. To obtain a specific delay, the appropriate counter output is connected to the input of U_{3-a} as shown in the figure.

When U_1 has counted the clock pulses required to produce the desired delay, the jumper-selected output of U_1 goes high and U_2 is reset. In turn, the output terminal, which had been high, then goes low. The circuit remains in this state until the low level is removed from the input.

Longer delays of any duration may be obtained by inserting an appropriate divider between the output of U_2 and clock input of U_1. □

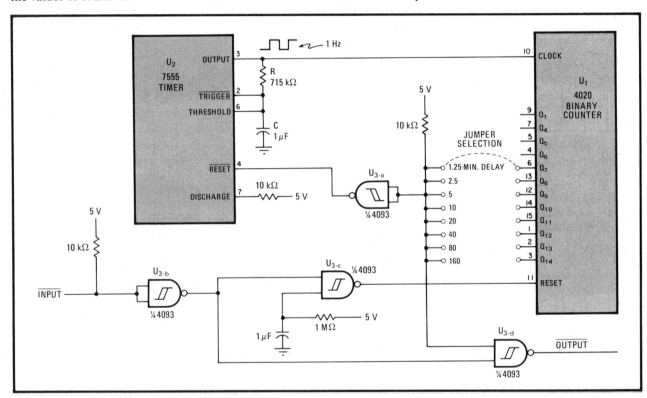

Prolonged. Using three chips, this circuit produces one-shot delays of up to 160 minutes with just a resistor and a capacitor as timing elements. Astable multivibrator U_2 generates clock pulses for counter U_1. The output of U_1 is tied to the appropriate delay input of U_{3-a}, and U_2 is reset when U_1's jumper-selected output goes high. This state change makes the output go low.

Synchronous one-shot has integral pulse width

by Robert D. Guyton
Mississippi State University, Starkville, Miss.

The pulse width of a one-shot multivibrator is usually adjusted with variable RC elements. However, this synchronous one-shot uses three integrated circuits to generate a pulse whose width is an integral multiple of the input clock period. The accuracy of the one-shot's pulse width depends on that of the input clock frequency.

The circuit uses switch S_1 to select pulse widths ranging from 1 to 10 clock periods wide. Starting at logic state 1111, synchronous 4-bit counter U_2 resumes counting when input pulse P(b) triggers the one-shot circuit.

The input pulse can be synchronized with the clock or may be asynchronous. The width of the asynchronous pulse must be about 1.5 times the clock's period. Because the logic is connected to the enable-p input of U_2, the counting continues until the input-to-load function of U_2 is set low. This low resets the counter to the 1111 state where it waits for another input.

The load input for U_2 is obtained from the selected output of the binary-coded-decimal–to–decimal decoder U_3, which in turn gives the selection of pulse widths ranging from 1 to 10 clock periods wide. The combination of waveforms Z and P, using gates U_{1-a} through U_{1-d}, enable the counter.

If only one output pulse width is required, the circuit may be modified by replacing U_3 and S_1 with a single NAND gate having the appropriate inputs. ☐

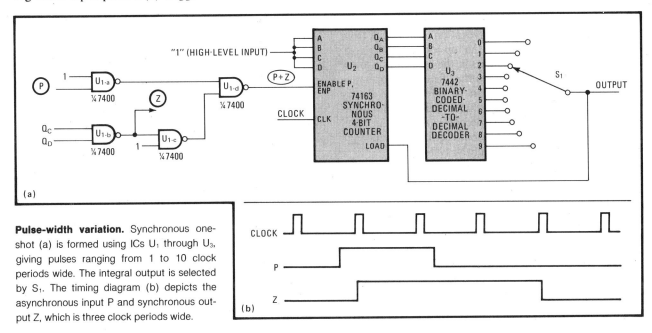

Pulse-width variation. Synchronous one-shot (a) is formed using ICs U_1 through U_3, giving pulses ranging from 1 to 10 clock periods wide. The integral output is selected by S_1. The timing diagram (b) depicts the asynchronous input P and synchronous output Z, which is three clock periods wide.

Synchronous pulse catcher snares narrow glitches

by Marian Stofka
Bratislava, Czechoslovakia

Positive and negative glitches are easily detected by this pulse catcher, which costs less than $5. It is especially useful in synchronous data systems, since it is capable of indicating if the glitch has been detected on the positive or negative level of the system clock. As a consequence of using clocked flip-flops, the unit snares pulses as narrow as 7 nanoseconds—a tenth of the width caught by conventional low-cost units built with monostable

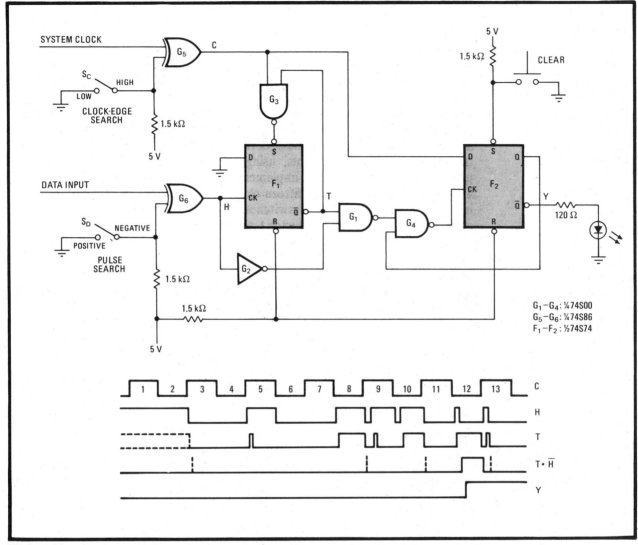

Glitch grabber. Synchronous pulse catcher detects narrow positive- or negative-going glitches on either 0-to-1 or 1-to-0 transition of system clock. Width of shortest pulse detected is determined by minimum width of clock needed to drive flip-flop F_1—in this case, 7 ns, which is well beyond the capabilities of conventional pulse catchers. LED serves as the output indicator. Timing diagram details circuit operation.

multivibrators and rudimentary logic.

Circuit operation is based on the principle that true data signals create only one input transition for a 0-to-1 or 1-to-0 system clock cycle; more than one data-input transition constitutes a glitch. Generally, data is stored in the 74S74 flip-flop, F_1, as shown, and compared with the input signal already stored at NAND gate G_1. If these data signals differ, flip-flop F_2 is triggered and the light-emitting diode at the output is turned on, indicating a glitch has been detected.

The operation of the circuit is subdivided into the search (C = 0) and prepare phases, which alternate with the logic state of the clock. The search and prepare phases can be inverted by selecting the initiating clock cycle's edge polarity that is desired (switch S_c). Similarly, a search for either positive or negative glitches can be selected (switch S_D).

As seen with the aid of the timing diagrams, whenever a positive data transition occurs during a 0-to-1 clock cycle, the Q output of F_1 is set high. Slightly before this time, the output of inverter G_2 has gone low, however, and so there is no logic-race condition at comparator G_2.

During the search phase, the set input of F_1 is released because G_3 is disabled by C = 0. The input data, once high (H = 1) must remain so within the present search phase. Otherwise the output of inverter G_2 will go high and the comparator G_1 will produce a pulse that triggers glitch-memory F_2. Selection of a search for positive and negative glitches by simple inversion of the input data is made possible by the fact that F_1 is edge-triggered. Gate G_4, connected to the clock input of F_2, prohibits loss of information once a glitch is detected. The LED will then light and remain on until the circuit is manually cleared.

During the preparatory phase, the ability to catch glitches is lost, but F_1 is set for next search because it is cleared during this time. F_1 clears itself through G_3, rather than directly. Thus there is no dead time during which a glitch might slip through. Short false-pulse indications may still appear if input-data transitions immediately follow the clock signal, however. To ignore these indications, the D input of F_2 is driven by G_5, which serves as a buffer for the clock. □

Dual one-shot keeps firmware on track

by Patrick L. McLaughlin
Teletech LaGuardia Inc., R&D Labs, Lafayette, Colo.

By noting the absence of pulses generated by status-reporting statements inserted in a running program, this missing-pulse detector reinitializes a microprocessor-based system when glitches on the power line or peripheral circuitry occur. The circuit provides more efficient system performance than a periodic reset timer and is much less expensive than installing line filters or isolators. Only one chip is required—a dual retriggerable monostable multivibrator.

Problems created by a power glitch—such as shuffling of information in the data registers and program jumps to undefined locations or to a location that gives rise to infinite loops—are conventionally solved by placing a timer in the system's reset line to initialize the system every 15 minutes or by using a brute-force power-line filter or even a dynamotor power isolator. A timer probably offers the best low-cost solution, but system speed is degraded by the unnecessary periodic interruptions.

A better solution is to provide a way for the program to report to the system hardware that it is running and on track. Using the 74123 dual one-shot, as shown in the figure, to monitor so-called report statements that are entered in the program's housekeeping loop automatically resets the microprocessor if and when the reports stop for longer than a specified period.

In general operation, both one-shots (one serving as the missing-pulse detector, the other as the output timer) trigger each other alternately in an astable, free-running mode, with $R_1 C_1$ setting the report window, t_w, and R_2 setting the reset time, t_s. On power up, pin 12 of the 74123 is low and the processor is kept at rest until both one-shots time out. Then pin 12 is brought high, enabling the processor. If no report is made before time t_w,

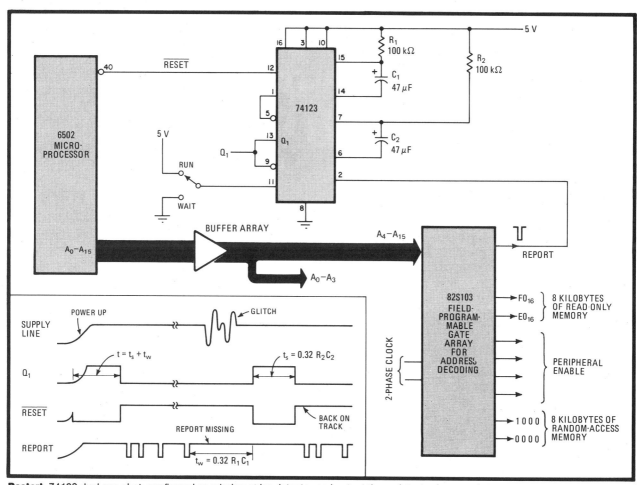

Restart. 74123 dual one-shot, configured as missing-pulse detector and output timer, detects absence of program report statements caused by power-line glitches in order to efficiently reinitialize microprocessor. Reports are entered as often as required in wait-for-data–type systems to ensure pulse rate falls within t_w window. Circuit accommodates static-type stop typical 8080/8085 wait instructions.

the cycle is repeated. An active-low series of report pulses made any time before t_w resets the missing-pulse detector (the output is Q_1), keeping pin 12 high and the processor running.

Usually, report statements are routinely entered before, after, or at both ends of the program's housekeeping loop and in most cases will be called frequently enough to fall within the t_w time window. In loops that may delay normal reporting, however, such as wait-for-data types, inclusion of additional report statements is advisable.

Note that if pin 11 is brought to ground, the one-shot at the output will be inhibited without resetting the processor. Thus, this circuit can accommodate static-type stops typical of the 8080/8085 wait instruction and is usable with slow-running programs and single-stepping arrangements.

The 74123 can be rewired to accept positive-going report pulses simply by introducing the report line to pin 1 of the chip and making pin 5 the output reset line. Pin 3 is then connected to 5 volts and pin 10 disconnected from the positive supply and connected to pin 4 instead. Finally, pin 2 is connected to pin 12. □

Parallel power MOS FETs increase circuit current capacity

by Herb Saladin and Al Pshaenich
Motorola Semiconductor Products Sector, Phoenix, Ariz.

A fast high-voltage and -current pulse is often needed to evaluate the characteristics of a switching power device. For such applications, the semiconductor switch must be much faster than the device under test. Power MOS field-effect transistors serve this function well, but are limited by their current-carrying capability. However, this current capability can be increased without altering the switching speed of the generated pulse by paralleling the transistors.

Power MOS FETs Q_1 through Q_{15} (Fig. 1) are connect-

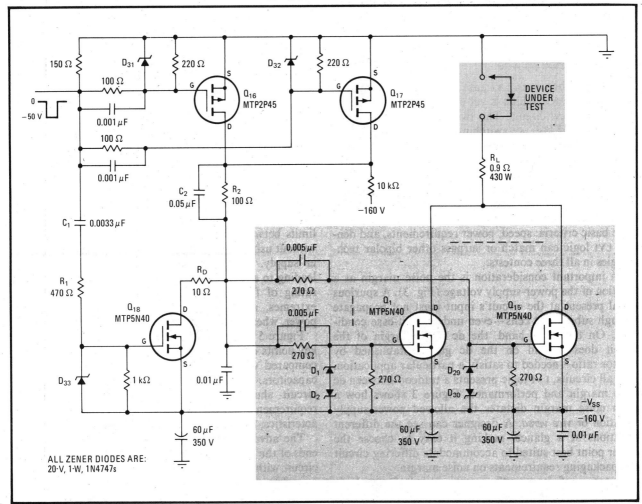

1. Power switcher. This high-speed and high-current semiconductor switch uses 15 n-channel power MOS FETs in parallel to achieve the circuit capability of 150 A of peak pulsed current. Unmatched FETs are used because the self-limiting ability of each FET tends to equalize the drain current. A low-duty-cycle, 50-V input drive pulse has a rise time of 10 ns.

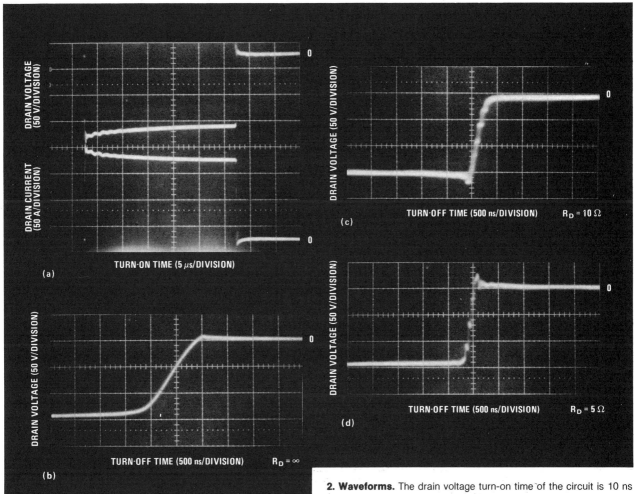

2. Waveforms. The drain voltage turn-on time of the circuit is 10 ns and drain current rises in about 250 ns due to reactive load. (a). The drain voltage turns off in about 1 μs (b) without clamp Q_{18}. With the clamp the turn-off time is reduced to 0.2 μs (c). The FET switch turns off faster as the value of resistance R_d is reduced (d).

ed in parallel to obtain 150 amperes of peak pulsed current for the system. Initially, the FET with the highest transconductance (g_{fs}) draws the largest drain current and consequently causes high dissipation. The resulting temperature rise increases the drain-source on-resistance $r_{ds(on)}$ and self-limits the drain current. This process will continue until all of the FETs in the circuit have equal drain currents.

Two p-channel FETs (Q_{16} and Q_{17}) are connected in parallel to provide the drive current for the MOS FET switch. These FETs are turned on by the negative-going input pulse. Limiting resistor R_2 and speed-up capacitor C_2 in the drain path of Q_{16} and Q_{17} feed the 15 gate circuits of Q_1 through Q_{15}. For simplicity, only the gate circuits of Q_1 and Q_{15} are shown, each formed by a directly coupled resistor, a speed-up capacitor, and back-to-back zener diodes for protection.

In the circuit, approximately 150 A at 140 volts is switched extremely fast, with voltage turn-on time being less than 10 nanoseconds and current rise time about 250 ns (Fig. 2a). The turn-off time for the power switch is improved with the n-channel FET clamp Q_{18}, which turns on at the input pulse's trailing edge and supplies a reverse gate voltage to the power switch. The drain voltage turn-off time of about 1 microsecond (Fig. 2b) is reduced to 0.2 μs with this clamp (Fig. 2c). Reducing resistor R_d further lessens turn-off time (Fig. 2d).

Extreme care must be exercised in the layout of the 15 parallel MOS FETs. Lead lengths must be kept as short as possible and rf bypass capacitors placed at several points along the source bus line to minimize reactive effects. A duty cycle of less than 1% is used to ensure safe operation. □

'Demultiplexing' pulses of different widths

by D. S. Jain
National Remote Sensing Agency, Hyderabad, India

Process-control instrumentation is only one of a multitude of applications that could use a circuit capable of sorting and counting pulses in accordance with their durations in a data stream. This pulse-width discriminator (a) accomplishes this feat by generating a pulse at one of nine outputs that corresponds to one of nine windows in which the input pulse width lies.

Monostables U_1 and U_2 make up an astable multivibrator that is triggered at the input pulse's leading edge. This astable circuit generates a square wave of period T, determined by RC, at output Q_1 that is counted by binary-coded-decimal counter U_5.

The trailing edge of the input pulse triggers monostable U_3, and a pulse is created at Q_3 clearing U_1 and U_2. As a result, astable multivibrator oscillation and BCD counting are forbidden. In addition, the data at the counter's output is transferred to U_6's output.

The trailing edge of the pulse at Q_3 triggers monostable U_4 and produces a narrow pulse at output Q_4. The counter and the latch is reset to zero by this pulse. As a

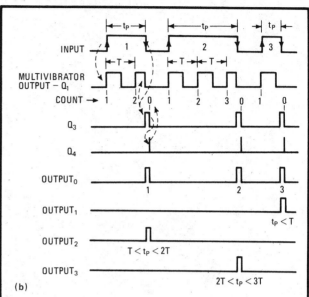

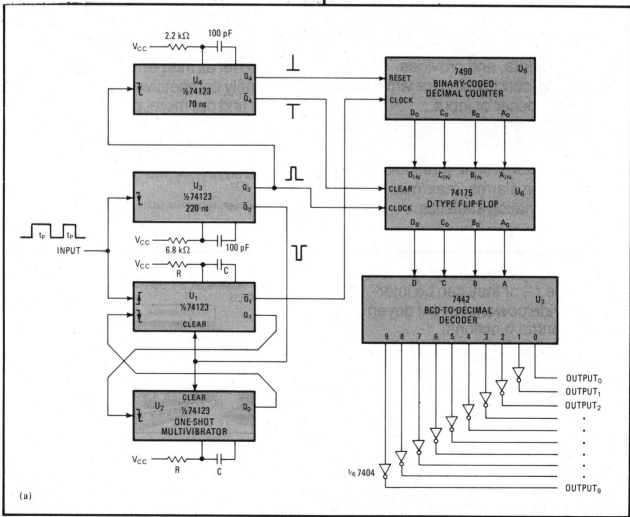

Discrimination. The circuit (a) sorts input pulses according to their durations and generates a positive pulse at the corresponding output. The number of pulses at $output_0$ gives the total number of input pulses. The waveforms (b) show outputs for three different input pulses.

result, the outputs go low, producing a positive pulse at $output_0$ and $output_N$; $output_N$ corresponds to the input pulse width.

The circuit's operation for three input pulses of different widths is illustrated in the timing diagram (b). A positive pulse is produced by the circuit at $output_1$, $output_2$, and $output_3$ for input pulse widths of $t_p < T$, $T < t_p < 2T$, and $2T < t_p < 3T$, respectively. A pulse is generated at $output_0$ for each input pulse, regardless of duration, thus counting the total number of input pulses.

The present range of the circuit is 0 to 9T. BCD data is decoded into decimal N, where N = 1, 2, ... with BCD-to-decimal decoder U_7. Cascading additional BCD counters, latches, and BCD-to-decimal decoders extends the number of windows, enabling the circuit to cover a wider range of inputs. □

Pulse-width discriminator eliminates unwanted pulses

by George Raffoul
Lockheed Engineering & Management Services Co., Houston, Texas

In many digital communications systems, nonreturn-to-zero data is converted into the biphase type before transmission because the receiver can then recover clock information embedded within the data through phase-locked-loop techniques. Unfortunately, the biphase data is noisy since the NRZ data is rarely (if ever) in phase with the transmitter clock (Fig. 1a). These unwanted chirps may be silenced by including a pulse-width discriminator in the receiver block (Fig. 1b).

The circuit (Fig. 2a) eliminates the undesired pulses by slightly delaying the biphase pulse. The 0-to-1 transition of the noisy data triggers one-shot U_{2-a}, which produces pulses at its Q and $\overline{Q}$ outputs, respectively. To compensate for U_{2-a}'s turn-on delay, the positive pulse is switched through exclusive-OR gates U_{3-a} and U_{3-b} and delayed by about 150 nanoseconds.

If the 0-to-1 transition in the input signal represents the leading edge of a pulse that is longer than U_{2-a}'s pulse width—400 ns for the components shown—the AND gate U_{4-b}'s output is a 150-ns pulse. However, if this 0-to-1 transition in the input represents the leading edge of a pulse that is shorter than U_{2-a}'s pulse width, U_{4-b} produces no output. Thus the output of U_{4-b} sets latch U_5

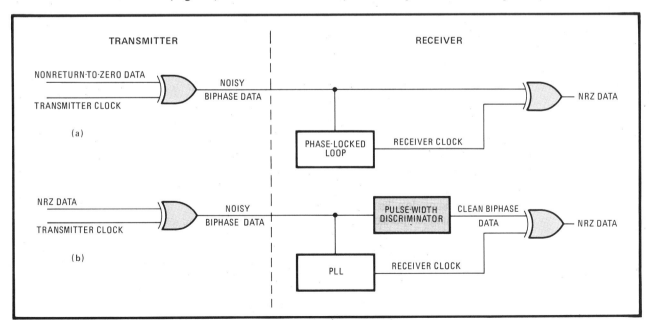

1. Biphase data. At the transmitter, biphase data is obtained by passing NRZ data and the transmitter clock signal through an exclusive-OR gate (a). This data is noisy and therefore requires the use of a pulse-width discriminator in the receiver (b) to eliminate unwanted pulses.

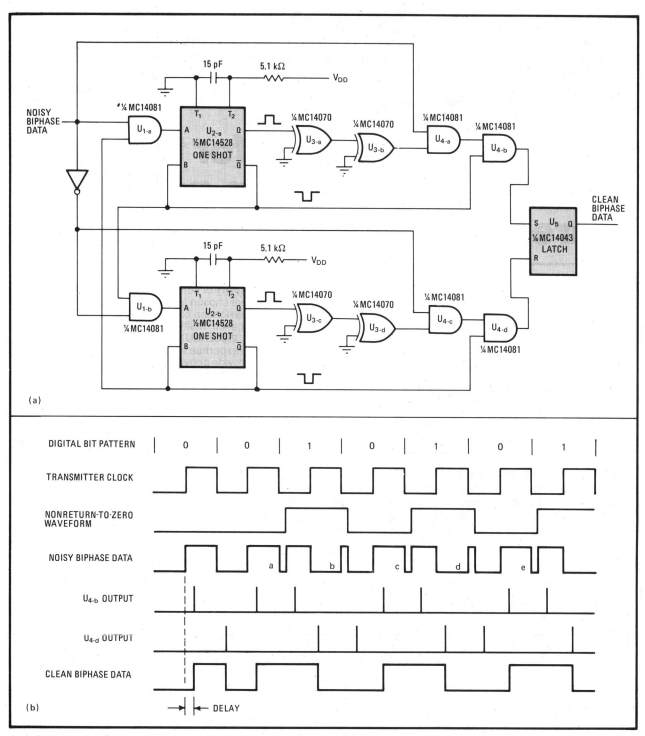

2. Refinement. This pulse-width discriminator circuit uses two one-shots, an RS latch, and a few exclusive-OR and AND gates to eliminate undesired pulses from biphase data (a). The timing diagram (b) shows that the unwanted pulses a, b, c, d, and e are the result of a delay between NRZ data and the transmitter clock signal. The clean output data is delayed.

whenever a pulse of desired width is detected.

Similarly, one-shot U_{2-b} is triggered by a 1-to-0 transition of the input, and gates U_{3-c} and U_{3-d} compensate for the turn-on delay of U_{2-b}. AND gate U_{4-c} samples the inverse of the input signal for a duration of 400 ns, and U_{4-d} produces a narrow pulse if the input signal is greater than $\overline{Q}$ output of U_{2-b}.

The narrow pulse output of U_{4-d} resets RS latch U_5. The timing diagram (Fig. 2b) shows narrow pulses at the outputs of U_{4-b} and U_{4-d} along with a clean biphase waveform at the Q output of U_5. The figure shows that the data is now free of undesired pulses a, b, c, d, and e, but delayed by an amount equal to the one-shot's pulse width. This width is adjusted by proper selection of the one-shot's RC time constant. □

Low-cost generator delivers all standard bit rates

by Robert E. Turner
Martian Technologies, Spring Valley, Calif.

Costing only $2, this generator can drive universal asynchronous receiver-transmitters and other RS-232 serial interface chips at any standard bit rate selected by the user. It is especially suitable for systems based on the Z80 microprocessor from which the generator can derive its 4-megahertz quartz-crystal time base.

As the figure shows, a 4-MHz input clock is divided by 13 by the 74LS393 counter (A_1) and a 74LS11 AND gate, thereby providing a 3.25-microsecond signal that is suitable for driving the CD4024 seven-stage counter, A_2. This signal is close to 16 times the maximum 19.2-kilobit/s output frequency. Smaller divisions are handled by the counter, which supplies 9,600-, 4,800-, 2,400-, 1,200-, 600-, 300-, 150-, and 110-b/s outputs.

The counter outputs are wired so that a bit rate may be selected for both channels of a Z80A serial input/output module or dual asynchronous receiver-transmitter serial interface chip. The output frequency of the generator is 16 times greater than the bit rate, so each serial data bit is sampled 16 times per bit period.

The bit-rate clock frequency is selected with either

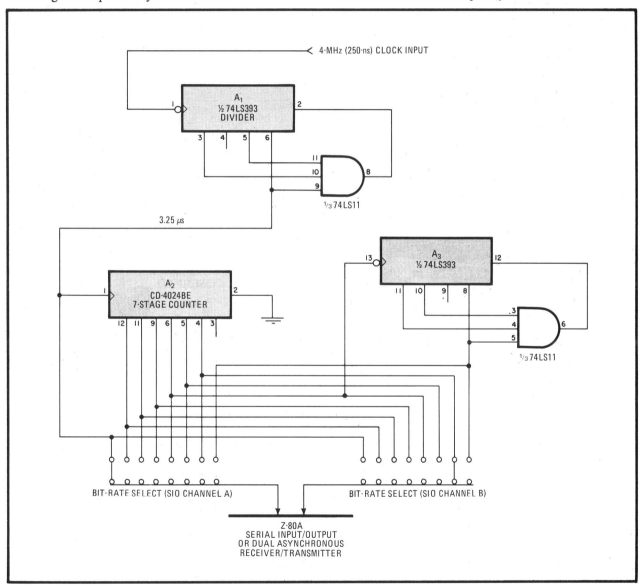

Trimmed taps. With a Z80 system clock trimming costs to $2, the four-chip generator delivers all standard bit rates for RS-232–based systems. The generated rate accuracy is high (see table). The system can be easily modified for older interfaces.

GENERATOR RESPONSE			
Period	Actual output (b/s)	Ideal output (b/s)	Error
52 µs	19,230.76	19,200	+0.16%
104 µs	9,615.38	9,600	+0.16%
208 µs	4,807.69	4,800	+0.16%
416 µs	2,403.84	2,400	+0.16%
832 µs	1,201.92	1,200	+0.16%
1.66 ms	600.96	600	+0.16%
3.33 ms	300.48	300	+0.16%
6.65 ms	150.24*	150	+0.16%
9.15 ms	109.26	110	−0.68%

*OUTPUT AVAILABLE, BUT NOT CONNECTED IN CIRCUIT SHOWN

printed-circuit-board jumpers or by small dual-in-line-packaged switches. The entire circuit is small enough to be mounted next to the Z80's DB-25 connectors that are mounted on its rear panel, making it easy for the end user of the RS-232 interface to select a bit rate. The dual-channel version of the circuit requires only five interface lines: the 4-MHz clock input, bit-rate clock A and B outputs, the 5-volt line, and logic ground.

As for the accuracy of the rates generated, they are well within the 1% timing variation standard required by the RS-232 interface (see table). The generator has been used with many different terminals and printers, and no operating difficulties have been encountered.

The circuit may also be used with most of the older serial interface chips like the 8251 and the 6850, if the user is willing to sacrifice the 19.2-kb/s output. In this case, the divide-by-11 counter would be driven by a 2-MHz clock, with the highest bit rate available becoming 9.6-kb/s. This signal drives the counter. The output of the counter that divides the 110-b/s signal by 11 (counter A_3) would then be connected to the new 1,200-b/s output of the seven-stage counter. □

Generator has independent duty cycle and frequency

by Harry H. Lamb
Richmond Hill, Ont., Canada

Using just a 555 timer, two transistors, and an operational-amplifier chip, this pulse generator provides pulses whose duty cycle and frequency of oscillation can be controlled independently. The frequency varies linearly with the gain of the amplifier, and the duty cycle is linearly controlled with a potentiometer.

Transistors Q_1 and Q_2 allow capacitor C to charge only when timer U_2 is not causing it to discharge. As a result, the sum of charge and discharge times remains constant. U_2 switches when the trigger input drops to $V_{cc}/3$ or the threshold input increases to $2V_{cc}/3$. Rearranging the above voltage levels reveals that the threshold-trigger input operates at $V_{cc}(1 \pm 1/3)/2$ and that, as a result, capacitor voltage V_c varies between $V_{cc}(1+1/3A)/2$ and $V_{cc}(1-1/3A)/2$, where A is the gain of the amplifier U_{1-a} and controls the time required for V_c to take the threshold-trigger input from one switching point to the other. U_{1-b} simply sets the reference $V_{cc}/2$ for U_{1-a}.

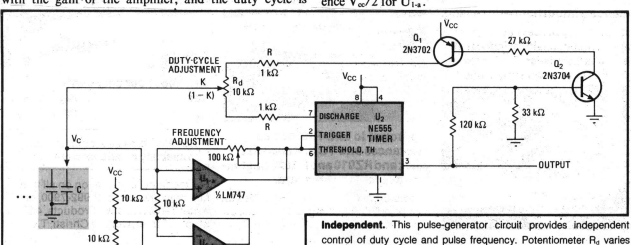

Independent. This pulse-generator circuit provides independent control of duty cycle and pulse frequency. Potentiometer R_d varies the duty cycle, and gain A of amplifier U_{1-a} controls the frequency. Timer 555 is connected in a free-running multivibrator mode.

During time interval t_{low}, the output of U_2 is low and the capacitor discharges from an initial value of $V_{cc}(1+1/3A)/2$. During the interval t_{high}, the output is high and C charges from an initial value of $V_{cc}(1-1/3A)/2$. The duty cycle for the circuit is given by $t_{high}/\tau = (R+R_dk)/(2R+R_d)$, where $\tau = t_{low}+t_{high}$ is the period of oscillation and k is a parameter controlled by the potentiometer R_d. The frequency of oscillation is expressed as $f = 1/\tau = 3A/2C(R+R_d)$. The above relations for duty cycle and frequency show that the two can be controlled independently. Parameter k varies the duty cycle and gain A controls the frequency. The error in the approximation is about 4% at $A = 1$ and decreases rapidly with gain. In addition, the duty cycle varies linearly with the parameter k. □

Thumbwheel switch programs retriggerable one-shot

by Dil Sukh Jain
National Remote Sensing Agency, Hyderabad, India

This programmable synchronous one-shot is a few jumps ahead of the rest by being able to generate a synchronized output pulse whose width can be varied through an externally controlled clock period or a programmable thumbwheel switch. In addition, the circuit is retriggerable with a provision for a clear input.

A narrow negative pulse applied at the trigger input loads the synchronous binary-coded-decimal down-up counter 74190 with the number (N) set on the thumbwheel switch and simultaneously sets Q_1 of flip-flop A_1 high. This loading in turn sets Q_2 of flip-flop A_2 high. The low level at $\overline{Q}_1$ enables the counter to count down from N on successive positive edges of the clock.

When the counter reaches zero, a negative pulse is produced at the ripple clock output of the counter 74190, which corresponds to the negative edge of the clock. This negative pulse, inverted by the 7404 chip, triggers A_1 to make Q_1 low and the enable ($\overline{Q}_1$) input high, which in turn disables the counter and inhibits the circuit. The

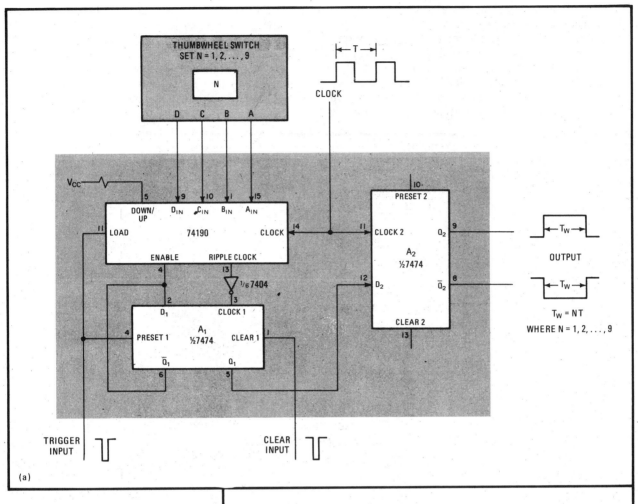

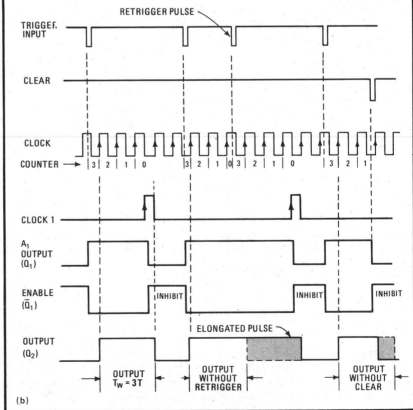

Programmable one-shot. The output pulse width of this synchronized one-shot (a) is programmed with a thumbwheel switch. The circuit uses a binary-coded–decimal counter 74190 and D type flip-flop 7474 to provide the retrigger and clear feature. The timing diagram (b) for $N = 3$ illustrates the control of retrigger and clear inputs on the output pulse.

low input level at D_2 terminates the output pulse whose width is given by $T_w = NT$, where T is the clock's period. The one-shot output pulse width when $N = 3$ is 3T(b).

A retrigger pulse applied while the counter is counting down reloads the circuit with set number N and begins a new countdown, resulting in a single stretched pulse at output Q_2. A negative pulse applied at the clear input (while the counter is counting down) terminates the output pulse at the clock's following positive edge. □

24. SOFTWARE/CALCULATORS

TI-59 program tracks satellites in elliptical orbits

by Janos Molnár
Siófok, Hungary

This versatile program for the TI-59 can find orbital parameters for man-made satellites that rotate in highly elliptical paths around the earth. Quickly solving an interrelated set of 26 equations based upon Kepler's laws of planetary motion, it will serve as an invaluable satellite tracking tool for the astronomy buff.

The program first determines the major-axis elliptical orbit, given the apogee (H_A) and perigee (H_P) of the satellite path, with the aid of formulas 1 and 2 from the equation set shown in the table. The rotating period of the satellite (τ) and its mean angular velocity (ω) depend only on the length of the major axis (a) and a few natural constants and so τ may be found.

Next, the time (t) relationship between the eccentric anomaly (E) and the time measured from the point of the perigee, (t_P) is solved using formulas 4 and 8. Then the relationship between the instantaneous central angle of the satellite (θ) measured with respect to its elliptical geometrical position is calculated by formulas 10 and 11.

The relationship between the axial rotation of the earth and its corresponding satellite equatorial-ascending position helps to ascertain the relative earth-based coordinates. These measurements are found with formulas 13, 14, and 15. Once the coordinates of the true observing point and the trace points (subsatellite) points are known, the spherical distance, elevation, and azimuth of the satellite may be found by applying formulas 12, 16, 17, 18, and 19.

Various properties of motion are also taken into consideration in this program. For instance, the attractive force from the earth pushes a satellite into an elliptical orbit and also rotates this orbit within a sphere. Slight path displacements that occur in the satellite's orbit are found by using formulas 21, 23, and 24. Perigee wandering and satellite nodal ascending time may be calculated with formulas 20, 22, 24, and 25.

An example illustrates the program's usefulness. Consider the case of an elliptical satellite in the Oscar series of amateur radio vehicles located over Budapest, Hungary, that has $\phi_O = 47.5°$ and $\lambda_O = 19.2°$. A tracking accuracy of $\delta = 0.005$ (n = 3) is desired. By entering an

	LIST OF SYMBOLS		
Symbol	**Definition**	**Symbol**	**Definition**
A	azimuth angle measured from North Pole in the horizontal plane (°)	T*	orbital period, presented (min)
a	major axis (km)	Δt	time between the ascending node and the satellite position (min)
B	parameter	t_p	time of perigee passage (min)
b	minor axis (km)	t_{po}	intital perigee passage time (min)
D	distance traveled by the satellite from the observation post (km)	t	time difference (min)
E	eccentric anomaly (rad)	δ	error limit of formula 9
E_p	eccentric anomaly at perigee (rad)	ϵ	eccentricity
e	elevation angle above the horizontal plane (°)	Δ	earth's central angle between the perigee and the observation post and the point of ground trace (°)
H_A	altitude of apogee (km)	ζ	parameter
H_P	altitude of perigee (km)	θ	orbital central angle between the perigee and the satellite position (°, rad)
i	inclination angle of the orbit that is related to the equatorial plane (°)	λ_E	longitude of an ascending node (°)
J	dimensionless term from the potential function	λ_{EO}	initial longitude of an ascending node (°)
k	parameter	λ_O	longitude of observation post (°)
M	mean anomaly-average angular velocity (rad/min)	λ_S	longitude of the subsatellite point (°)
M_P	angular velocity at perigee (rad/min)	μ	earth's gravitational constant (km³/s²)
n	root's precious number of formula 9	Σ	orbital central angle between the ascending node and the satellite (°)
R	orbital radius (km)	Σ_P	amount of perigee (°, rad)
R_F	mean radius of earth (km)	Σ_{PO}	initial perigee argument (°, rad)
R_{FQ}	equatorial radius of earth (km)	τ	time measured from perigee passage (min)
T	time interval between two successive perigee passes, anomalistic period, calculated (min)	ϕ_O	latitude of observation post (°)
T_E	actual time of equatorial ascending node (h, min, s)	ϕ_P	latitude of the perigee (°)
T_N	nodal orbital period (min)	ϕ_S	latitude of the subsatellite point (°)
T_λ	rotation period of orbital plane (min)	ω	mean angular velocity (rad/min)
T_Σ	rotations period of perigee in orbital plane (min)	ω_F	angular velocity of the earth (°, min)

INTEGRATED EQUATION SET FOR SATELLITE-ORBIT ANALYSIS

(1) $a = \dfrac{H_A + H_P}{2} + R_F$

(2) $\epsilon = \dfrac{H_A - H_P}{2a}$

(3) $\omega = \dfrac{2\pi}{T} = 60\sqrt{\mu a^{-3}}$

(4) $\Sigma'_P = \sin^{-1}\left\{\dfrac{\sin\phi_P}{\sin i}\right\}$

$\Sigma_P = \Sigma'_P$ for $0° \leq \xi < 90°$
$\quad\;\; = 180° - \Sigma'_P$
$\quad\quad$ for $90° \leq \xi < 270°$
$\quad\;\; = 360° + \Sigma'_P$
$\quad\quad$ for $270° \leq \xi < 360°$

where $\xi = (\lambda_P - \lambda_E)\,\text{sign}(\cos i)$

(5) $E'_P = \cos^{-1}\left\{\dfrac{\epsilon + \cos\Sigma_P}{1 + \epsilon\cos\Sigma_P}\right\}$

$E_P = E'_P$ for $0 \leq \phi_P$
$\quad\;\; = 2\pi - E'_P$ for $0 \geq \phi_P$

(6) $t_P = \dfrac{E_P - \epsilon\sin E_P}{\omega} = \dfrac{M_P}{\omega}$

(7) $t = K\,\Delta t = T^* - T_E$

(8) $M = \omega(t - t_p) = \omega\tau$

$\tau = \tau$ for $0 \leq \tau < T$
$\;\; = \tau - T$ for $T \leq \tau$

(9) $0 = \epsilon\sin E - E + M$

(10) $\theta' = \cos^{-1}\left\{\dfrac{\cos E - \epsilon}{1 - \epsilon\cos E}\right\}$

$\theta'' = \theta'$ for $0 \leq E \leq \pi$
$\quad\;\; = -\theta' + 2\pi$ for $\pi < E \leq 2\pi$

where $\theta = \theta''\,(180/\pi)\,\text{sign}(M)$

(11) $R = a(1 - \epsilon\cos E)$

(12) $\zeta = \dfrac{R_F}{R}$

(13) $\Sigma = \Sigma_P + \theta$

Constants

$R_F\; = 6{,}371.0$ km
$R_{Fq} = 6{,}378.2$ km
$J\; = 1.627 \times (10^{-3})$
$\mu\; = 398{,}603$ km^3/s^2
$\omega_F \cong 0.25°$/min

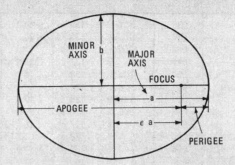

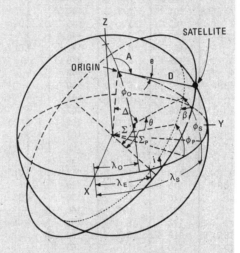

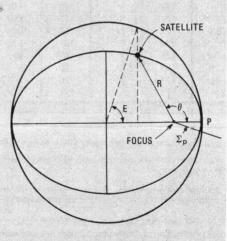

(14) $\phi_S = \sin^{-1}\{\sin i \sin\Sigma\}$

(15) $\lambda_S = \dfrac{\cos i}{|\cos i|}\cos^{-1}\left\{\dfrac{\cos\Sigma}{\cos\phi_S}\right\} - \omega_F t + \lambda_E$

(16) $\Delta = \cos^{-1}\{\sin\phi_O \sin\phi_S + \cos\phi_O \cos\phi_S \cos(\lambda_S - \lambda_O)\}$

(17) $e = \tan^{-1}\left\{\dfrac{\cos\Delta - \zeta}{\sin\Delta}\right\}$

(18) $|A| = \cos^{-1}\left\{\dfrac{\sin\phi_S - \sin\phi_O\cos\Delta}{\sin\Delta\cos\phi_O}\right\}$

(19) $D = R\sqrt{1 + \zeta^2 - 2\zeta\cos\Delta}\;\text{sign}(e)$

(20) $\Sigma_P = \Sigma_{PO} + 360(t - t_{po})/T_\Sigma$

(21) $T_\Sigma = \dfrac{T}{B(5\cos^2 i - 1)/2}$

(22) $\lambda_E = \lambda_{EO} \pm 360(t - t_{po})/T_\lambda$

(23) $T_\lambda = \dfrac{T}{B\cos i}$

(24) $\dfrac{J}{4}\left(\dfrac{1}{\frac{H_A}{R_{FQ}} + 1} + \dfrac{1}{\frac{H_P}{R_{FQ}} + 1}\right)^2 = B$

(25) $T_N = T\left[1 - J\left(\dfrac{5\cos^2 i - 1}{2}\right)\left(\dfrac{R_F}{a}\right)^2\right]$

(26) $\delta = \tfrac{1}{2}10^{-n} \geq |f(E_O) - f(E)|$

PRINTER OUTPUT FOR TI-59 ELLIPTICAL SATELLITE TRACKING PROGRAM

TIM (t)	F(ϕ_S)	L(λ_S)	Δ	e	A	D	Σ	R
0.00	0.00	0.00	50.35	−9.33	154.72	−6,502.06	0.01	8,332.85
100.00	52.65	96.61	47.51	31.00	53.41	23,573.59	108.58	27,404.14
200.00	38.82	98.36	55.86	25.44	67.59	34,876.05	131.63	38,050.33
300.00	27.28	85.23	54.42	27.88	86.94	38,684.83	146.88	42,042.81
629.50	−56.80	105.10	126.19	−56.45	137.41	−14,849.06	266.12	10,168.44
659.50	9.50	−159.09	122.98	−56.57	2.01	−13,354.96	11.35	8,770.73

PRINTER LISTING: TI-59 PROGRAM FOR ELLIPTICALLY ORBITING SATELLITES

Location	Key									
000	LBL	043	7	086	RCL	129	STO	172	06	215	+
001	A'	044	2	087	18	130	00	173	INV	216	RCL
002	(	045	4	088	SUM	131	|X|	174	GE	217	26
003	STO	046	3	089	30	132	EXC	175	CE	218	=
004	09	047	0	090	RCL	133	00	176	π	219	SBR
005	SIN	048	SBR	091	30	134	OP	177	×	220	DEG
006	×	049	(	092	RTN	135	10	178	2	221	STO
007	RCL	050	2	093	LBL	136	STO	179	÷	222	37
008	20	051	1	094	DEG	137	26	180	RCL	223	SIN
009	−	052	SBR	095	÷	138	CLR	181	27	224	×
010	RCL	053	(	096	3	139	STO	182	=	225	RCL
011	09	054	2	097	6	140	01	183	STO	226	14
012	+	055	7	098	0	141	RAD	184	27	227	SIN
013	RCL	056	SBR	099	=	142	PGM	185	LBL	228	=
014	00	057	(	100	INV	143	08	186	CE	229	INV
015	)	058	7	101	INT	144	E	187	RCL	230	SIN
016	RTN	059	5	102	×	145	COS	188	27	231	STO
017	LBL	060	SBR	103	3	146	×	189	×	232	31
018	(	061	(	104	6	147	RCL	190	1	233	RCL
019	OP	062	5	105	0	148	20	191	8	234	37
020	04	063	4	106	=	149	−	192	0	235	COS
021	RC*	064	SBR	107	RTN	150	1	193	÷	236	÷
022	28	065	(	108	LBL	151	=	194	π	237	RCL
023	OP	066	1	109	C	152	+/−	195	=	238	31
024	06	067	3	110	STO	153	STO	196	PRD	239	COS
025	OP	068	SBR	111	30	154	38	197	26	240	=
026	00	069	(	112	÷	155	÷	198	DEG	241	INV
027	CLR	070	1	113	RCL	156	(	199	RCL	242	COS
028	1	071	6	114	17	157	RCL	200	21	243	STO
029	SUM	072	SBR	115	=	158	06	201	PRD	244	32
030	28	073	(	116	INV	159	COS	202	38	245	1
031	LBL	074	7	117	INT	160	−	203	RCL	246	8
032	=	075	7	118	×	161	RCL	204	19	247	0
033	RTN	076	SBR	119	RCL	162	20	205	÷	248	X⇌T
034	LBL	077	(	120	17	163	=	206	RCL	249	RCL
035	D'	078	3	121	−	164	1/X	207	38	250	37
036	3	079	5	122	RCL	165	INV	208	=	251	INV
037	0	080	SBR	123	24	166	COS	209	STO	252	GE
038	STO	081	(	124	=	167	STO	210	39	253	)
039	28	082	ADV	125	×	168	27	211	LBL	254	RCL
040	FIX	083	RTN	126	RCL	169	π	212	Σ+	255	32
041	02	084	LBL	127	22	170	X⇌T	213	RCL	256	+/−
042	3	085	X⇌T	128	=	171	RCL	214	12	257	+

Instructions

- Key in program, first partitioning calculator for 40 data registers by entering (4) OP* 17

- Enter earth coordinates of observation point and accuracy index to which orbital parameters are to be found:
  (ϕ_O), A, (λ_U), R/S, (n), R/S

- Specify orbital parameters of apogee and perigee (km), inclination of orbit relative to equatorial plane (°), argument of perigee (°), longitude of ascending node (°), and actual time equatorial ascending (if known), in hours (h), minutes (min), seconds (s):
  (H_A), R/S, (H_P), R/S, (i), R/S, (Σ_P), R/S, (λ_E), R/S, (T_E), R/S
  Calculator will display satellite's orbital time, T, in minutes, between two successive perigee passes after entry of H_P

- Press B' to find initial coordinates of satellite's perigee point as observed from specified earth location:
  t, ϕ_S, λ_S, Δ, e, A, D, Σ, and R are printed

- Select intervals of time at which satellite's position is to be found from its ascending node and time from its ascending node to the final satellite coordinates of interest:
  (Δt), B, (t), C
  Program finds orbital parameters from T_E to t, in steps of Δt

- Alternatively, specify the actual (real) time in from which the orbital parameters are to be found in steps of new Δt:
  (Δt), B, (t*), D

- Press R/S RST CLR to terminate program

#	Code	#	Code	#	Code	#	Code	#	Code	#	Code
258	3	322	RCL	386	00	450	STO	514	=	578	24
259	6	323	39	387	=	451	10	515	STO	579	+/−
260	0	324	=	388	NOP	452	R/S	516	22	580	+
261	=	325	÷	389	SBR	453	STO	517	1/X	581	2
262	STO	326	RCL	390	X⇌T	454	11	518	×	582	×
263	32	327	33	391	IFF	455	R/S	519	π	583	π
264	LBL	328	SIN	392	01	456	STO	520	×	584	=
265	)	329	=	393	DMS	457	08	521	2	585	STO
266	RCL	330	INV	394	GTO	458	+/−	522	=	586	24
267	32	331	TAN	395	C	459	INV	523	STO	587	LBL
268	×	332	STO	396	LBL	460	LOG	524	17	588	1/X
269	RCL	333	34	397	DMS	461	÷	525	R/S	589	RCL
270	25	334	OP	398	INV	462	2	526	STO	590	24
271	=	335	10	399	FIX	463	STO	527	14	591	−
272	+	336	STO	400	÷	464	03	528	COS	592	RCL
273	RCL	337	36	401	6	465	=	529	OP	593	20
274	13	338	1	402	0	466	EXC	530	10	594	×
275	−	339	+	403	+	467	08	531	STO	595	RCL
276	RCL	340	RCL	404	RCL	468	R/S	532	25	596	24
277	30	341	39	405	23	469	STO	533	RCL	597	SIN
278	×	342	X^2	406	=	470	15	534	14	598	=
279	.	343	−	407	÷	471	STO	535	R/S	599	÷
280	2	344	2	408	2	472	20	536	STO	600	RCL
281	5	345	×	409	4	473	STO	537	12	601	22
282	0	346	RCL	410	=	474	21	538	×	602	=
283	6	347	39	411	INV	475	R/S	539	π	603	STO
284	8	348	×	412	INT	476	STO	540	÷	604	24
285	4	349	RCL	413	×	477	16	541	1	605	DEG
286	=	350	28	414	2	478	INV	542	8	606	6
287	SBR	351	=	415	4	479	SUM	543	0	607	
288	DEG	352	√X	416	=	480	20	544	=	608	3
289	STO	353	×	417	INV	481	SUM	545	STO	609	STO
290	32	354	RCL	418	DMS	482	21	546	24	610	02
291	−	355	38	419	LBL	483	2	547	RAD	611	RCL
292	RCL	356	=	420	D	484	INV	548	COS	612	12
293	11	357	PRD	421	STF	485	PRD	549	+	613	R/S
294	=	358	36	422	01	486	20	550	RCL	614	STO
295	COS	359	RCL	423	FIX	487	INV	551	20	615	13
296	×	360	31	424	04	488	PRD	552	=	616	RTN
297	RCL	361	SIN	425	PRT	489	21	553	÷	617	STO
298	31	362	−	426	STO	490	6	554	(	618	23
299	COS	363	RCL	427	29	491	3	555	1	619	DMS
300	×	364	28	428	DMS	492	7	556	+	620	EXC
301	RCL	365	×	429	−	493	1	557	RCL	621	23
302	10	366	RCL	430	RCL	494	STO	558	20	622	RTN
303	COS	367	10	431	23	495	19	559	×	623	LBL
304	+	368	SIN	432	=	496	SUM	560	RCL	624	B'
305	RCL	369	=	433	CP	497	21	561	24	625	STF
306	31	370	÷	434	GE	498	RCL	562	COS	626	00
307	SIN	371	RCL	435	STF	499	21	563	=	627	GTO
308	×	372	10	436	+	500	INV	564	INV	628	Σ+
309	RCL	373	COS	437	2	501	PRD	565	COS	629	LBL
310	10	374	÷	438	4	502	20	566	STO	630	B
311	SIN	375	RCL	439	=	503	Y^x	567	24	631	STO
312	=	376	33	440	LBL	504	1	568	1	632	18
313	STO	377	SIN	441	STF	505		569	8	633	INV
314	28	378	=	442	×	506	5	570	0	634	STF
315	INV	379	INV	443	6	507	+/−	571	X⇌T	635	00
316	COS	380	COS	444	0	508	×	572	RCL	636	CLR
317	STO	381	STO	445	=	509	3	573	12	637	RTN
318	33	382	35	446	GTO	510	7	574	INV		
319	RCL	383	SBR	447	C	511	8	575	GE		
320	28	384	D'	448	LBL	512	8	576	1/X		
321	−	385	IFF	449	A	513	1	577	RCL		

apogee of 35,786 kilometers, a perigee of 1,500 km, and the previously mentioned parameters, λ_o and ϕ_o, the satellite's orbital time is discovered to be 656.195 minutes.

Now entering the angle of inclination relative to the equatorial plane (57°). the argument of the perigee (329.74°), the longitude of the ascending node (0°), and the actual time of equatorial ascending (13.1530), given an interval of time equal to 100 min at which the satellite's position is to be known, the set of data displayed in the table on page 147 is obtained by the program. The first four rows of the table is for the stipulated condition. The last two rows represent Δt being changed to 30 min and T* changed to 23.4500. □

Calculator plots time response of inverse Laplace transform

by Hank Librach
Fairfield, Conn.

Combining the benefits of two previous notes[1,2] in this series, this HP-41C calculator note solves the inverse Laplace transform and then displays the results on the calculator's miniprinter-plotter.

Using the Gaver algorithm[3] to approximate the time-domain representation of a Laplace transfer function, the Hewlett-Packard calculator and program form a portable tool for visualizing the time-domain response of a system given the Laplace transform. The user is prompted by the program to choose the bounds of the time axis as well as the time increments. Then by redefining the start and end points of the axes, the user can zoom in on smaller time increments.

Step 59 (LBL A) in the program shown is where the transfer function is entered—in this case, it is $F(s) = 1/(s+2)$. Then the corresponding time-varying function—namely, $f(t) = e^{-2t}$—is plotted along the axes specified by the user.

The prompts that specify the plot coordinates are shown in the table. The resulting printout, for a range of 0 to 2 seconds in 0.1-s increments, is shown in the figure. The Gaver Approximation makes conventional assumptions regarding the time functions—namely that f(t) is bounded and well-behaved. Finally, an extra memory module is needed to run the printer-plotter. □

References
1. Kin-chu Woo, "TI-59 inverts Laplace transforms for time-domain analysis," *Electronics*, Oct. 9, 1980, p. 178.
2. Michael A. Wyatt, "Home computer displays inverse Laplace transforms," *Electronics*, March 24, 1981, p. 163.
3. D. P. Gaver, "Observing Stochastic Processes and Approximate Transform Inversion," Operational Research, Vol. 14, No. 3, 1966, pp. 444–459.

PROMPTS FOR SETTING X AND Y AXES			
NAME ?		AXIS ?	
ILT	RUN	0.00	RUN
Y MIN ?		X MIN ?	
0.00	RUN	0.001	RUN
Y MAX ?		X MAX ?	
1.00	RUN	2.00	RUN
		X INC ?	
		.10	RUN

Miniplotting. The plotted curve of the inverse Laplace transform is generated from the computed results by an HP-41C calculator. Axes are set up according to the user's response to series prompts that are part of the overall program for calculating the inverse transform. The X and Y coordinates are rotated 90° clockwise from the standard Cartesian representation in the arrangement, right, which depicts the actual plot as it emerges from the calculator's miniprinter.

```
       PLOT OF ILT
      X (UNITS = 1.) ↓
      Y (UNITS = 1.) →
      0.00           1.00
      0.00
      +---------------+
0.00:                 x
0.10:               x
0.20:              x
0.30:            x
0.40:           x
0.50:         x
0.60:        x
0.70:      x
0.80:     x
0.90:    x
1.00:   x
1.10:  x
1.20: x
1.30: x
1.40: x
1.50: x
1.60: x
1.70: x
1.80: x
1.90: x
```

```
HP-41C PROGRAM LISTING FOR INVERSE
          LAPLACE TRANSFORM

PRP "ILT"

01   ♦LBL "ILT"        33      STO 12
02    STO 26           34      STO 24
03    .083333333
04    STO 13           35   ♦LBL 02
05    -32.0833333      36      1
06    STO 14           37      ST+ 12
07    1279.000076      38      ST+ 25
08    STO 15           39      RCL 23
09    -15623.66689     40      RCL 12
10    STO 16           41      *
11    84244.16946      42      XEQ A
12    STO 17           43      RCL IND 25
13    -236957.5129     44      *
14    STO 18           45      ST+ 24
15    375911.6923      46      RCL 12
16    STO 19           47      10
17    -340071.6923     48      X=Y?
18    STO 20           49      GTO C
19    164062.5128      50      GTO 02
20    STO 21
21    -32812.50256     51   ♦LBL C
22    STO 22           52      RCL 24
                       53      RCL 23
23   ♦LBL B            54      *
24    12               55      RTN
25    STO 25           56      GTO B
26    RCL 26           57      "ENTER FUNCTION"
27    2                58      PROMPT
28    LN
29    X<>Y             59   ♦LBL A
30    /                60      2
31    STO 23           61      +
32    0                62      1/X
                       63      RTN
                       64      END
```

HP-41C generates a pseudorandom sequence

by Ian Patterson, *Chemical Engineering Department, Ecole Polytechnique, Montreal, Quebec, Canada*

In much the same, time-honored way as a shift register whose output bit is exclusive-ORed with a selected bit in the register and fed back as input, this HP-41C program will generate a pseudorandom binary sequence that is useful for statistical communications analysis. This one provides a sequence ranging in length from 3 (2^2-1) to 1,023 ($2^{10}-1$) bits.

The program is based upon determining the value of the register bit to be exclusive-ORed and fed back from the relationship:

$$n_f = \text{INT}(R/2^{l-n}) \text{ MOD } 2$$

where
l = the register length
n = the bit position of the bit to be exclusive-ORed in the feedback loop
R = the register value (the contents, decimal equivalent, of the register).

Only the register length and a seed (the initial value in the register) must be provided. The program is executed by keying XEQ PRBS. The user then responds to the prompt, "No. of bits?" with the register length (from 2 to 10), and then with the seed, which can be any positive integer. The feedback bit position, n, is calculated by subroutine 10 (LBL 10) in the program, using the method detailed by Davies[1].

Having the required information, the program then derives the sequence, with the calculator then generating a high tone when each stepped output bit is a 1 and a low tone for 0. The tone output may be suppressed and the value displayed by deleting instructions 68 to 71 and substituting a pause (PSE) or STOP instruction in their place. □

HP-41C PROGRAM FOR PSEUDO RANDOM BINARY SEQUENCE GENERATION

Line	Code							
01	LBL T PRBS	30	8	60	*	90	T BAD LENGTH
02	CF 10	31	−	61	RCL 02	91	AVIEW
03	T NO. OF BITS?	32	x = 0?	62	2	92	GTO T PRBS
04	PROMPT	33	SF 10	63	/	93	LBL 08
05	STO 01	34	LBL 03	64	INT	94	RCL 02
06	2	35	FS? 10	65	+	95	4
07	RCL 01	36	GTO 08	66	STO 02	96	/
08	y ↑ x	37	RCL 02	67	RCL 04	97	INT
09	1	38	2	68	7	98	2
10	−	39	RCL 01	69	*	99	MOD
11	INT	40	RCL 03	70	TONE IND X	100	RCL 02
12	T SEED?	41	−	71	RCL 04	101	8
13	PROMPT	42	y ↑ x	72	GTO 03	102	/
14	LBL 01	43	/	73	LBL 10	103	INT
15	x ≤ y?	44	INT	74	3481121110	104	2
16	GTO 02	45	2	75	RCL 01	105	MOD
17	2	46	MOD	76	1	106	+
18	/	47	LBL 04	77	−	107	RCL 02
19	INT	48	RCL 02	78	CHS	108	16
20	GTO 01	49	2	79	10 ↑ X	109	/
21	LBL 02	50	MOD	80	*	110	INT
22	x ≤ 0?	51	+	81	10	111	2
23	XEQ 10	52	2	82	/	112	MOD
24	STO 02	53	MOD	83	FRC	113	+
25	XEQ 10	54	STO 04	84	10	114	GTO 04
26	FIX 0	55	2	85	*	115	END
27	RCL 03	56	RCL 01	86	INT		
28	x = 0?	57	1	87	STO 03		
29	GTO 09	58	−	88	RTN		
		59	y ↑ x	89	LBL 09		

References
W. D. T. Davies, "System identification for self-adaptive control," John Wiley & Sons, New York, 1970.

Macroinstruction for Z80 guarantees valid output data

by Daniel Ozick
Irex Medical Systems, Ramsey, N. J.

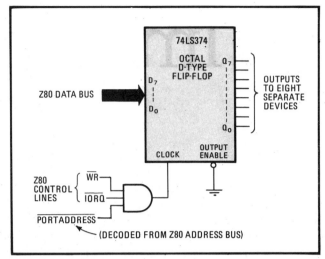

Output port. A standard 74LS-series octal latch is used as an output-only port in a Z80 microprocessor system. The Z80 instruction sequence guarantees that the output port is always in a valid state. The OUTI instruction latches data into the flip-flop.

Microprocessor-based systems often utilize 74LS-series octal latches as output-only ports (see figure). With this Z80 instruction sequence, a single output line can be made to change its state without affecting the states of the others. The program guarantees valid output data.

The standard practice has been to maintain an up-to-date image copy of the states of the output lines in the memory. When the state of an output line is to be changed, the image is fetched from the memory, updated with the desired bit change, transferred to the output, and stored back in the memory again. This procedure is adequate for systems without interrupts, but is unsafe when interrupt and background (noninterrupt) portions of a program share the same output port.

The conflict between interrupt and background can be avoided by using the safe macroinstruction shown in the table. In this sequence, the Z80 SET or RES instruction modifies the port image in memory in one uninterruptible instruction, ensuring that the port image is always valid. The output-and-increment (OUTI) instruction transfers the port image to the output port hardware in another indivisible instruction, ensuring that the image always matches the state of the port hardware. Thus, neither image nor port is left in an invalid state.

This sequence works well with Z80 systems, where interrupts cannot be disabled, and does not impose constraints on the assignment of devices to ports. □

SAFE PORT-OUTPUT MACROINSTRUCTION FOR THE Z80 MICROPROCESSOR	
Source statement	Comments
PortOut MACRO, PortAddress, BitNum, BitVal	; microsoft macro definition
	; input arguments are the address of the port (value ; from 0_{16} to $0FF_{16}$), the number of the bit to be ; modified, (values from 0 to 7), and the new bit ; value (0 or 1)
	; example of macro invocation using symbolic ; constants for port address and bit number: ; PortOut LightPort, ErrorLight, 1
PUSH BC PUSH HL	; save the registers to be used in the instruction ; sequence (optional)
LD HL, PortImage	; use HL as a pointer to the port image
IF BitVal EQ 0	; assemble the next line if the bit to be outputted ; has a value of 0
RES BitNum, (HL)	; in one indivisible instruction, reset the bit in the ; port image pointed to by HL
ELSE	; assemble the next line if the bit to be outputted ; has a value of 1
SET BitNum, (HL)	; in one indivisible instruction, set the bit in the ; port image pointed to by HL
ENDIF	; end of the conditional assembly
LD C, PortAddress OUTI	; (C) gets (HL), HL gets HL+1, B gets B−1 in one ; indivisible instruction, the port image held in ; memory is transferred to the port
POP HL POP BC	; restore the pushed registers ;
ENDM PortOut	; end of the PortOut macro definition

HP-67 calculates maximum nonlinearity error

by Kyong Park
Kavlico Corp., Chatsworth, Calif.

Although Hewlett-Packard Co. offers an HP-67 program for plotting a straight line through a random set of points using the least-squares method, the maximum nonlinearity error of the line must be laboriously calculated separately. This program eliminates this chore and, in addition, allows the user to find point values that will best fit into the set of input points rather than on the calculated line. As a result, the line's nonlinearity error is minimized.

PRINTER LISTING: HP-67 PROGRAM FOR CALCULATING NONLINEARITY ERROR OF TRANSDUCERS

Location	Key								
001	*LBL a	027	RCL (i)	054	X	081	R↓	108	GTO 5
002	CL REG	028	RCL 0	055	−	082	RC I	109	RCL 1
003	DSP 4	029	−	056	RCL 9	083	X<0	110	GSB 6
004	P⇌S	030	ABS	057	÷	084	GTO 2	111	STO 0
005	CL REG	031	P⇌S	058	STO A	085	R↓	112	RTN
006	0	032	STO 3	059	-X-	086	GTO 1	113	*LBL E
007	ST I	033	RCL 8	060	RCL B	087	RTN	114	GSB 6
008	P⇌S	034	RCL 4	061	-X-	088	*LBL 2	115	RTN
009	0	035	RCL 6	062	P⇌S	089	X<0	116	*LBL 6
010	RTN	036	X	063	RC I	090	R↓	117	STO E
011	*LBL A	037	RCL 9	064	1	091	P⇌S	118	RCL A
012	GSB 0	038	÷	065	+	092	RCL 3	119	RCL B
013	X⇌Y	039	−	066	ST I	093	P⇌S	120	RCL E
014	GSB 0	040	ENTER	067	GSB 5	094	÷	121	X
015	Σ+	041	ENTER	068	RCL C	095	1	122	+
016	RTN	042	RCL 4	069	ST I	096	0	123	RTN
017	*LBL 0	043	X²	070	*LBL 1	097	0	124	*LBL D
018	ST (i)	044	RCL 9	071	RCL (i)	098	X	125	STO E
019	ISZ	045	÷	072	ABS	099	RTN	126	RCL B
020	RTN	046	RCL 5	073	RC I	100	*LBL 5	127	1/X
021	*LBL C	047	X⇌Y	074	2	101	RCL (i)	128	RCL A
022	RC I	048	−	075	−	102	GSB 6	129	RCL E
023	2	049	÷	076	ST I	103	DSZ	130	X⇌Y
024	−	050	STO B	077	R↓	104	STO-(i)	131	−
025	STO C	051	RCL 6	078	RCL (i)	105	DSZ	132	X
026	ST I	052	RCL 4	079	ABS	106	RC I	133	RTN
		053	RCL B	080	X≤Y	107	X>0		

Instructions

- Key in program
- Initialize: f a
- Enter all P and V values:
  P_1 enter V_1, press A
  ⋮
  P_n enter V_n, press A for n ≤ 5
- Find a, b, and nonlinearity error in percent: press C
- Enter known P, press E for theoretical V
- Enter known V, press D for theoretical P

TEST DATA	
Absolute pressure, lb/in.²	Output, V dc
20.0000	3.313
30.0000	3.982
40.0000	4.655
50.0000	5.328
60.0000	6.005

CALCULATOR PROCEDURE		
Key entry	Display	Comments
f a	0.0000	
20 Enter	20.0000	
3.313 A	1.0000	
30 Enter	30.0000	
3.982 A	2.0000	
40 Enter	40.0000	
4.655 A	3.0000	
50 Enter	50.0000	
5.328 A	4.0000	
60 Enter	60.0000	
6.005 A	5.0000	
C	1.9646	a: intercept
	0.0673	b: slope
	0.0892	nonlinearity error in percent
45 E	4.9931	output expected at 45 lb/in.²
3.5 D	22.8143	pressure expected for 3.5 V dc output

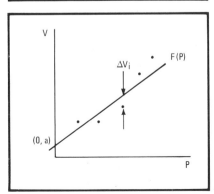

Nonlinear. The intercept (a) and the slope (b) of straight line f(P) are calculated with this HP-67 program, which minimizes the sum of the squares of the deviations (ΔV_i).

If a pair of measurements are linearly related by the equation $V = f(P)$, the deviation of point (P_i, V_i) on line $f(P)$ is defined as $\Delta V_i = V_i - f(P_i) = V_i - (a + bP_i)$, where a is the intercept on the V axis and b is the slope of the line (see figure). The values of a and b are chosen by minimizing the sum of the squares of ΔV_i. Thus the best fitting straight line will result.

The equation for summing the square of ΔV_i for n pairs of experimental points is:

$$\sum_{i=1}^{n}(\Delta V_i^2) = \sum_{i=1}^{n}V_i^2 + na^2 + b^2\sum_{i=1}^{n}P_i^2 - 2a\sum_{i=1}^{n}(V_i - bP_i) - 2b\sum_{i=1}^{n}P_iV_i$$

Appropriate values for a and b, which result in a minimum, are calculated by partially differentiating the above equation with respect to a and b, setting each equation equal to zero, and solving the simultaneous equations. As a result:

$$a = [\sum_{i=1}^{n}P_i\sum_{i=1}^{n}P_iV_i - \sum_{i=1}^{n}P_i^2\sum_{i=1}^{n}V_i] / [(\sum_{i=1}^{n}P_i)^2 - n\sum_{i=1}^{n}P_i^2]$$

and $b = [\sum_{i=1}^{n}P_i\sum_{i=1}^{n}V_i - n\sum_{i=1}^{n}(P_iV_i)] / [(\sum_{i=1}^{n}P_i)^2 - n\sum_{i=1}^{n}P_i^2]$

The maximum nonlinearity error in percent is now calculated from $[(\Delta V_i)_{max}/(V_n - V_i)] \times 100\%$. However, because of the limited register availability of the HP-67, the calculation is accurate for only up to five pairs of data entries. A numerical example (see table) illustrates the programming procedure. In addition, this program may be easily adapted to the HP-41C calculator, which can have higher values for n. □

Codec program compands samples for μ-law simulation

by Clive McCarthy
Northern Telecom Inc., Santa Clara, Calif.

Simulating the operation of a pulse-code-modulation encoder that conforms to the industry-standard μ-255 companding law, this TI-59 program will be useful to telecommunications engineers who design or test codec circuits. The 476-step program has been designed so that it is especially easy to use.

Given a sinusoidal input signal of any amplitude, phase, and frequency, the program finds the corresponding binary output data from the following standard approximation of a logarithmic compression curve:

$$F(x) = Sgn(x)\frac{\ln(1+\mu|x|)}{\ln(1+\mu)} \qquad 0 \leq |x| \leq 1$$

where x = the input signal, $Sgn(x)$ = the sign of the input signal, and $\mu = 255$ for the 15-segment curve, each of whose segments has 16 steps.

A sampling rate of 8 kilohertz is assumed for solving the equation. The corresponding output data, displayed as a logic 1 or 0, with the sign bit to the left of a decimal point and the chord and step bits to the right, may be examined byte by byte or, alternatively, as a continuous stream of data if the calculator's optional PC-100 printer is available.

As an example, consider the case where an input signal at 2,600 hertz having an amplitude of 0 dBmo (the zero reference digital-milliwatt level for the data channel) and a phase angle of 117° is to be sampled. Keying

TI-59 PRINTER LISTING: µ255 PULSE-CODE-MODULATION ENCODER

Location	Key																				
000	LBL	052	PRD	106	EXC	160	GTO	214	STO	268	GE	322	0	376	X	430	÷
001	A	053	06	107	OP	161	LST	215	19	269	|X|	323	0	377	RCL	431	2
002	STO	054	2	108	27	162	LBL	216	1	270	÷	324	=	378	04	432	0
003	01	055	INV	109	LBL	163	√x	217	1	271	2	325	+	379	÷	433	=
004	X	056	PRD	110	PRD	164	−	218	GTO	272	5	326	RCL	380	2	434	INV
005		057	09	111	−	165	1	219	LST	273	6	327	07	381	÷	435	LOG
006	0	058	GTO	112	1	166	2	220	LBL	274	=	328	=	382	2	436	X
007	4	059	DMS	113	=	167	8	221	RCL	275	STO	329	FIX	383	√x	437	8
008	5	060	LBL	114	GE	168	=	222	−	276	19	330	07	384	=	438	1
009	=	061	−	115	X⇌T	169	GE	223	1	277	CLR	331	RTN	385	INV	439	5
010	EXC	062	RCL	116	1	170	1/X	224	0	278	LBL	332	LBL	386	FIX	440	9
011	01	063	09	117	6	171	÷	225	2	279	LST	333	E	387	RTN	441	=
012	INV	064	INV	118	+/−	172	8	226	4	280	STO	334	OP	388	LBL	442	EXC
013	FIX	065	SUM	119	STO	173	LBL	227	=	281	20	335	25	389	C'	443	02
014	R/S	066	08	120	19	174	STO	228	GE	282	(	336	(	390	X	444	INV
015	LBL	067	GTO	121	1	175	19	229	SUM	283	0	337	RCL	391	2	445	FIX
016	E'	068		122	1	176	1	230	÷	284	0	338	05	392	√x	446	R/S
017	INT	069	LBL	123	1	177	0	231	6	285	0	339	INT	393	X	447	LBL
018	STO	070	=	124	GTO	178	1	232	4	286	1	340	X	394	2	448	C
019	00	071	RCL	125	LST	179	GTO	233	=	287	STO	341	RCL	395	÷	449	STO
020	INV	072	20	126	LBL	180	LST	234	STO	288	06	342	01	396	RCL	450	02
021	FIX	073	÷	127	X⇌T	181	LBL	235	19	289	4	343	)	397	04	451	X
022	R/S	074	1	128	−	182	1/X	236	1	290	STO	344	+	398	=	452	2
023	LBL	075	0	129	3	183	−	237	0	291	0	345	RCL	399	LOG	453	√x
024	A'	076	0	130	0	184	2	238	GTO	292	8	346	03	400	X	454	X
025	STO	077	0	131	=	185	5	239	LST	293	STO	347	=	401	2	455	1
026	03	078	=	132	GE	186	6	240	LBL	294	08	348	SIN	402	0	456	6
027	INV	079	+	133	X²	187	=	241	SUM	295	(	349	X	403	=	457	3
028	FIX	080	RCL	134	÷	188	GE	242	−	296	1	350	RCL	404	+	458	1
029	R/S	081	07	135	2	189	STO	243	2	297	÷	351	02	405	3	459	8
030	LBL	082	−	136	=	190	÷	244	0	298	9	352	=	406	.	460	÷
031	DMS	083	RCL	137	STO	191	1	245	4	299	0	353	NOP	407	1	461	RCL
032	RCL	084	06	138	19	192	6	246	8	300	0	354	SBR	408	7	462	04
033	08	085	=	139	1	193	=	247	=	301	0	355	IFF	409	=	463	=
034	EQ	086	FIX	140	1	194	STO	248	GE	302	)	356	PAU	410	INV	464	EXC
035	=	087	07	141	1	195	19	249	Y^x	303	SUM	357	PRT	411	FIX	465	02
036	GE	088	RTN	142	GTO	196	1	250	÷	304	07	358	DSZ	412	RTN	466	INV
037	−	089	LBL	143	LST	197	0	251	1	305	RCL	359	00	413	LBL	467	FIX
038	RCL	090	D	144	LBL	198	0	252	2	306	19	360	E	414	D'	468	R/S
039	09	091	RCL	145	X²	199	GTO	253	8	307	+	361	R/S	415	STO	469	LBL
040	SUM	092	02	146	−	200	LST	254	=	308	1	362	LBL	416	04	470	|X|
041	08	093	LBL	147	6	201	LBL	255	STO	309	6	363	B'	417	INV	471	RCL
042	RCL	094	IFF	148	4	202	STO	256	19	310	=	364	−	418	FIX	472	07
043	06	095	X⇌T	149	=	203	−	257	1	311	INT	365	3	419	R/S	473	FIX
044	INV	096	CLR	150	GE	204	5	258	GTO	312	X⇌T	366	.	420	LBL	474	07
045	SUM	097	STO	151	√x	205	1	259	LST	313	CLR	367	1	421	B	475	RTN
046	07	098	07	152	÷	206	2	260	LBL	314	INV	368	7	422	STO	476	0
047	LBL	099	X⇌T	153	4	207	=	261	Y^x	315	EQ	369	=	423	02		
048		100	GE	154	=	208	GE	262	−	316	DMS	370	÷	424	−		
049	1	101	EXC	155	STO	209	RCL	263	4	317	RCL	371	2	425	3		
050	0	102	+/−	156	19	210	÷	264	0	318	20	372	0	426	.		
051	INV	103	GTO	157	1	211	3	265	9	319	÷	373	=	427	1		
		104	PRD	158	1	212	2	266	6	320	1	374	INV	428	7		
		105	LBL	159	0	213	=	267	0	321	0	375	LOG	429	=		

Instructions

- Key in program
- Enter frequency of modulating signal to be encoded:
  (f), A
- Specify the signal's amplitude (dBmo or V_{rms}) and phase angle (°):
  (A), B (φ), A'
  If amplitude is expressed in V_{rms}, the peak-to-peak reference level must also be entered in register D'. Values of dBmo placed in register B' will be converted into V_{rms} values. Conversely, V_{rms} quantities placed in register C' will be converted into dBmo values.
- Calculate the digital value of each byte:
  Press E' for each byte (assuming no PC-100 printer). Press E', E to allow PC-100 to print the entire data steam automatically. PCM format has sign bit to the left of the decimal point and the chord and step bits to the right.
- Press D to find the peak value of the PCM input signal

Sine-wave coding

0.0001101
0.0110011
1.0001010
0.0010000
0.0100100
1.0001010
0.0010101
0.0011011
1.0001001
0.0011011
0.0010101
1.0001010
0.0100100
⋮

$f = 2{,}600$ Hz
$f_s = 8{,}000$ Hz
$\phi = 117°$

in the required data as specified in the instructions yields the data tabulated to the right of the instructions in the program listing. Note that with a sampling rate of 8 kHz, modulating frequencies should be kept below 4 kHz in order to meet the Nyquist criteria.

As seen by inspection, or alternately by pressing the D key, the peak positive value of the signal is 1.001001. Note that the chord-step information is expressed in inverted binary code, which is the industry-accepted standard. □

HP-41C calculator analyses resistive attenuators

by Albert E. Hayes Jr.
Albert Hayes & Associates, Fullerton, Calif.

Formulas for designing resistive attenuators (a) usually flow from the pages of standard textbooks. However, when these same pages are tapped for ways of calculating the attenuation of an such a circuit, given the exact values of its elements, only an informational desert is found. This HP-41C calculator program is designed to eliminate this problem.

The attenuation in decibels of a resistive T attenuator is:

$$20 \log \left[1 + \frac{R_3}{R_L} + \left(\frac{R_3}{R_2 R_L} + \frac{1}{R_2} + \frac{1}{R_L} \right)(R_S + R_1) \right]$$
$$- 20 \log \left(\frac{R_L + R_S}{R_L} \right)$$

For the pi configuration, the attenuation is:

$$20 \log \left\{ 1 + \frac{R_2}{R_3} + \frac{R_2}{R_L} + R_S \left[\frac{1}{R_1} + \frac{R_2}{R_1} \left(\frac{1}{R_3} + \frac{1}{R_L} \right) \right] \right.$$
$$\left. + R_S \left(\frac{1}{R_3} + \frac{1}{R_L} \right) \right\} - 20 \log \left(\frac{R_L + R_S}{R_L} \right)$$

The last term in each of the above equations represents the attenuation introduced by the voltage divider that is formed by the source and load resistance (R_S and R_L). This term is present even when the three-element attenuator is removed. A pi or T configuration is selected by pressing 1 or 0 respectively after initialization, as detailed in the instruction table. Selecting 0 at step 5 causes flag 1 to be set. This selection causes the program to automatically branch to the T subroutine at step 23

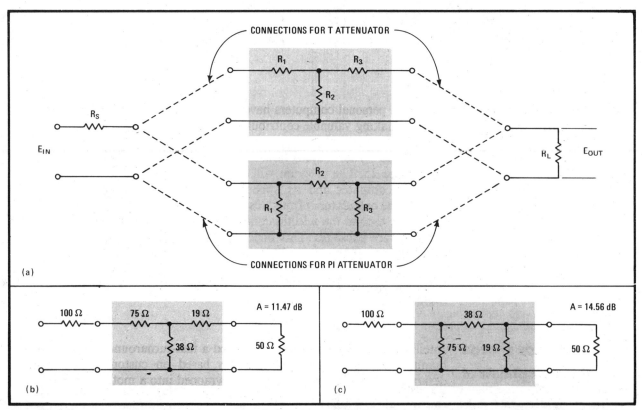

Attenuator. This program can calculate the attenuation of a T or pi network (a). R_S is the source and R_L the load resistance. Networks (b) and (c) show the usefulness of the program. The program calculates attenuation in decibels.

after the element values are loaded. For example, the attenuation of a T-configuration attenuator (b) is found by entering RTN, R/S, 0, R/S, 100, A, 75, B, 38, C, 19, D, 50, E. The attenuation value (11.47 dB) appears on the calculator's display.

Similarly for the pi network (c), which has the same element values, the computaton is the same as before except after initialization, 1 must be entered to call in the pi subroutine. The displayed attenuation value will be 14.56 dB. □

PRINTER LISTING: HP-41C PROGRAM TO COMPUTE THE ATTENUATION OF A PI-OR T-CONFIGURATION RESISTIVE ATTENUATOR

Location	Key											
01	*LBL -ATTEN-	18	STO 03	36	/	54	RCL 00	72	20	90	/
02	CLX	19	STOP	37	+	55	RCL 03	73	*	91	+
03	CF 01	20	*LBL E	38	RCL 00	56	/	74	RCL 06	92	RCL 02
04	STOP	21	STO 04	39	RCL 02	57	+	75	LOG	93	/
05	X = 0?	22	FS? 01	40	*	58	RCL 00	76	20	94	RCL 04
06	SF 01	23	GTO 01	41	RCL 01	59	RCL 04	77	*	95	1/X
07	STOP	24	1	42	/	60	/	78	−	96	+
08	*LBL A	25	ENTER↑	43	RCL 03	61	+	79	FIX 2	97	RCL 00
09	STO 00	26	RCL 02	44	/	62	*LBL 02	80	STOP	98	RCL 01
10	STOP	27	RCL 03	45	+	63	STO 05	81	*LBL 01	99	+
11	*LBL B	28	/	46	RCL 00	64	RCL 04	82	1	100	*
12	STO 01	29	+	47	RCL 02	65	RCL 00	83	RCL 03	101	+
13	STOP	30	RCL 02	48	*	66	+	84	RCL 04	102	GTO 02
14	*LBL C	31	RCL 04	49	RCL 04	67	RCL 04	85	/	103	END
15	STO 02	32	/	50	/	68	/	86	+		
16	STOP	33	+	51	RCL 01	69	STO 06	87	1		
17	*LBL D	34	RCL 00	52	/	70	RCL 05	88	RCL 03		
		35	RCL 01	53	+	71	LOG	89	RCL 04		

Instructions

- Key in program
- Initialize by pressing RTN, R/S
- Select network
    If T network is used, enter 0, then press R/S
    If PI network is used enter 1, then press R/S
- Enter appropriate circuit values in ohms
    For R_S, enter value and press A
    For R_1, enter value and press B
    For R_2, enter value and press C
    For R_3, enter value and press D
    For R_L, enter value and press E
- Read attenuation in decibels

TI-59's reverse-Polish routine simplifies complex arithmetic

by John Bunk
University of Pittsburgh, Johnstown, Pa.

Solving complex arithmetic equations by hand is both tedious and subject to errors. However, by using a system of reverse-Polish notation (RPN) similar to that used in the Hewlett-Packard series of calculators, this Texas Instruments-59 program performs this task easily. A six-register stack stores intermediate results automatically, and equations can therefore be solved with a minimum of keystrokes. The real and imaginary components of the equation are displayed separately.

The 224 step program is a collection of simple subroutines that are joined to perform the complex mathematical calculations. The subroutines, which perform all the mathematical operations required for the calculations, are the addition, subtraction, multiplication, division, exponentiation, and inverse functions. The functions corresponding to their subroutine names are listed in the table.

Calculators using the RPN system require a specific memory-stack structure. Each stack register occupies two data registers that store the real and imaginary parts of any input number. The arithmetic is solved with the aid of both the A and Z registers, which contain the data that is keyed into the stack by means of the SBR =

TI-59's OPERATIONAL ROUTINES FOR COMPLEX MATH

Subroutine	Function
SBR +	A + Z
SBR −	A − Z
SBR ×	A × Z
SBR ÷	A ÷ Z
SBR Y^x	A^Z
SBR =	enter display/t-register into stack
SBR CLR	clear stack
SBR lnx	e^Z
SBR 1/X	1/Z

TI-59 PRINTER LISTING: SIMPLIFYING COMPLEX MATH USING REVERSE-POLISH NOTATION

Location	Key
000	RCL
001	20
002	STO
003	01
004	RCL
005	21
006	STO
007	02
008	RCL
009	18
010	STO
011	03
012	RCL
013	19
014	STO
015	04
016	CP
017	STF
018	01
019	RTN
020	STO
021	20
022	X⇌T
023	STO
024	21
025	NOP
026	NOP
027	RCL
028	29
029	EXC
030	27
031	EXC
032	25
033	EXC
034	23
035	EXC
036	21
037	STO
038	19
039	X⇌T
040	NOP
041	NOP
042	RCL
043	28
044	EXC
045	26
046	EXC
047	24
048	EXC
049	22
050	EXC
051	20
052	STO
053	18
054	X⇌T
055	RTN
056	LBL
057	=
058	EXC
059	19
060	EXC
061	21
062	EXC
063	23
064	EXC
065	25
066	EXC
067	27
068	STO
069	29
070	NOP
071	NOP
072	X⇌T
073	EXC
074	18
075	EXC
076	20
077	EXC
078	22
079	EXC
080	24
081	EXC
082	26
083	STO
084	28
085	NOP
086	NOP
087	RCL
088	18
089	X⇌T
090	RCL
091	19
092	RTN
093	LBL
094	CLR
095	1
096	9
097	STO
098	18
099	0
100	ST*
101	18
102	RCL
103	18
104	X⇌T
105	2
106	9
107	EQ
108	01
109	16
110	1
111	SUM
112	18
113	GTO
114	00
115	99
116	0
117	STO
118	18
119	RTN
120	RCL
121	19
122	STO
123	02
124	X⇌T
125	RCL
126	18
127	STO
128	01
129	STF
130	01
131	CP
132	RTN
133	LBL
134	+
135	SBR
136	00
137	00
138	PGM
139	04
140	B
141	SBR
142	00
143	20
144	RTN
145	LBL
146	−
147	SBR
148	00
149	00
150	PGM
151	04
152	B'
153	SBR
154	00
155	20
156	RTN
157	LBL
158	×
159	SBR
160	00
161	00
162	PGM
163	04
164	C
165	SBR
166	00
167	20
168	RTN
169	LBL
170	÷
171	SBR
172	00
173	00
174	PGM
175	04
176	C'
177	SBR
178	00
179	20
180	RTN
181	LBL
182	Y^x
183	SBR
184	00
185	00
186	PGM
187	04
188	E'
189	PGM
190	04
191	D
192	SBR
193	00
194	20
195	RTN
196	LBL
197	LNX
198	SBR
199	01
200	20
201	PGM
202	05
203	B'
204	STO
205	18
206	X⇌T
207	STO
208	19
209	RTN
210	LBL
211	1/X
212	SBR
213	01
214	20
215	PGM
216	05
217	E
218	STO
219	18
220	X⇌T
221	STO
222	19
223	RTN

Instructions

- Key in program
- Given the algebraic expression, key in the variables corresponding to their real and imaginary magnitudes and their mathematical operations

    Place all real quantities in the t register

    Enter all imaginary quantities into the stack by means of the SBR = function

    If multiplication or division of two terms is necessary, press SBR × or SBR ÷ keys after entering the last term

    If addition or subtraction is desired, press the SBR + or SBR − keys after entering the last term

    If quantities must be multiplied exponentially, press either the y^x or lnx key, depending on the application

    If the inverse of an algebraic quantity is required, press the 1/x key

    Stack may be cleared by pressing the SBR CLR keys

- Imaginary quantities that result from a given algebraic operation will be displayed directly. Real quantities may be recovered from the t register (press the x ⇌ t key)

operation. After the data is entered, the complex equation is solved by pressing the appropriate function keys. In this program all stack entering is manual.

As an example of the program's usefulness, consider $x + jy = (2 + j2)(3 + j5) - (9 - j15) + (1.4 + j3)$. The solution to this equation may be found by entering: 2, x ⇌ t, 2, SBR =, 3, x ⇌ t, 5, SBR =, SBR ×, 9, x ⇌ t, 15, ±, SBR =, SBR −, 1.4, x ⇌ t, 3, SBR =, SBR +. The imaginary part of the result, y = 34, appears on the display. The real part of the result, x = −11.6, may be found by pressing the x ⇌ t key. □

25. SOFTWARE/COMPUTERS

Interface program links a-d chip with microprocessor

by M. F. Smith
Reading University, Reading, England

When combined with a peripheral interface adapter, this microprocessor program eases interfacing an ICL7135 analog-to-digital converter with the MC6800 family of microprocessors. Through placing the 16-bit MC6821 interface chip between the a-d converter and processor interface, all digital data and control signals may be funneled through the parallel interface.

The abbreviated program does not check for polarity or input range. In this routine, the digit strobe of U_1 is used to interrupt the microprocessor through the CB_1 input to U_2. Binary-coded–decimal data from the a-d converter is stored in a temporary buffer, labeled BUFFER TEMP, and the digit status in the buffer is determined by digit output D_1 through D_5 of U_1 (see figure). When the last digit D_1 is sent out, the temporary buffer is moved to a final buffer labeled BUFFER-ADC. Subroutine ADC-INIT initializes MC6821 and allows interrupts before data can be sent from the a-d converter.

This software allows the ICL7135 chip to operate at full speed and without any possibility of misreading the multiplexed digits. The software may be synchronized with the converter, and the final buffer may be provided by a software flag, instead of a hardware strobe. □

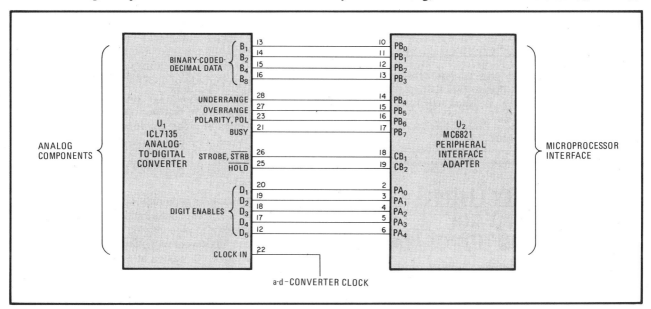

Microprocessor interface. Intersil's analog-to-digital converter ICL7135 is linked to Motorola's 6800 microprocessor family with peripheral interface adapter MC6821 and the associated software. This method funnels digital data and control signals through the interface and eliminates the need for additional circuitry. The digit strobe of U_1 interrupts the processor through CB_1.

\	\	\	INTERFACE PROGRAM LINKING 6800 MICROPROCESSOR WITH ICL7135 ANALOG-TO-DIGITAL CONVERTER	\	\
Location	Object code	Line	Source statement		Comments
		1	"6800"		
		2			
		3			; ICL7135 4-1/2 DIGIT a-d CONVERTER
		4			
		5			
	(8100)	6	PIA_DIGIT	EQU 8100H	; SIDE FOR DIGIT NUMBER
	(8102)	7	PIA_BCD	EQU 8102H	; SIDE FOR BINARY-CODED-DECIMAL DATA ERRORS
	(8103)	8	ADC_STROBE	EQU 8103H	; STROBE FOR a-d CONVERTER
		9			
		10			

Push-pop program aids 6800's register swapping

by P. R. Apte
Tata Institute of Fundamental Research, Bombay, India

Unfortunately, the popular 6800 microprocessor lacks the push and pull stack instructions that its counterpart, the 8080, uses so efficiently to exchange index registers and accumulators. With the addition of a simple routine and appropriate hardware for modifying the system's nonmaskable interrupt service routine, however, implementation of these functions is easy.

In the unmodified system, it is necessary to store the index register in memory in order to access two or more tables in a program, because the 6800's code is nonreentrant and thus is inefficient for register swapping. This program and three chips coordinate the data transfer much as is done in the 8080, with little complexity and at low cost.

As seen from the program and with the aid of the figure, the routine first checks for the occurrence of an active interrupt request on lines IRQA and IRQB by polling ports CRA and CRB of the 6820 peripheral interface adapter. The PSHX and PULX routines are called in accordance; otherwise, the system's usual nonmaskable interrupt routine is executed.

The TST codes are used for ordering the PSHX and PULX routines, with addresses D_1 and D_2 on the 6800 bus being decoded and presented to the 6820. Upon the arrival of a pulse at CA_1, the interrupt bit in the control and status register is set, making IRQA low. The timing diagram shows how this signal combines with the system-generated NMI output so as to interrupt the microprocessor.

Flip-flop A_3 is set by the NMI system input. One-shot A_2 generates a negative pulse when either IRQA or IRQB goes low. As one-shot A_1 is triggered only when there is no IRQ signal from the 6820, A_1 is not triggered during the execution of the PSHX and PULX codes, making these routines noninterruptible.

The PSHX and PULX routines take the IRQA or IRQB signal high just before returning from the service routine. The maximum delay for responding to a system NMI signal is 158 clock cycles. The PSHX and PULX require 150 and 158 cycles, respectively. However, it is possible to reduce this time to 98 cycles by allowing the system's condition codes to be modified. □

REENTRANT ROUTINE FOR 6800 MICROPROCESSOR

Label	Source statement		Comments
NMI	LDAA	CRB	
	BMI	PSHX	Test PSHX interrupt
	LDAA	CRA	
	BMI	PULX	Test PULX interrupt
	BRA	RNMI	
PSHX	DES		Push X register on stack, shift
	DES		present 7-byte status up stack by 2 bytes
	TSX		X register = SP + 1
	LDAA	2,X	
	STAA	0,X	Move CCR
	LDAA	3,X	
	STAA	1,X	Move ACC-B
	LDAA	4,X	
	STAA	2,X	Move ACC-A
	LDAA	5,X	
	STAA	3,X	Move X-H
	LDAA	6,X	
	STAA	4,X	Move X-L
	LDAA	7,X	
	STAA	5,X	Move PC-H
	LDAA	8,X	
	STAA	6,X	Move PC-L
	LDAA	3,X	Put X register on stack
	STAA	7,X	in the 2 empty byte locations
	LDAA	4,X	
	STAA	8,X	Move X register in place
	TST	DRA	Remove IRQA
	RTI		Return from PSHX
PULX	TSX		Pull X register from stack
	LDAA	7,X	transfer from stack to status on stack
	STAA	3,X	
	LDAA	8,X	
	STAA	4,X	Shift X register
	LDAA	6,X	Move 7-byte status down stack 2 bytes
	STAA	8,X	Move PC-L
	LDAA	5,X	
	STAA	7,X	Move PC-H
	LDAA	4,X	
	STAA	6,X	Move X-L
	LDAA	3,X	
	STAA	5,X	Move X-H
	LDAA	2,X	
	STAA	4,X	Move ACC-A
	LDAA	1,X	
	STAA	3,X	Move ACC-B
	LDAA	0,X	
	STAA	2,X	Move CC-REG
	INS		
	INS		New SP
	TST	DRB	Remove IRQB
	RTI		Return (regular NMI routine following)

Exchange. Hardware and software for implementing push and pop functions on data registers of 6800 microprocessor is simple. Timing diagram (figure) details logic sequences required to generate interrupts. System, which generates the reentrant routine, coordinates data transfers efficiently. Table shows locations of stack pointer before and after push-pop operations.

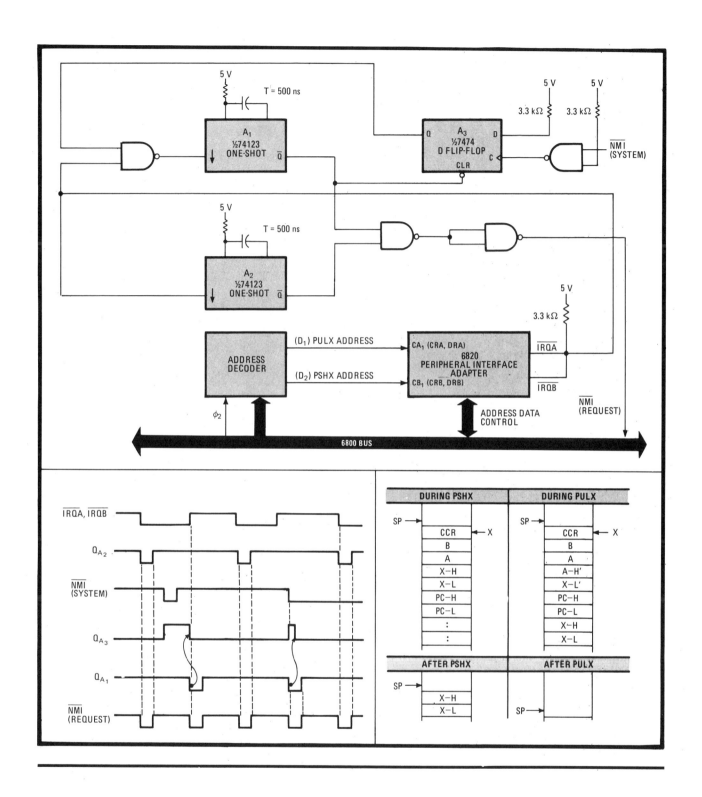

Nystrom integration gives dynamic system's response

by David Eagle
Lear-Siegler Inc., Grand Rapids, Mich.

The response of many dynamic systems, described by second-order vector differential equations that are written in the form $\ddot{\vec{X}} = \vec{f}(\vec{X}, \dot{\vec{X}}, t)$, are determined by numerical methods—a popular approach is the fourth-order Runge-Kutta technique. However, this program uses the fourth-order Nystrom-integration method to solve systems of second-order differential equations and can be a faster solution for some dynamic systems.

An integration subroutine beginning at line 1,000 and a driver routine that shows how the program works are the software's main components. In addition, the derivatives are evaluated in the subroutine on line 3,000. The program example demonstrates the usefulness of Nystrom integration and integrates the single second-order differential equation defined by: $\ddot{X}(X,t) = 0.5X^2(1+(1+t)^2 X)$ where $X(0) = 1$ and $\dot{X}(0) = 0.5$ are the initial conditions.

The above differential equation has the exact solution $X(X,t) = 4/(4-2t-t^2)$ and $\dot{X}(X,t) = 0.5(1+t)X^2$. The example prints the time and values for X and $\dot{X}$, and exact solutions for X and $\dot{X}$ are also included for comparison with the Nystrom-integration method. □

BASIC SUBROUTINE FOR INTEGRATING SECOND-ORDER VECTOR DIFFERENTIAL EQUATIONS

```
1000    * FOURTH-ORDER NYSTROM-INTEGRATION SUBROUTINE
1010    LET T2 = T1
1020    FOR I = 1 TO N1
1030    LET X2(I) = X1(I)
1040    LET V2(I) = V1(I)
1050    NEXT I
1060    GOSUB 3010
1070    *
1080    LET T2 = T1 + A2 * S1
1090    FOR I = 1 TO N1
1100    LET D1(I) = S1 * E1(I)
1110    LET X2(I) = X1(I) + S1 * (A2 * V1(I) + A1 * D1(I))
1120    LET V2(I) = V1(I) + A2 * D1(I)
1130    NEXT I
1140    GOSUB 3010
1150    *
1160    LET T2 = T1 + B2 * S1
1170    FOR I = 1 TO N1
1180    LET D2(I) = S1 * E1(I)
1190    LET X2(I) = X1(I) + S1 * (B2 * V1(I) + A6 * D1(I) + A7 * D2(I))
1200    LET V2(I) = V1(I) - A8 * D1(I) + A9 * D2(I)
1210    NEXT I
1220    GOSUB 3010
1230    *
1240    LET T2 = T1 + S1
1250    FOR I = 1 TO N1
1260    LET D3(I) = S1 * E1(I)
1270    LET X2(I) = X1(I) + S1 * (V1(I) + B3 * D1(I) + B4 * D2(I) + B5 * D3(I))
1280    LET V2(I) = V1(I) + B6 * D1(I) - B7 * D2(I) + B8 * D3(I)
1290    NEXT I
1300    GOSUB 3010
1310    *
1320    LET T1 = T1 + S1
1330    FOR I = 1 TO N1
1340    LET D4(I) = S1 * E1(I)
1350    LET X1(I) = X1(I) + S1 * (V1(I) + A3 * D1(I) + A4 * D2(I) + A5 * D3(I))
1360    LET V1(I) = V1(I) + A3 * (D1(I) + D4(I)) + B1 * (D2(I) + D3(I))
1370    NEXT I
1380    RETURN
1390    *
2000    * INTEGRATOR-COEFFICIENTS SUBROUTINE
2010    LET A1 = .045: A2 = .3: A3 = 13/126: A4 = 5/18: A5 = 5/42: A6 = 7/600
2020    LET A7 = 7/30: A8 = 7/15: A9 = 7/6: B1 = 25/63: B2 = .7: B3 = 19/78
2030    LET B4 = 35/312: B5 = 15/104: B6 = 64/39: B7 = 70/39: B8 = 15/13
2040    RETURN
2050    *
3000    * DIFFERENTIAL-EQUATION SUBROUTINE
3010    LET A = (1 + T2) * (1 + T2)
3020    LET B = X2(1) * X2(1)
3030    LET E1(1) = .5 * B * (1 + A * X2(1))
3040    RETURN
```

PROGRAM COMPARING NYSTROM-INTEGRATION METHOD WITH EXACT SOLUTION

```
0010    * PROGRAM "NYM4"
0020    *
0030    * A DEMONSTRATION OF THE FOURTH-ORDER NYSTROM METHOD
0040    * FOR INTEGRATING SYSTEMS OF SECOND-ORDER
0050    * VECTOR DIFFERENTIAL EQUATIONS
0060    *
0070    * SET PRINTER LINE LENGTH
0080    LINE = 110
0090    * DEFINE SIZE OF SYSTEM AND DIMENSION ARRAYS
0100    LET N1 = 1
0110    DIM X1(N1), X2(N1), V1(N1), V2(N1), E1(N1), D1(N1)
0120    *
0130    * COMPUTE INTEGRATOR COEFFICIENTS
0140    GOSUB 2010
0150    *
0160    * INTEGRATION STEP SIZE
0170    LET S1 = 0.05
0180    * INITIAL TIME
0190    LET T1 = 0.0
0200    * INITIAL X
0210    LET X1(1) = 1.0
0220    * INITIAL X DOT
0230    LET V1(1) = 0.5
0240    * FINAL TIME
0250    LET T2 = 0.5
0260    *
0270    * CLEAR SCREEN AND PRINT HEADER
0280    HOME :PRINT TAB(5); "PROGRAM NYM4" :PRINT
0290    PRINT TAB(5); "TIME", "COMPUTED", "COMPUTED", "EXACT", "EXACT"
0300    PRINT TAB(3); "(SECONDS)"; TAB(20); "X"; TAB(35); "X DOT";
0310    PRINT TAB(51); "X"; TAB(65); "X DOT"
0320    *
0330    * INTEGRATE SYSTEM OF EQUATIONS
0340    FOR I1 = 1 TO T2/S1
0350    GOSUB 1010
0360    *
0370    * COMPUTE EXACT SOLUTION
0380    LET X3 = 4/(4-2*T1-T1*T1)
0390    LET V3 = .5*(1+T1)*X1(1)*X1(1)
0400    *
0410    * PRINT RESULTS
0420    PRINT :PRINT TAB(5); T1, X1(1), V1(1), X3, V3
0430    *
0440    NEXT I1
0450    END
0460    *
```

Time (s)	Computed X	Computed Ẋ	Exact X	Exact Ẋ
0.05	1.0262989	0.552976961	1.0262989	0.552976949
0.1	1.05540895	0.612638448	1.05540897	0.612638426
0.15	1.08769542	0.680271785	1.08769544	0.68027176
0.2	1.12359547	0.757480102	1.1235955	0.757480066
0.25	1.16363632	0.846280972	1.16363636	0.846280925
0.3	1.20845916	0.94924286	1.20845921	0.949242799
0.35	1.25885122	1.06967691	1.25885129	1.06967681
0.4	1.31578937	1.2119113	1.31578947	1.21191116

Pocket computer solves for LC resonance using Basic

by Cass R. Lewart
System Development Corp., Eatontown, N.J.

The pocket computers that were recently introduced by several manufacturers represent a new breed of calculator. A unit manufactured by Sharp Corp. and distributed by Radio Shack (catalog No. 26-3501), for example, executes a comprehensive set of Basic statements, thus allowing easy writing, editing, and debugging of programs. Here, a coil-resonance program illustrates some of the computer's strong and weak points as compared with an advanced programmable calculator.

To make programming it easier, the pocket computer has alphanumeric prompts, alphanumeric variables, and dimensioned variables. Programming in Basic also has the inherent advantage over reverse Polish notation (RPN) or arithmetic operating-system (AOS) notation of keeping the display register free—the display is not explicitly used by the computer for computation (except in the INPUT statement). Thus, a running program can be interrupted at any point, its variables reviewed or modified, and the program then restarted without affecting the computation process.

The program evaluates a resonant circuit having a lossless, single-layer airwound coil and a capacitor. The circuit is fully described by two equations—an approxi-

COIL RESONANCE PROGRAM

```
  5: "S"CLEAR: USING"##.## ^": PRINT "COIL RESONANCE PROGRAM"
 10: "Z" Z=1: PAUSE "COMPUTE": END
 15: "C" AREAD C: IF Z=0 GOTO 20
 17: Z=0: C=1/(2πF)^2/L
 20: PRINT "CAPACITANCE=";C
 25: "L" AREAD L: IF Z=0 GOTO 30
 27: Z=0: L=1/(2πF)^2/C
 30: PRINT "INDUCTANCE=";L
 35: "F" AREAD F: IF Z=0 GOTO 40
 37: Z=0: F=1/2π/√(LC)
 40: PRINT "FREQUENCY=";F
 45: "D" AREAD D: IF Z=0 GOTO 50
 47: Z=0: D=(9L+√(81LL+40LANN* EXP-6))/NN* EXP6
 50: PRINT "DIAMETER=";D
 55: "N" AREAD N: IF Z=0 GOTO 60
 57: Z=0: GOSUB 100: N=√(XL)/D
 60: PRINT "# TURNS=";N
 65: "A" AREAD A: IF Z=0 GOTO 70
 67: Z=0: A=((DN* EXP-3)^2-18DL)/40L
 70: PRINT "LENGTH=";A
 75: "X" Z=0: GOSUB 100
 77: L=NNDD/X: GOTO 30
100: X=(18D+40A)* EXP6: RETURN
```

	Shift S	clear all variables, start computation
⟨value⟩	Shift F	enter frequency in hertz
Shift Z	Shift F	compute F from C and L
⟨value⟩	Shift C	enter capacitance in farads
Shift Z	Shift C	compute C from F and L
⟨value⟩	Shift L	enter coil inductance in henrys
Shift Z	Shift L	compute L from F and C
Shift Z	Shift X	compute L from A, D, and N
⟨value⟩	Shift A	enter coil length in inches
Shift Z	Shift A	compute A from L, D, and N
⟨value⟩	Shift D	enter coil diameter in inches
Shift Z	Shift D	compute D from L, A, and N
⟨value⟩	Shift N	enter number of turns
Shift Z	Shift N	compute N from L, A, and D

To review value of any parameter, key its name followed by ENTER; for example, to find the currently stored frequency, press F, then ENTER.

LC RESONANCE EXAMPLES

A single-layer coil 1.5 in. long, 0.6 in. in diameter, with 14 turns, resonates with an unknown capacitor at 10 MHz. What parallel capacitor value is required to lower the resonant frequency to 9 MHz?

Key entry	Display	Remarks
Shift S	COIL RESONANCE PROGRAM	start, clear variables
Exp 7 Shift F	FREQUENCY= 1.00 E07	enter frequency
1.5 Shift A	LENGTH= 1.50 E00	enter coil length
0.6 Shift D	DIAMETER= 6.00 E−01	enter coil diameter
14 Shift N	# TURNS = 1.40 E01	enter number of turns
Shift Z Shift X	COMPUTE INDUCTANCE= 9.96 E−07	find L from A, D, and N
Shift Z Shift C	COMPUTE CAPACITANCE= 2.54 E−10	find C from F and L
T=C Enter	2.54 . . . E−10	save C temporarily
9 Exp 6 Shift F	FREQUENCY= 9.00 E06	enter new frequency
Shift Z Shift C	COMPUTE CAPACITANCE= 3.13 E−10	find new capacitor
C−T Enter	5.96 . . . E−11	an extra 60 pF is required

Instead of adding a capacitor, how many turns could be added to the coil?

Key entry	Display	Remarks
C=T Enter	2.54 . . . E−10	restore previous C
T=N Enter	14.	save old N temporarily
Shift Z Shift L	COMPUTE INDUCTANCE= 1.23 E−06	find new inductance
Shift Z Shift N	COMPUTE # TURNS= 1.55 E01	find number of turns equals 15.5
N−T Enter	1.55	add 1.5 turns

mation formula for the coil inductance together with the resonance equation of a resonant parallel or series LC circuit: $L = [(ND)^2 \times 10^{-6}]/(18D + 40A)$ and $(2\pi F)^2 LC = 1$, where F is the frequency in hertz, C is the capacitance in farads, L is the coil inductance in henrys, A is the coil length in inches, D is the coil diameter in inches, and N is the number of turns.

When running the program, the user either stores a parameter value or computes it by means of an appropriate linear or quadratic equation that can be derived from the above formulas.

For example, once the program on page 189 has been listed, the computer should be placed into the define (DEF) mode. The parameters F, C, L, A, D, or N may then be entered, computed, or reviewed in any order by pressing the appropriate keys, as designated in the table's lower portion. Then to review the value of any parameter, its name is keyed and followed by an enter statement.

Typical examples are illustrated above. Suppose a single-layer 14-turn coil 1.5 in. long and with a diameter of 0.6 in. resonates at 10 megahertz when paired with a capacitor of unknown value, and it is desired to lower that resonance to 9 MHz by adding in parallel another capacitor. By following the steps in the first of the examples, the program will indicate an additional 60 picofarads is required. Alternatively, the problem could have been solved by adding turns to the coil. The second of the examples indicates the coil would require an additional 1.5 turns.

This program was originally prepared for the Hewlett-Packard HP-67 calculator. For the unit, a 112-step program occupied half of the calculator's program memory and took many hours to write. Though it is simple to use, the calculator program is difficult to modify and also difficult to follow.

In contrast, the Basic program took less than an hour to write, occupies only 40% of a pocket computer's program memory, has good prompts, and is easy to follow and modify for other personal computers that can be programmed in Basic. Because of this and other programs, the Basic calculator has about 1.2 to 2 times the program capacity of the HP-67, but executes programs at speeds that are 20% slower.

The weak points of the hand-held Basic computer, as compared with a desktop or larger computer, are its lack of string-manipulation commands and DATA statements, which often lead to long and repetitive programs. However, it is possible to enter repetitive statements quickly into program memories by assigning a string of program text in the reserve mode to a single key.

Finally, although a hand-held computer can store programs and data on an external cassette recorder, the process is somewhat more inconvenient than the use of the magnetic strips of calculators. However, the unit's complementary-MOS memory retains programs and data even when the computer is turned off. □

Bibliography
C. Lewart, "Coil Resonance Program for HP67/97," Microwaves, December 1978, pp. 86–88.

TRS-80 program helps to load cassette data twice as fast

by H. Lee
Dover Heights, New South Wales, Australia

Radio Shack's TRS-80 level 2 personal computer is becoming increasingly popular for simple control tasks such as monitoring instruments that measure various physical quantities. Unfortunately, however, users who lack the expensive disk operating system must store and call data on or from the standard cassette tape source, and the machine's 500-bit-per-second read and write speeds are often too slow to satisfy the average engineer, especially if the program is 12-K bytes long or longer. But the short program shown here allows machine-language software to run on the computer at twice the speed (1,000 b/s) without any change having to be made to the hardware.

The increase in speed is achieved by altering the machine's standard read and write formats. The original

TRS-80 PRINTER LISTING FOR FAST TAPE LOADING

Location	Label	Mnemonic		Comments
00010	START	EQU	5000H	; start of program, may be changed
00020	NOBYTS	EQU	2000H	; no. of bytes in user's program
00030		ORG	7F60H	; location of 1000 bauds drivers
00040	FASWR	LD	HL,START	; start of user's program
00050		LD	DE,NOBYTS	; no. of bytes to write
00060		XOR	A	; zero accumulator
00070		CALL	212H	; switch on tape recorder
00080		CALL	287H	; write out tape leader
00090	W1	LD	A,(HL)	; put byte into accumulator
00100		CALL	WRITE	; write byte at 1000 bauds to tape
00110		INC	HL	; next byte
00120		DEC	DE	; decrease count
00130		LD	A,E	; check if completed
00140		OR	D	
00150		JR	NZ,W1	; all bytes completed ?
00160		CALL	1F8H	; yes, switch off tape recorder
00170		JP	6CCH	; go to Basic interpreter
00180	;			
00190	; WRITE ROUTINE			
00200	;			
00210	WRITE	PUSH	HL	; save registers HL
00220		PUSH	DE	; and DE
00230		LD	C,9	; transfer 9 bits (1 synch and 8 bits)
00240		LD	D,A	; save byte in register D
00250	W2	CALL	PULSE	; write pulse on tape
00260	W3	DEC	C	; decrease count
00270		JR	Z,W4	; all 9 bits done ?
00280		LD	A,D	; restore data byte
00290		RLCA		; rotate data byte
00300		LD	D,A	; save accumulator
00310		JR	C,W2	; if carry then it's 1
00320		LD	B,54H	; delay here for 0 bit
00330		DJNZ	$	
00340		JR	W3	; return for more bits
00350	W4	POP	DE	; restore registers DE
00360		POP	HL	; and HL
00370		RET		; subroutine ends
00380				
00390	PULSE SUBROUTINE			
00400				
00410	PULSE	PUSH	AF	; save registers AF
00420		LD	HL,0FC01H	; set a positive pulse to tape
00430		CALL	221H	; output to tape
00440		LD	B,5	
00450		DJNZ	$	; for about 50 microseconds
00460		LD	HL,0FC02H	; set a negative pulse to tape
00470		CALL	221H	; output to tape
00480		LD	B,5	
00490		DJNZ	$	; for about 50 microseconds
00500		LD	HL,0FC00H	; set tape output to zero level

system inserts 8 sync-reference bits for every byte of data. Here, only 1 sync bit is used for each byte. The bit time has also been shortened slightly. These fast tape drivers have worked reliably with the original tape recorder supplied with the machine, provided Radio Shack's XRX-111 hardware fix for tape loading has been implemented.

Users need only specify the program with the starting memory address of interest and the length of the routine they wish to convert. Providing a jump-to-execute address is optional, but the default address will initiate a return to the Basic interpreter. As for the drivers, they may be either loaded using a monitor program such as T-BUG or entered using the poke command. The drivers are location-independent so they may be loaded anywhere in free memory.

The tape can be saved or loaded using a jump to FASWR or FASRD, respectively. It would also be advantageous to note the location of the software driver for the fast read (FASRD). This would facilitate reliable power startup. □

```
00510            CALL    221H
00520            LD      B,32H                ; delay after each pulse
00530            DJNZ    $
00540            POP     AF                   ; restore A
00550            RET
00560    ;
00570    ; 1000 BAUDS LOAD DRIVER
00580    ;
00590    FASRD   LD      HL,START             ; start of program
00600            LD      DE,NOBYTS            ; no. of bytes
00610            XOR     A                    ; zero accumulator
00620            CALL    212H                 ; switch on tape recorder
00630            CALL    296H                 ; look for tape leader
00640    R1      CALL    READ                 ; read 1 byte from tape
00650            LD      (HL),A               ; store it in memory
00660            INC     HL                   ; point to next memory
00670            DEC     DE                   ; decrease count
00680            LD      A,E
00690            OR      D
00700            JR      NZ,R1                ; continue if not completed
00710            JP      6CCH                 ; go to Basic interpreter
00720    ;
00730    ; READ SUBROUTINE
00740    ;
00750    READ    PUSH    HL                   ; save registers HL

00760    R2      IN      A,(0FFH)             ; get byte from tape
00770            RLA                          ; rotate until synch bit appears
00780            JR      NC,R2                ; synch bit ?
00790            LD      B,0AH                ; delay
00800            DJNZ    $
00810            CALL    21EH                 ; reset input flip-flop
00820            LD      B,3AH                ; delay after first bit
00830            DJNZ    $
00840            LD      C,8                  ; read remaining 8 bits
00850    R3      PUSH    AF                   ; save accumulator
00860            LD      B,21H                ; delay for next byte
00870            DJNZ    $
00880            IN      A,(0FFH)             ; read next byte from tape
00890            LD      B,A                  ; save bit in register B
00900            CALL    21EH                 ; reset input flip-flop
00910            POP     AF                   ; restore byte
00920            RL      B                    ; add bit from carry
00930            RLA
00940            LD      B,2AH                ; delay before next byte
00950            DJNZ    $
00960            DEC     C                    ; decrease bit count
00970            JR      NZ,R3                ; again if not completed
00980            POP     HL                   ; restore registers HL
00990            RET
01000            END
```

TRS-80 program simplifies design of PROM decoders

by Gideon Gimlan
Loral Electronic Systems, The Bronx, N. Y.

Coding data into programmable read-only memories is tedious when done by hand. This TRS-80 Basic program takes the drudgery out of the effort. Although performing a myriad of tasks, its major advantages are simplifying the design of PROM decoders by providing documentation of the design, simulating sequential-state machines, and preparing the data for immediate PROM burn-in.

To understand the program and its relation to character-generation redundancy or the sequential-state machine on which this program is based, consider the typical video-display circuit shown at the right. On it, the address lines are designated as inputs and the data lines form the outputs. Each input combination has a unique output, but in general practice, redundancy occurs frequently because of the so-called "don't care" condition.

In this example, a 5-bit ASCII code of the character to be displayed is placed on the least significant address lines of the PROM, as shown. The 3 most significant bits are connected to a row counter that increments from 0 to

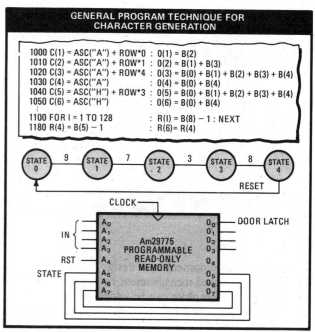

Basic burn-in. Writing data into a PROM with TRS-80 Basic, as grasped with the aid of a video-display scheme and the general programming technique (top), is relatively simple for state machines because of their inherent character-generation redundancy properties. The technique is easily extended to practical programs and the one shown (immediately above) is a combination lock that opens when the sequence 9-7-3-8 is entered.

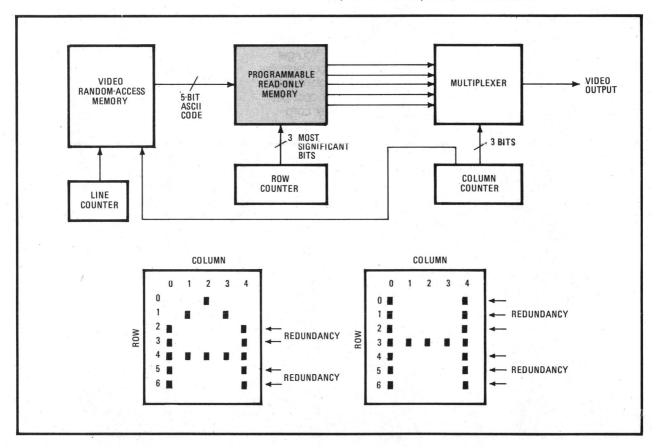

TRS-80 PROGRAM FOR READ-ONLY-MEMORY GENERATION OF SEQUENTIAL-STATE MACHINES

```
10 '-- PROM CAD
60 CL5:DEFINTA-Z
70 CLEAR 1000
80 DIM C(255),R(255),O(255),WD(255)
90 X$=CHR$(23):N$=CHR$(28):EL$=CHR$(30):ES$=CHR$(31)
91 ' 32/LINE  64/LINE   ERASE LINE   ERASE SCREEN
92 UB$=CHR$(18):UF$=CHR$(19):UX$=CHR$(8)
93 ' CURSOR BACK, FORWARD, BACK & ERASE CHAR
94 DIM B(15):FOR I=0TO15: B(I)=2↑I  :NEXT

100 PRINTX$:PRINT"PROM BASED STATE MACHINE"
110 PRINT STRING$(24,"-"):PRINT:PRINT
120 PRINT"THIS PROGRAM GENERATES THE DATA"
130 PRINT"FOR A SEQUENTIAL STATE MACHINE."
140 PRINT
150 PRINT"THE CONDITION STATEMENTS START"
160 PRINT"AT LINE 1000"
170 PRINT"C(I)= 1'S & 0'S  R(I)=REQ'D BITS"
180 PRINT"ENTER 'L' TO LIST THE CONDITIONS "
190 PRINT"ENTER 'R' TO GENERATE TABLE"
200 I$=INKEY$
210 IF I$="L" THEN CLS:LIST 990-1005
220 IF I$="R" THEN CLS:PRINT"EXECUTING":GOTO 1000
230 GOTO 200

500 FOR A=0 TO AX:' (PROCESS ENTIRE ADDRESS SPACE)
505 PRINT@50,A
510 I=0
520 I=I+1:IF R(I)=0 THEN WD(A)=OX :GOTO540
530 IF (A AND R(I))=C(I) THEN WD(A)=O(I) ELSE 520
540 NEXT A
550 PRINT"NOW FINISHED. LAST ADDRESS=";AX
560 INPUT"DO YOU WISH TO LIST PROM CONTENTS (Y/N)";I$
564 IF I$="N" THEN 580
570 FOR A=0TOAX:PRINTUSING"###";A;:PRINT") ";WD(A)," ":
    NEXT A

580 PRINT"DO YOU WISH TO SIMULATE THE MACHINE?"
590 PRINT" 'Y' = YES  'N' = NO"
600 I$=INKEY$
610 IF I$="N" THEN LIST 1000-
620 IF I$="Y" THEN 800
630 GOTO 600

800 CLS:PRINT"WE START AT STATE NUMBER 0":PRINT:A=0:SP=0:ST=0
805 PRINT"STATE=";SP;TAB(20)"OUTPUT=";WD(A);TAB(50)"INPUT=";
810 INPUT IN
815 A=ST+(IN AND (B(0)+B(1)+B(2)+B(3)+B(4)))
820 ST=WD(A) AND (B(5)+B(6)+B(7))
821 SP=ST/B(5)
825 IF IN>255 THEN STOP ELSE 805

990 CONDITION STATEMENTS BEGIN AT LINE 1000
991 CHANGE THESE LINES TO CREATE A NEW SYSTEM
992 C(I)=CONDITION BITS, UP TO 16 BITS
994 R(I)=REQUIRED BITS, THOSE OF C(I) THAT ARE NOT
955     DON'T CARE BITS AND MUST MATCH THE PROM ADDRESS
996 O(I)=PROM OUTPUT WORD. IF A CONDITION MATCH IS FOUND
997 OX =DEFAULT OUTPUT IF NO MATCH IS FOUND
999          ---- CONDITIONS ----
1000 IN=B(0) : RST=B(4) : STATE=B(5) :    -(DEFINE INPUT BITS
1001 NXT=B(5) : LATCH=B(1) : '            -(DEFINE OUT BITS
1002 N=8 : AX=B(N)-1 : ALL=AX : '         -(DEFINE NO OF BITS
1003 '
1004 C(1) = RST  :  R(1) = RST  : O(1)=STATE*0: ' -(IF RESET
1005 C(2) = 9*IN + STATE*0    : O(2)=STATE*1: ' -(IF IN=9
1006 C(3) = 7*IN + STATE*1    : O(3)=STATE*2: ' -(IF IN=7
1007 C(4) = 3*IN + STATE*2    : O(4)=STATE*3: ' -(IF IN=3
1008 C(5) = 8*IN + STATE*3    : O(5)=STATE*4 + LATCH
1009 C(6) =     STATE*4  : O(6)=STATE*4 :R(6)=STATE*?
1010       R(2)=ALL : R(3)=ALL : R(4)=ALL : R(5)=ALL
1020 OX=STATE*0 + B(2) :' DEFAULT TO STATE 0
2000 GOTO 500
2001 END
```

6, keeping track of which row of a five-by-seven-dot matrix the system is on.

The data word to be burned into each PROM location is determined by a set of condition equations. For each input combination, the program scans the condition-equation list for a match. If the input condition is met, then the associated output word is placed in the PROM's desired memory location.

For example, to generate the dot matrix shown for the character A presented at the inputs of the PROM, the equations in the figure commencing at line 1000 and ending at line 1030 are entered. The C(I) matrix indicates the bits that must be at logic 1 in the input data, O(I) is the associated output word, and B(N) is the binary value of each data bit and is equal to 2^N.

For the character H, only two lines of data are needed—at locations 1040 and 1050 in the general programming technique—because of the redundancy in the required output data. That is, rows 0, 1, 2, 4, 5, and 6 of the H are all the same. Only row 3 is important. Bits that are required to meet any condition, be they 1s or 0s, are designated by the R(I) matrix. Those input bits that are not required are designated "don't care" bits, as seen at lines 1100 and 1180 in the TRS-80 program.

With this information, the practical implementation of ROM generation for a sequential-state machine that functions as a combination lock may be considered (see program listing). In this example, the lock opens a latch when the sequence 9-7-3-8 is entered, as shown in the bottom half of the figure. Otherwise the sequencer returns to state number 0. The hardware implementation is shown at the bottom of the program. □

8048 program transmits messages in ASCII

by Mark I. Bresler
Rehabilitation Engineering Center, University of Virginia, Charlottesville, Va.

The physically handicapped can still communicate in writing with the aid of this subroutine, which is the output section of a communicator. The omission of both a fixed memory allocation and a starting-address look-up table shortens the message table and maximizes message flexibility. Also, the program uses the most significant bit to indicate the end of a message of any length. Although the code presented is for 8048 microcontroller,

MESSAGE SUBROUTINE FOR A COMMUNICATOR

SUBROUTINE MESSAGE

THIS IS AN 8048 SUBROUTINE TO BE USED IN ASCII MESSAGE TRANSMISSION, 8251A COMMUNICATIONS INTERFACE CHIP IS CONNECTED TO THE 8048 AS FOLLOWS: P26=RESET, P25=CHIP SELECT P24=CONTROL/DATA. THE 8251A IS ASSUMED TO BE RESET AND INITIALIZED WITH THE PROPER MODE AND COMMAND DATA. R0 IS THE MESSAGE POINTER, R1 IS THE CHARACTER POINTER, R6 IS TEMPORARY CHARACTER STORAGE, F1 WHEN SET UPON SUBROUTINE CALL INDICATES LOWER CASE TO BE PRINTED

Address	First instruction byte	Second instruction byte	Label	Mnemonic	Operands	Comments
0200				ORG	0200H	
0200	B9	00		MOV	R1,#000H	;CLEAR CHARACTER POINTER
0202	F8			MOV	A,R0	;JUMP IF FIRST MESSAGE DESIRED
0203	C6	0D		JZ	FSTMSG	
0205	F9		NXTCHR	MOV	A,R1	;MOVE ADDRESS TO ACCUMULATOR
0206	19			INC	R1	;INCREMENT R1 TO NEXT CHARACTER
0207	E3			MOVP3	A,@A	;MOV CHAR TO ACCUMULATOR
0208	37			CPL	A	;CHECK IF MESSAGE END IF NOT LOOP TO NXTCHR
0209	F2	05		JB7	NXTCHR	
020B	E8	05		DJNZ	R0,NXTCHR	;CHK IF R0 AT DESIRED MESSAGE IF NOT LOOP TO NXTCHAR
020D	F9		FSTMSG	MOV	A,R1	;MOV CONTENTS OF NEXT ADDRESS TO ACCUMULATOR
020E	E3			MOVP3	A,@A	
020F	AE			MOV	R6,A	;MOVE CHAR TO TEMPORARY STORAGE
0210	B5			CPL	F1	;JUMP IF LOWER CASE NOT DESIRED
0211	76	22		JF1	PRINT	
0213	37			CPL	A	;JUMP IF MESSAGE IS MULTICHARACTER
0214	F2	22		JB7	PRINT	
0216	37			CPL	A	;RECOMPLEMENT ACCUMULATOR
0217	03	25		ADD	A,#025H	;ADD 37_{10} TO ACCUMULATOR, JMP IF ACCUMULATOR > ASCII CHAR CAPITAL Z
0219	F6	22		JC	PRINT	
021B	03	1A		ADD	A,#01AH	;ADD 26_{10} TO A JMP IF ACCUMULATOR < ASCII CHAR CAP A
021D	E6	22		JNC	PRINT	
021F	03	E1		ADD	A,#0E1H	;ADD 241_{10} TO ACCUMULATOR FOR A TOTAL ADDITION FROM START TO HERE OF 32_{10} (IGNORING CARRY), CHANGING UPPER TO LOWER CASE
0221	AE		CHAR	MOV	R6,A	;MOV CHAR TO TEMP
0222	54	A0	PRINT	CALL	USART	
0224	F2	2B		JB7	END	;JMP IF MESSAGE END
0226	19			INC	R1	;INC CHAR POINTER
0227	F9			MOV	A,R1	;MOV POINTER TO ACCUMULATOR
0228	E3			MOVP3	A,@A	;MOV CHAR TO ACCUMULATOR
0229	44	21		JMP	CHAR	;LOOP TO PRINT CHAR
022B	A5		END	CLR	F1	;CLR LOWER CASE FLAG
022C	93			RETR		;RETURN
022D			USART	EQU	02A0H	

USART SUBROUTINE TO MESSAGE SUBROUTINE

Address	First instruction byte	Second instruction byte	Label	Mnemonic	Operands	Comments
02A0				ORG	2A0H	
02A0	9A	80		ANL	P2,#080H	;ENABLE 8251A
02A2	8A	10		ORL	P2,#010H	;CHECK STATUS
02A4	80		STATUS	MOVX	A,@R0	;LOOP UNTIL READY FOR NEXT CHARACTER
02A5	37			CPL	A	
02A6	12	A4		JB0	STATUS	
02A8	9A	80		ANL	P2,#080H	;8251A INTO DATA MODE
02AA	FE			MOV	A,R6	;MOV CHARACTER TO 8251A
02AB	90			MOVX	@R0,A	
02AC	8A	20		ORL	P2,#020H	;DISABLE 8251A
02AE	93			RETR		;RETURN
DONE						

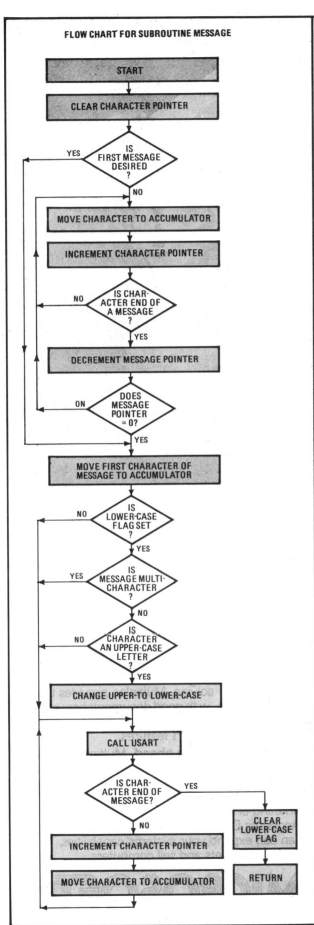

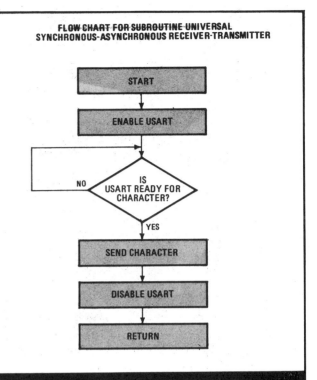

SAMPLE TABLE LISTING			
Message No.	Character	Address$_H$	Hexadecimal code
00	A N D	300 301 302	41 4E C4
01	A	303	C1
02	B	304	C2
03	T H E	305 306 307	54 48 C5

the algorithm can be used on any other microcomputer

The listing for the message subroutine, opposite, shows that the ASCII code for the first character of message 00_H is entered into bit 0 to 6 of the first table byte. The remaining characters are entered into sequential memory bytes, with the MSB set for the last character to be printed. Thus if a message is a single character, the first character is also the last character to be printed, so its bit 7 is set. Subsequent messages are entered in the same manner, progressing sequentially until all of them have been entered. The subroutine is initiated by clearing character pointer R1 and, when called, the number of the message to be printed is contained in register R0.

The flow chart presented on the left outlines the selection process for the message desired. The search for the correct message uses only six machine code instructions and takes about 22.5 microseconds to complete a loop. Setting flag F1 before calling the subroutine changes the output to lower-case. This output is printed

Pocket computer tackles classical queuing problems

by Cass R. Lewart
System Development Corp., Eatontown, N. J.

This program, written in Basic, permits pocket computers such as the Radio Shack and the Sharp PC-1211 to tackle problems in classical waiting-line (queuing) theory that has proved so useful in solving the tradeoffs that have to be made between utilization and capacity in telecommunications systems. The program can easily be translated to work on other machines.

The classical queuing considerations assume an expo-

SOLUTION OF QUEUING EQUATION IN RADIO SHACK/SHARP BASIC

```
10  "Z" CLEAR: USING "##.##^"
15  PAUSE "QUEUING PROG. C.R.LEWART"
20  INPUT "ARRIVAL RATE?";L: IF L <= 0 GOTO 45
25  INPUT "SERVICE RATE?";M: IF M <= 0 GOTO 45
30  INPUT "# SERVERS?";S: IF (S <> INT S)+(S <= 0) GOTO 45
40  X=L/M: U=X/S: V=1-U: IF (V > 0)*((S-1)*LOGX < 100)*
    (S < 70) GOTO 50
45  BEEP 2: GOTO 20
50  Y=S: GOSUB 200
55  T=Z: FOR I=0 TO S-1: Y=I: GOSUB 200
60  P=P+X^I/Z: NEXT I
65  Y=S: GOSUB 200
70  P=1/(P+X^S/T/V): B=X^S*P/T/V
75  W=B/SMV: R=W+1/M: BEEP 1: PRINT "READY"
80  "B" PRINT "P(ALL BUSY)=";B
85  "X" PRINT "UTILIZATION=";U
90  "A" PRINT "P(0)=";P
95  "S" PRINT "AV. WAIT=";W
100 "V" PRINT "AV. RESP.=";R: D=√(2B-BB+SSVV)*W/B
105 PRINT "ST.DEV.(TR)=";D
110 "N" D=LR: PRINT "AV. IN SYS.=";D
115 "D" D=LW: PRINT "AV. IN QUEUE=";D
120 "F" INPUT "# ITEMS IN SYSTEM?";N: D=P*X^N:
    IF N < S GOTO 130
125 E=D/TS^(N-S): GOTO 140
130 Y=N: GOSUB 200
135 E=D/Z
140 PRINT "P(N=";N;")=";E
145 "C" INPUT "TIME?";H
150 F=B*EXP -SMVH: A$="W"
155 PRINT USING "##.#^";"P(T";A$;">";H;")=";
    USING "##.##^";F
160 G=V-1/S: A$="R": D=EXP -MH: IF G=0 GOTO 170
165 F= D*(1+B/S*(1-EXP -MSGH)/G): GOTO 155
170 F=D*(1+BMH): GOTO 155
200 Z=1: FOR J=1 TO Y: Z=JZ: NEXT J: RETURN
```

| CALCULATION PROCEDURE FOR SAMPLE PROBLEM |||
Key entry	Display	Remarks
SHIFT Z	QUEUING PROGRAM C.R.LEWART	CLEAR ALL VARIABLES
	ARRIVAL RATE?	
16 ENTER	SERVICE RATE?	
4 ENTER	# SERVERS?	
5 ENTER	(BEEP TONE) READY	INITIALIZATION COMPLETED
SHIFT X	UTILIZATION = 8.00E−01	UTILIZATION FACTOR
SHIFT B	P(ALL BUSY) = 5.54E−01	
SHIFT A	P(0) = 1.29E−02	P (OF NO JOBS IN SYSTEM)
SHIFT F	# ITEMS IN SYSTEM?	
2 ENTER	P(N = 2.00E00) = 1.03E−01	P (2 ITEMS IN SYSTEM)
SHIFT S	AV. WAIT = 1.38E−01	WAIT IN QUEUE
SHIFT V	AV. RESP. = 3.88E−01	WAIT + SERVICE
ENTER	ST. DEV. (TR) = 3.35E−01	STANDARD DEVIATION OF ABOVE
SHIFT C	TIME?	P (WAIT > T)
0.25 ENTER	P(TW > 2.5E−01) = 2.03E−01	P (QUEUE WAIT > 0.25)
ENTER	P(TR > 2.5E−01) = 5.71E−01	P (TOTAL WAIT > 0.25)
SHIFT D	AV. IN QUEUE = 2.21E00	QUEUE LENGTH
SHIFT N	AV. IN SYSTEM = 6.21E00	
RUN 30 ENTER	# SERVICES?	RECOMPUTE FOR S = 6
6 ENTER	(BEEP TONE) READY	
SHIFT S	AV. WAIT = 3.55E−02	

nential distribution of customer arrival rates and serving times, identical servers, and a first-in, first-out order of service. Exponential distribution implies independence of events: customers seek service independent of the queue length and the servers operate at a steady rate independent of the load and queue length. Such assumptions usually result in safe estimates of waiting times and other queuing parameters and can be turned into simple equations that can be solved in a reasonable time on this pocket computer.

Given the customer's average arrival rate, r, the average serving rate, m, and the number of servers, s, the program first finds the system's utilization factor, u, from $u = r/sm$. It then finds the probability of finding all servers busy from:

$$B = \sum_{n=s}^{\infty} P(n) = (r/m)^s P(0)/[s!(1-u)]$$

where

$$P(0) = \left[(r/m)^s/s!(1-u) + \sum_{j=0}^{s-1}[(r/m)^j/j!] \right]^{-1}$$

Following this, the probability of finding n items already in the system is calculated from:

$$P(n) = P(0)(r/m)^n(1/n!) \qquad n < s$$

or

$$P(n) = P(0)(r/m)^n[1/(s!s^{n-s})] \qquad n \geq s$$

Next, the average waiting time in the queue is computed from $T_w = B/sm(1-u)$, with the average response time being $T_r = T_w + 1/m$. The standard deviation of T_r is then found by:

$$S_{TR} = (T_w/B)[B(2-B) + s^2(1-u)^2]^{1/2}$$

and from this, the probabilities of $T_w > T$ and $T_r > T$:

$$P(T_w > T) = B \exp[-smT(1-u)]$$
$$P(T_r > T) = \exp(-mT) \times \{1 + [1 - \exp(-msTK)] \times (B/SK)\}, \qquad K \neq 0$$
$$P(T_r > T) = \exp(-mT) \times (1 + BmT), \qquad K = 0$$

where $K = 1 - u - (1/s)$.

Finally, the average queue length is determined from $Q = rT_w$ and the average number of jobs in the system from $N = rT_r$. The 33-line program for finding all desired queuing parameters is shown in Table 1.

Consider the case where a computer having five terminals averages 16 customers per hour who arrive at random intervals, with each job taking an average of 15 minutes (four per hour). The aim is to ascertain several queuing parameters, as well as the reduction in waiting time if a sixth terminal is installed.

If the program is initialized as illustrated in Table 2 (the average initialization time will vary from 10 seconds for s = 1 to 18 minutes for s = 69), it will find that u = 0.8, B = 0.554, P(0) = 0.0129, P(n = 2) = 0.103, T_w = 0.138 hour, T_r = 0.388 h, S_{TR} = 0.335 h, P(T_w>0.25 h) = 0.203 h, and P(T_r>0.25 h) = 0.571 h. With s = 6, T_w is reduced to 0.036 h. □

Interface, software form smart stepper controller

by Robert Ward
McPherson College, McPherson, Kansas

Although a stepper motor may be driven by a few logic bits on a microprocessor [*Electronics*, July 28, 1982, p. 119], many real-time applications cannot handle the attendant timing loops and software. In addition, if the timing loops generate stepper pulses while real-time activity is controlled through interrupt routines, the motor may overrun inaccurately. This inexpensive programmable stepper-motor controller (a) overcomes these limitations and greatly simplifies the task of writing software through an intelligent interface.

The motor controller is realized with chips U_1 through U_4 (b). Through simple input/output instructions (Table 1), a programmer may specify the characteristics of the

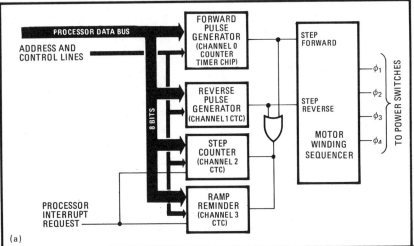

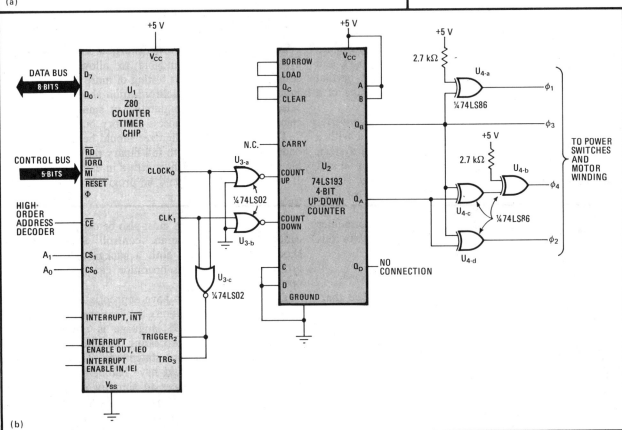

Smart interface. This intelligent programmable stepper-motor controller (a) is realized with chips U_1 through U_4 (b). The clear and load lines are used to shorten the count sequence of U_4, thereby enabling it to function as a 2-bit counter. Simple input/output instructions specify the direction, rate, and length of the motor's motion.

TABLE 1: Z80 ASSEMBLY CODE

```
CHAN0    EQU    CARD ADR  ; FORWARD PULSE GENERATOR
CHAN1    EQU    CHAN0+1   ; REVERSE PULSE GENERATOR
CHAN2    EQU    CHAN0+2   ; STEP-DOWN COUNTER
CHAN3    EQU    CHAN0+3   ; RAMP REMINDER
                          ; DOWN COUNTER
STOP     EQU    03H       ; SET-RESET IN
                ; CHANNEL CONTROL WORD

; INTERRUPT VECTOR TABLE
; USES Z-80 MODE II INTERRUPTS
         ORG A WORD BOUNDARY
         JP ERROR
         JP ERROR
         JP DONE
         JP RAMP
FREE     DS 1  ; FLAG: 0 = BUSY, 1 = FREE
; NOTE THIS ALLOWS MOVES ONLY
; 256 STEPS.
DONE:
    PUSH   AF              ; WHEN DONE WE
    LD     A, STOP         ; STOP THE PULSES
    OUT    (CHAN0),A       ; ON BOTH
    OUT    (CHAN1),A       ; CHANNELS
    LD     A, 1            ;
    LD     (FREE),A        ; MARK CONTROLLER
    POP    AF              ; STATUS FREE
    RETI   ; CLEAR INTERRUPT, RETURN
; IMPLEMENTS MORE COMPLEX RAMP
; FUNCTION THAN TEXT DISCUSSES.
; BEFORE BEGINNING MOTION, LOADS
; UP TO FOUR RATE CHANGES AND TRAVEL
; DIRECTION. RAMP ADJUSTS THE
; PULSE RATE DYNAMICALLY. "ERROR"
; ROUTINE NOT DEFINED HERE
DIR      DS 1      ; FIRST BYTE IS
NLEFT    DS 1      ; NEW RATE. SECOND
LAST     DS 2      ; IS COUNT TO
THIRD    DS 2      ; NEXT
SECOND   DS 2      ; REMINDER
FIRST    DS 2
RAMP:
    EX     AF, AF'         ; WHOLESALE SAVE
    EXX                    ; OF PROCESSOR STATE
    LD     A, (NLEFT)      ; MORE RATE
    OR     A               ; CHANGES LEFT?
    JP     Z, ERROR        ; SOMETHING'S WRONG
    LD     (NLEFT),A       ; SAVE NEW COUNT
    LD     HL,NLEFT+1      ; COMPUTE
    LD     D,0             ; ADDRESS OF
    LD     E,A             ; NEW RATE
    ADD    HL,DE
    ADD    HL,DE
    LD     B, (HL)         ; NEW RATE IN B
    INC    HL
    LD     C, (HL)         ; NEXT REMINDER
                           ; INTERVAL IN C
    LD     A, (DIR)        ; FORWARD = 0: REVERSE = 1
    OR     A
    JP     Z, DOREV
```

```
;CHANGE RATE IN FORWARD PULSE GENERATOR
    LD      A,NEWRATE   ;WARN FORWARD GENERATORS
    OUT     (CHAN0), A  ;NEW RATE COMING
    LD      A, B        ;NOW WE SET THE RATE
    OUT     (CHAN0), A
    JP      SETRMP      ;SET NEW INTERVAL
DOREV:
    LD      A, NEWRATE  ;CHANGE RATE
    OUT     (CHAN1), A  ;IN REVERSE PULSE
    LD      A, B        ;GENERATOR
    OUT     (CHAN1), A
;SET RAMP REMINDER
SETRMP:
    LD      A, (NLEFT)  ;MORE STOPS?
    DEC     A           ;MARK THIS INTERVAL
    LD      (NLEFT), A  ;SAVE NEW COUNT
    JP      Z, EXIT     ;QUIT IF NO
    LD      A, NXRMP    ;WARN OF NEW COUNT
    OUT     (CHAN3), A
    LD      A, C        ;GET NEW COUNT
    OUT     (CHAN3), A
EXIT:
    EXX                 ; RESTORE CENTRAL-PROCESSING-UNIT STATE
    EX      AF, AF'
    RETI                ; CLEAR INTERRUPTS, RETURN
```

TABLE 2: STEPPER-MOTOR-CONTROL PROGRAM IN C

```
/*The programmer works with these commands:
  set_ramp (r1, s1, r2, s2, r3, s3, r4, s4)
      He sets four ramp intervals (in steps) and subsequent
      step rates. A zero rate ends the sequence.
  move_mtr (dir,rmp_rate,steps)
      Direction and length of travel are set. The speed
      changes in set_ramp may be overridden by
      specifying rmp_rate. In addition, a new set_ramp command must
      precede each move_mtr command.
  status ( )
      Non-zero when interface is ready.
  mtr_pos ( )
      Returns the number of steps left in a currently
      executing command (0 if motor is not busy).
                    Initialization
  Assumes mode II interrupts. The interrupt vector
  register should be loaded with the upper byte of
  the interrupt vector table address. */

#define     BUSY    0
#define     FORWARD 0
#define     CHAN0           /*must match assembly listing*/
#define     CHAN1   CHAN0+1
#define     CHAN2   CHAN0+2
#define     CHAN3   CHAN0+3
#define     DIR     /*these must match*/
#define     NLEFT   /*the assembly values*/
#define     LAST
#define     FREE
#define     VTAB    /*interrupt vector cable address */
#define     MODE1   0x2B    /* counter timer chip control words */
#define     MODE2   0x2B
#define     MODE3   0xC3
#define     MODE4   0xC3
#define     NEWRATE 0x2F
#define     NEWCOUNT 0xC7
```

```c
char *dir, *last_ramp, *free, *nleft; /*globals */

initialize ( ) {
    outp (CHAN0,MODE1); /*set mode of each counter timer chip */
    outp (CHAN1, MODE2); /* channel */
    outp (CHAN2, MODE3);
    outp (CHAN3, MODE4);
    outp (CHAN0, VTAB);       /*set interrupt vector */
    outp (CHAN1, VTAB + 2); /*for each channel */
    outp (CHAN2, VTAB + 4);
    outp (CHAN3, VTAB + 6);
    dir = DIR; nleft = NLEFT; last_ramp = NLEFT + 1;
    free = FREE; *free = !BUSY; }

set_ramp (r1, s1, r2, s2, r3, s3, r4, s4)
int r1, r2, r3, r4, s1, s2, s3, s4;
{char *data;
    while (!status ( ) );     /*wait for stepper */
    *nleft = 4;               /*to be free */
    data = last_ramp;
    if (r4 == 0) { data = last_ramp; *nleft = 3; }
        else { *data = r4; data += 1; *data = s4; data += 1; }
    if (r3 == 0) { data = last_ramp; *nleft = 2; }
        else { *data = r3; data += 1; *data = s3; data += 1; }
    if (r2 == 0) { data = last_ramp; *nleft = 1; }
        else { *data = r2; data += 1; *data = s2; data += 1; }
    if (r1 == 0) { data = last_ramp; *nleft = 1; }
        else { *data = r1; data += 1; *data = s1; data += 1; } }

move_mtr (dr, rmp_rate, steps)
int dr, rmp_rate, steps;
{int begin_rate;
    while (!status ( ) ); /*wait for motor to be free */
    *dir = dr; /*record it in case ramp needs it */
    if (rmp_rate == 0) {
        begin_rate = * (last_ramp + 6); /*from ramp info */
        outp (CHAN3, NEWCOUNT); /*set up ramp reminder */
        outp (CHAN3, *(last_ramp + 7); /*first step interval */
        } /*in auto-start mode, counter is now active */
    else begin_rate = rmp_rate;
    outp (CHAN2, NEWCOUNT); /*now set up */
    outp (CHAN2, steps); /*the step counter */
    if (dr == FORWARD) { /*now start appropriate */
        outp (CHAN0, NEWRATE); /*pulse generator */
        outp (CHAN0, begin_rate); }
    else { outp (CHAN1, NEWRATE);
        outp (CHAN1, begin_rate); } }

int status {return *free; }

int mtr_pos ( )
{ if (!status ( ) ) return 0; return inp (CHAN2); }
```

signal that will control the motor. In addition, the motor's movement can be observed by the microprocessor without disturbing the programmed motion sequence. When the preset motor movement is complete, the controller interrupts the processor to indicate that it is ready for the next instruction.

To reduce software overhead in applications where the motor is ramped up to its slew rate, a spare counter is wired so that it can be used as a ramp reminder. This counter may be programmed to interrupt the processor while the stepper motor rotates. The processor then responds by loading a new rate constant into the forward- and reverse-pulse generators. Listed in the language C (Table 2), the program illustrates the use of high-level motor commands.

With this interface and software, other real-time activities may be handled in the background on an interrupt basis with little effect on pulse timing. However, proper ramp intervals must be selected for such activities. □

GPIB software helps to provide automatic test switching

by Mike Black
Texas Instruments Inc., Dallas, Texas

This automatic tester, which provides computer-controlled test switching, can examine a dual-channel, 60-megahertz intermediate-frequency amplifier. It combines a voltmeter with Hewlett-Packard's 9825 computer and 8660C frequency synthesizer, which uses the General-Purpose Interface Bus.

The table shows the subroutine that enables the computer to sequentially control the input switch, output switch, and bandwidth switch. Also, the circuit eliminates the need for a separate input/output bus adapter.

The radio-frequency output of the frequency synthesizer (Fig.1a) is fed into the switch controller, which is realized by the circuit in Fig. 1b. This circuit consists of high-Q–tuned amplifier Q_1, detector D_1, comparator U_1, one-shot U_2, and shift register U_3. Q_1 is set to 80 MHz by inductor L_1 and capacitors C_4 and C_5. Resistor R_1 gives the circuit a high input impedance and also sets its operating level. When the output of D_1 exceeds U_1's threshold, set by R_8, R_9, R_{10}, and R_{11}, the comparator

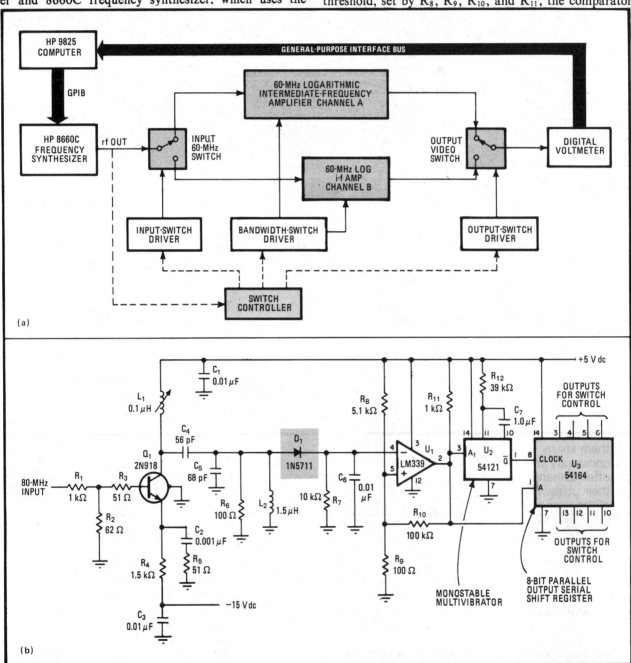

1. GPIB-based ATE. This automatic test system uses GPIB-based frequency synthesizer HP 8660C, a 9825 computer, and a DVM to test a dual-channel 60-MHz amplifier (a). The computer sequentially controls the input, output, and bandwidth switches. The output of the switch controller (b) is utilized to control the three switch drivers. The 8-bit output of U_3 represents the 8-bit code entered in string variable B$.

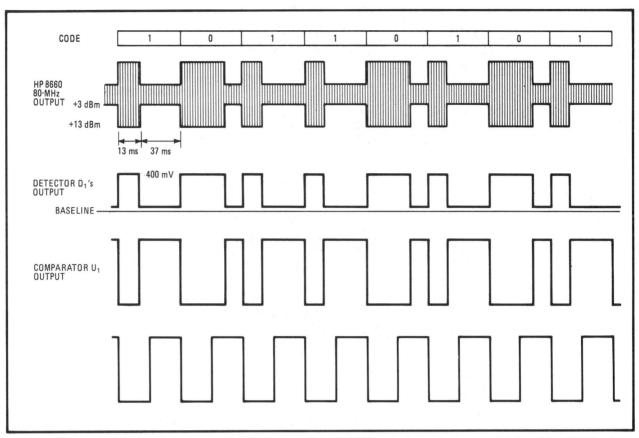

2. Timing. Since the switch controller is tuned to 80 MHz, program lines 4 and 5 enable the synthesizer to switch from 60 to 80 MHz when a change of switch settings is required. The synthesizer output is pulse-amplitude-modulated, and the code sequence entered in B$ provides the pulse-width modulation. The total length of a cycle is set at 50 milliseconds.

output switches. For the values shown, this switching point is set at a +10-dBm input. When the synthesizer is set to 80 MHz with the output amplitude modulated above and below +10 dBm, the comparator output is a TTL-compatible replica of the modulation (Fig. 2).

The program in the table, written in a high-level programming language for the 9825 computer, allows the output of shift register U_3 to control the three switch drivers. An 8-bit sequence of 1s and 0s is entered in the string variable, B$. This sequence represents the data required to set the three switches. Only 3 bits are needed for the job—the other 5 may be either ignored at the output or used for parity checking. In this example, the 3 least significant bits are used for switch control.

Lines 4 and 5 of the program enable the synthesizer to switch from 60 to 80 MHz when a change of switch settings is needed. Using the for/next loop in line 6, the synthesizer is pulse-amplitude–modulated from +13 to +3 dBm by lines 9 and 11. As for the total cycle length, it is set to 50 milliseconds. The sequence in B$ determines the pulse-width modulation at the synthesizer output. The detected pulses are restored by U_1 whose output is used to trigger one-shot U_2. In turn, the $\overline{Q}$ output of U_2 clocks shift register U_3.

With each successive clock, the number stored is serially shifted through U_3. After eight shifts, the output of U_3 represents B$, and these outputs are now used to control the appropriate switch driver. The switches now change to the states set by the computer, the frequency synthesizer returns to 60 MHz, and the subroutine returns to its control program. □

HP 9825 SUBROUTINE FOR PULSE-WIDTH MODULATION OF FREQUENCY SYNTHESIZER HP 8660C

```
"10110101" -> B$
0: "SWITCHING"
1: "SUBROUTINE FOR PULSE-WIDTH ENCODING":
2: fmt 9, 3f.0
3: rem 7
4: wrt 719, 10, "C"
5: wrt 719.9, 800, "("
6: for N = 1 to 8
7: if val (B$ [N,N]) = 1; 13 -> C
8: if val (B$ [N,N]) = 0; 37 -> C
9: wrt 719.9, 0, "C"
10: wait C
11: wrt 719.9, 10 "C"
12: wait 50 - C
13: next N
14: wrt 719.9, 600, "("
15: ret
```

Interactive software controls data-acquisition process

by Carl D. Slater and William S. Wagner
Northern Kentucky University, Highland Heights, Ky.

Using a 6820 peripheral interface adapter, an 8-bit digital-to-analog converter, a comparator, and an operational amplifier, this circuit links an Apple II Plus personal computer and a heathkit MC680-based microcomputer trainer to provide a data-acquisition system capable of handling 250 data points per second. The trainer, which does the analog-to-digital conversion, requires an input between 0 and 5 volts. In addition, the 8-bit d-a converter yields a resolution of 0.02 v.

This process is controlled by an interactive Basic (Applesoft) program (Table 1) that requires three inputs: the title, the number of points, and the interval time between points. The data-acquisition rate may be selected through software and ranges from 2 to 250 points per second. Once the input parameters are set, control is transferred to a machine-language subroutine that initiates highly accurate data collection. The hardware interface (see figure) shows that handshaking between the Apple and the Heathkit units is via the

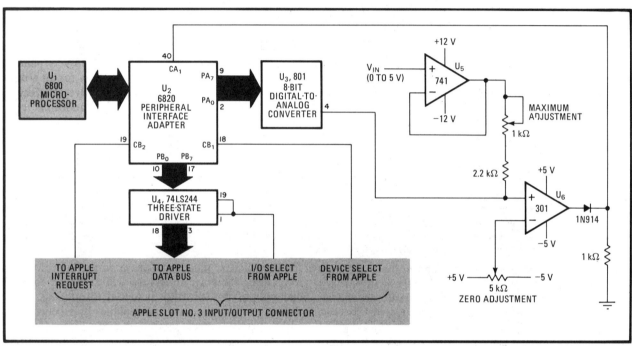

Data Acquisition. This interface hardware and software combine Motorola's MC6800 microprocessor and an Apple II Plus computer to produce a data-acquisition system. U_1 performs a-d conversion and places data at U_2's port B. This port acts as an input for driver U_4 whose output is connected to the Apple data bus. The Apple and the 6800 are connected through the Apple I/O connector slot No. 3.

INTERACTIVE BASIC PROGRAM FOR DATA ACQUISITION

```
10    REM PROGRAM DACQ2
12    HIMEM: 8190
15    TEXT : HOME
20    PRINT "DACQ2. THE TIME BETWEEN POINTS MUST BE A MULTIPLE OF 2 BETWEEN
         4 AND 512.": PRINT : PRINT
30    PRINT "INPUT THE TITLE OF THIS RUN": INPUT T$: PRINT : PRINT
40    PRINT "INPUT THE NUMBER OF DATA POINTS": INPUT N: PRINT : PRINT
50    PRINT "INPUT NO. OF MSECS BETWEEN POINTS": INPUT TT
55    TI = (TT - 2) / 2:N1 = INT (N / 255) : R1 = N1 * 255:R = N - R1
60    POKE 1022, 79: POKE 1023,101: POKE 25632,98: POKE 25605,TI
62    IF N1 ) = 1 GOTO 66
64    POKE 25601,R: POKE 25602,1: GOTO 70
```

```
66    POKE 25600,R: POKE 25601,255: POKE 25602,3: POKE 25608,N1 + 1
70    REM EXECUTE OBJECT CODE LOADER.
80    DIM BYTE(8): READ FIRST: READ LAST
90    FOR LINE = FIRST TO LAST
100   GOSUB 200
110   NEXT LINE
120   PRINT "TO PROCEED HIT RETURN": INPUT G$
130   CALL 25823: REM TAKE DATA
135   CALL 64477: REM DATA IN, RING BELL
140   PRINT "TO STORE DATA ON DISK ENTER 1": INPUT C1
145   IF C1 < > 1 GOTO 170
150   REM OUTPUT WILL BE STORED IN "DATA1" FILE.
160   GOSUB 400
170   PRINT : PRINT "TO PLOT DATA ON MONITOR ENTER 1": INPUT C2
180   IF C2 < > 1 GOTO 195
190   GOSUB 300
195   END
200   READ A:SUM = A: FOR J = 1 TO 8: READ TEMP:BYTE(J) = TEMP
210   SUM = SUM + BYTE(J): NEXT J
220   READ CHECK
230   IF SUM < > CHECK THEN 260
240   FOR J = 1 TO 8: POKE A + J - 1,BYTE(J): NEXT J
250   RETURN
260   PRINT "CHECKSUM ERROR IN DATA LINE ",LINE
270   PRINT "START ADDRESS GIVEN IN BAD DATA LINE IS ",A
280   END
300   HGR2 : HPLOT 0,32 TO 0,160: HPLOT 0,160 TO 255,160
310   IF N1 > = 1 GOTO 350
315   F = INT (256 / N)
320   FOR I = 1 TO N:X = I * F:P = PEEK (26112 + I) + 1
330   Y = 160 - P / 2: HPLOT X,Y: NEXT I
340   RETURN
350   K = 25856:F = INT (255 / (N1 + 1))
360   FOR J = 1 TO N1:K = K + 256
370   FOR I = 1 TO 255:X = INT (I / (N1 + 1)) + J * F -F
380   P = PEEK (K + I) + 1:Y = 160 - P / 2: HPLOT X,Y: NEXT I
390   NEXT J:K = K + 256
392   FOR I = 1 TO R:X = INT (I / (N1 + 1)) + F * N1
394   P = PEEK (K + I) + 1:Y = 160 - P / 2: HPLOT X,Y: NEXT I
396   RETURN
400   D$ = CHR$ (4): PRINT D$; "OPEN DATA1": PRINT D$; "WRITE DATA1"
410   PRINT T$: PRINT : PRINT N: PRINT : PRINT TT: PRINT
420   IF N1 > = 1 GOTO 450
430   FOR I = 1 TO N: PRINT PEEK (26112 + I): NEXT
440   GOTO 500
450   K = 25856
460   FOR J = 1 TO N1:K = K + 256
470   FOR I = 1 TO 255: PRINT PEEK (K + I): NEXT
480   NEXT :K = K + 266
490   FOR I = 1 TO R: PRINT PEEK (K + I): NEXT
500   PRINT D$; "CLOSE DATA1"
510   RETURN
900   DATA 1000
910   DATA 1015
1000  DATA 25823, 160, 0, 200, 88, 169, 16, 141, 48, 26645
1001  DATA 25831, 100, 173, 176, 192, 173, 48, 100, 201, 26994
1002  DATA 25839, 9, 176, 249, 169, 0, 141, 21, 100, 26704
1003  DATA 25847, 238, 21, 100, 162, 0, 232, 234, 234, 27068
1004  DATA 25855, 234, 234, 234, 236, 32, 100, 144, 245, 27314
1005  DATA 25863, 169, 3, 141, 69, 100, 206, 69, 100, 26720
1006  DATA 25871, 173, 69, 100, 201, 1, 234, 176, 245, 27070
1007  DATA 25879, 173, 21, 100, 205, 5, 100, 144, 216, 26843
1008  DATA 25887, 204, 1, 100, 144, 189, 169, 1, 24, 26719
1009  DATA 25895, 109, 84, 101, 141, 84, 101, 169, 1, 26685
1010  DATA 25903, 205, 2, 100, 176, 24, 206, 8, 100, 26724
1011  DATA 25911, 169, 1, 205, 8, 100, 144, 161, 173, 26872
1012  DATA 25919, 0, 100, 141, 1, 100, 169, 1, 141, 26572
1013  DATA 25927, 2, 100, 76, 223, 100, 96, 0, 0, 26524
1014  DATA 25935, 173, 0, 195, 153, 0, 102, 169, 6, 26733
1015  DATA 25943, 141, 48, 100, 64, 0, 0, 0, 0, 26296
```

6800 PROGRAM FOR ANALOG-TO-DIGITAL CONVERSION

Address	Op code	Mnemonic		Comment
		Initialization		
0010–	7F	CLR	$8001	
0013–	7F	CLR	$8003	
0016–	CE	LDX	#$FF04	
0019–	FF	STX	$8000	A side is output
001C–	CE	LDX	#$FF34	
001F–	FF	STX	$8002	B side is output
0022–	C6	LDAB	#$3C	
0024–	F7	STAB	$8003	Set CB_2
		Wait for control signal from Apple		
0027–	F6	LDAB	$8003	Control register B bit No. 7 is set on $CB_1\downarrow$
002A–	2A	BPL	$FB	(CB_1 is device-select from Apple computer)
		Analog-to-digital conversion by successive approximation		
002C–	7F	CLR	$8000	
002F–	7F	CLR	$00A0	
0032–	0D	SEC		
0033–	76	ROR	$00A0	When conversion is finished,
0036–	25	BCS	$2F	take branch to "Send data to Apple"
0038–	B6	LDAA	$8000	
003B–	9B	ADDA	$A0	
003D–	B7	STAA	$8000	Input to digital-to-analog converter
0040–	CE	LDX	#$0002	
0043–	09	DEX		
0044–	26	BNE	$FD	Time delay for comparator to react
0046–	7D	TST	$8001	Control-register A bit No.7 set on $\downarrow CA_1$
0049–	2B	BMI	0B	
004B–	01	NO	OP	
004C–	01	NO	OP	
004D–	CE	LDX	#$0003	
0050–	09	DEX		Time delay to make
0051–	26	BNE	$FD	all conversions equal time
0053–	7E	JMP	$0033	Go back to rotate right
0056–	B6	LDA	$8000	Clears control-register A bit No. 7
0059–	90	SUB	$A0	
005B–	B7	STAA	$8000	Input to digital-to-analog converter
005E–	CE	LDX	#$0002	
0061–	09	DEX		
0062–	26	BNE	$FD	Time delay for comparator to react
0064–	7E	JMP	$0033	Go back to rotate right
		Send data to Apple		
0067–	B7	STAA	$8002	Data is stored at peripheral-interface-adapter output
006A–	CE	LDX	#$000A	
006D–	09	DEX		Time delay to achieve a desired
006E–	26	BNE	$FD	digital-to-analog conversion time
0070–	C6	LDAB	#$B4	
0072–	F7	STAB	$8003	$CB_2\downarrow$ (CB_2 goes to Apple interrupt request)
0075–	F6	LDAB	$8002	Clears control-register B bit No.7
0078–	7E	JMP	$0022	Go back and wait for signal from Apple

Apple input/output connector slot 3.

Initially, the programs are loaded and the trainer is placed in a wait loop (above). After the Apple receives an interactive input, a call to the machine-language subroutine is initiated through a single keystroke on the Apple. The Apple then pulses the device-select line that is connected to PIA control line CB_1, and the Apple goes into a wait loop. The Heathkit performs a-d conversion and places the data byte on PIA port B, which is the input to the octal three-state driver U_4. U_1 then pulses PIA control line CB_2, which is connected to the Apple's interrupt-request line IRQ. A pulse is then generated on the I/O select line, thereby enabling driver U_4, whose output is connected to the Apple data bus. The data is withheld from the Apple bus until it is called by the I/O select pulse, which is synchronous with the data-reading operation. The data is then transferred to an appropriate storage location, counters are updated, and the process is repeated. ∎

Tracing out program bugs for Z80A processor

by U. K. Kalyanaramudu and G. Aravanan
Bharat Electronics Ltd., Bangalore, India

Advanced microprocessors like Motorola's MC68000 assist in program debugging by providing instruction-to-instruction tracing. This ability, which most 8-bit processors lack, is granted to all microprocessors with this logic circuit. In the case shown, it uses three NAND gates and a D-type positive-edge–triggered flip-flop to create a trace mode for a Z80A microprocessor so as to aid program development.

Trace-mode operation is selected and program 1 is executed once switch S_1 is closed. This program saves the contents of the refresh counter and loads it with value 7DH. For each fetch cycle $\overline{M}_1$, the refresh counter increments automatically until it reaches zero while executing the first user instruction—the one that follows RET. In addition, while the first user instruction is being executed, $\overline{M}_1$ resets flip-flop U_2.

Program 2 is executed as soon as the current instruction is over. This routine interrupts the user program and takes the Z80A's central processing unit to location 0038H, the point from where the trace program begins. During this interrupt routine, the refresh counter value is

Location	Object code	Mode statement	Source statement		Comments
		1	*H TRACE PROGRAMS 1 & 2		
		2			; PROGRAM 1 : THIS PROGRAM IS
		3			; EXECUTED TO ENTER IN TRACE MODE
		4			
0100		5		ORG 100H	
0100	F3	6		DI	
0101	ED56	7		IM 1	; SET INTERRUPT IN MODE 1
0103	E5	8		PUSH HL	; HL–>STARTING ADDRESS OF USER PROGRAM
0104	F5	9		PUSH AF	; STORE STATUS
0105	ED5F	10		LD A, R	; READ REFRESH COUNTER
0107	320040	11		LD (RS), A	; SAVE REFRESH COUNTER VALUE
010A	3E7D	12		LD A, 7DH	; LOAD REFRESH COUNTER
010C	ED4F	13	RFSH	LD R, A	
010E	F1	14	R7D	POP AF	
010F	FB	15	R7E	EI	
0110	C9	16	R7F	RET	; ENTER USER PROGRAM
		17			
		18			
		19			
		20			; PROGRAM 2 : INTERRUPT SERVICE
		21			; PROGRAM
0038		22		ORG 38H	; MODE 1 JUMP ADDRESS
0038	F5	23		PUSH AF	; SAVE STATUS
0039	3A0040	24		LD A, (RS)	; RESTORE REFRESH COUNTER
003C	ED4F	25		LD R, A	; LOAD IN REFRESH REGISTER
		26			;
		27			; INCLUDE TRACE PROGRAM
		28			;
003E	ED5F	29		LD A, R	
0040	320040	30		LD (RS), A	; SAVE REFRESH COUNTER
0043	3E7D	31		LD A, 7DH	
0045	ED4F	32		LD R, A	
0047	F1	33	RC7D	POP AF	
0048	FB	34	RC7E	EI	
0049	C9	35	RC7F	RET	
		36			; ENTER USER PROGRAM
		37	RS	EQU 4000H	; RANDOM-ACCESS-MEMORY LOCATION
		38		END	

PROGRAM LISTING FOR TRACE MODE AND INTERRUPT SERVICE

Tracing. The circuit uses the Z80 processor's machine cycle $\overline{M}_1$ and refresh counter to provide a powerful trace mode for program debugging. Flip-flop U_2 is set when refresh and read signals are low and address bit A_6 is high. It is reset when $\overline{M}_1$ is low.

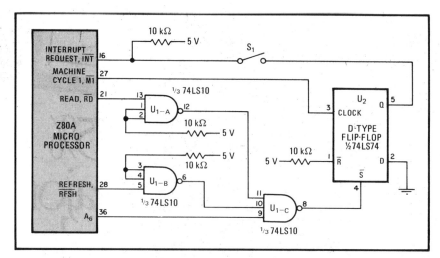

restored for proper refresh operation.

Trace-mode operation continues as long as switch S_1 is closed. A few special Z80 processors using 2-byte operating-code instructions need more than two $\overline{M}_1$ cycles for successful completion. These cycles depend on the result and the BC register count. □

Writing relocatable code for 8-bit microprocessors

by Richard L. Riggs
Sangamo Weston Inc.; Energy Management Division, Atlanta, Ga.

Position-independent code has proven itself in high-level languages as the way to move programs from system to system with little modification to the original software. But writing position-independent assembly code for first-generation 8-bit microprocessors is not always so easy or straightforward. Here is a way of doing it for a common look-up table.

The program at the top of page 131 shows a tradition-

TYPICAL PROGRAM FOR A POSITION-DEPENDANT LOOKUP TABLE FOR THE 6800

```
200   0000   CE   0019    A    LOOKUP   LDX    #TABLE   ⎫
210   0003   FF   0017    A             STX    TEMP     ⎪
220   0006   FB   0018    A             ADDB   TEMP+1   ⎬  calculate the address of the desired data item
230   0009   F7   0018    A             STAB   TEMP+1   ⎪
240   000C   24   03 0011                BCC    EXIT     ⎪
250   000E   7C   0017    A             INC    TEMP     ⎭
260   0011   FE   0017    A    EXIT     LDX    TEMP
270   0014   E6   00      A             LDAB   0,X           data item to B register
280   0016   39                         RTS
290                                     ****************
300   0017        0002    A    TEMP     RMB    2
310   0019        04      A    TABLE    FCB    4
320   001A        22      A             FCB    34
330   001B        1D      A             FCB    $1D
340   001C        03      A             FCB    3
350                                     *
360                                     *      ETC
370                                     *
380   001D        0087    A    LAST     FDB    $87
390
```
• Enter with B register containing offset. • Exits with table data in B register. • Uses X register.

TYPICAL PROGRAM FOR A RELOCATABLE LOOKUP TABLE FOR THE 6800

```
420   001F   8D   11   0032   RLOOK    BSR    PSHTAB
430   0021   E6   00      A            LDAB   0,X             data item to B register
440   0023   39                        RTS
450                                    ********************
460   0024   36              REXIT     PSHA    ⎫
470   0025   37                        PSHB    ⎬  set up the stack to look as though an
480   0026   07                        TPA     ⎪  interrupt occurred
490   0027   36                        PSHA    ⎭
500   0028   30                        TSX     ⎫
510   0029   EB   04      A            ADDB   4,X   ⎬ calculate the desired data item
520   002B   E7   04      A            STAB   4,X   ⎭
530   002D   24   02   0031            BCC    REXIT2  ⎱ address in the stack X register
540   002F   6C   03      A            INC    3,X     ⎰
550   0031   3B              REXIT2    RTI                    return to RLOOK+1
560                                    ************
570   0032   8D   F0   0024  PSHTAB    BSR    REXIT           push RTABLE on stack
580   0034        02      A  RTABLE    FCB    2
590   0035        04      A            FCB    4
600                                    *
610                                    *      ETC
620                                    *
630   0036        87      A  RLAST     FCB    $87
640                                    END
```

al look-up table for the 6800 microprocessor. However, the four instructions starting on line 200, as well as those at lines 250 and 260, make this routine position-dependent. Also, 2 bytes of temporary, read/write storage are required, shown here on line 300.

The second program, however, implements a look-up table for the 6800 that is relocatable, uses no read/write memory other than 7 bytes of stack, and works in read-only memory. The trick is on line 570.

For position independence, the branch-to-subroutine instruction on line 570 must be located at the beginning of the table. It pushes the address of the next instruction onto the stack. In this case, the next instruction is not really an instruction but the table base address.

That, along with the push instructions on lines 460 through 490, sets up the stack as though an interrupt had occurred, leaving the address of the table on the stack in the location where the X register would have been pushed had a true interrupt occurred. The balance of the routine adjusts the X register on the stack to point to the desired data item, restores the registers, and obtains the data. This technique works equally well for the 8080 family by using the call instruction. ∎

Universal E-PROM controller eases computer linkup

by Ralph Tenny
George Goode & Associates Inc., Dallas, Texas

A software-based controller lets an erasable programmable read-only memory programmer serve many different microprocessor systems, regardless of the individual microprocessors used. Programming the E-PROMs to meet any system format increases the system's capability, especially in transferring code between machines, in control of one computer by another, and in the direct transfer of data between an on-line computer and a word-processing system.

Here, TI's popular TM 990/189 University board is converted into a terminal that can receive an ASCII data stream, store it in memory, and then produce it on a cassette tape with a format acceptable to a TI loader. The tape input thus becomes the entree to any computer in the 990 family. With such a controller driving any of the many E-PROM programmers of standard design,[1,2] the memories can be programmed for any system.

The basic premise of this program is that almost any microprocessor's debugging routine contains a utility sequence that generates a memory dump. In the case of ASCII characters (figure), the program produces a continuous data stream that contains no control codes (CR and LF would be the most common extra characters). With this modification, the data stream becomes a serial version of the program data and can be recorded in consecutive memory locations. This serial data, when placed on a tape cassette, is then transferred to the machine through the E-PROM programmer.

The board is especially suited to communication between systems. It is one of the few computers with an on-board terminal that, under software control, supports full resource sharing between on-board and external terminals. In contrast, many systems derive both control and interaction from an external terminal and may not support two terminals simultaneously without additional interfacing.

A buffer at location 36_{16}, which identifies the terminal in use—on-board or external—makes dual-terminal operation possible. Thus, at line 170 of the program, the clear command is initially used to order external-terminal operation. Similarly, at line 640, the INV statement writes in a nonzero value, and the on-board terminal and keyboard resumes control.

With this control feature, the utility bus can either be used in a passive mode (as in the listing) or serve as a terminal for another computer. Here, the bus sends a series of commands that initialize the slave computer for data reception. Of course, the slave must have the right program to accept both data and commands, but such a program is usually easy to implement.

Program operation is straightforward. Once the system is initialized (up through line 250), the command XOP R4, 11 puts the utility bus in a wait loop until a string of serial data is transmitted. Note that the communications port of most computer systems is configured as a transmitter. That is, data in RS-232 format is put out on pin 3 of the port and incoming data is received on pin 2, so it is best to install a cable that lets these two lines be interchanged by a switch.

An incoming character, assumed to be hexadecimal ASCII, is then tested to see if it is numeric or alphabetic. After suitable adjustment, the program then deposits it in one of four buffers in read/write memory (figure), which shows the buffer configuration after four characters have been received and stripped to a hexadecimal nibble.

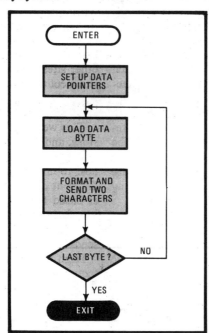

Bit forms. Flow chart outlines method for maintaining a continuous stream of ASCII characters, which represents a memory image from the transmitting computer. Technique described in text permits TI's TM 990 University board to reconfigure memory image for the purpose of programming an E-PROM to meet almost any system format.

After the fourth character arrives, the routine beginning at line 380 recombines the four original data nibbles. The SOC @BUFR(R1), R5 command is a classical OR function that accesses the four memory locations via indexed addressing, with the result being placed in register R_4. Finally, at line 420, the data word (which may be two characters, one 16-bit data word, or one word of 9900 machine code) is stored in the memory buffer using auto-increment addressing, and the buffer is tested to see if it is full. In normal operation, only 1 kilobyte of code will be transferred, and the utility routine will be terminated manually by the operator. However, if the memory buffer overflows, the routine terminates reception and returns to on-board terminal operation, issuing an audible warning and relighting the on-board display.

Larger quantities of data may be transferred and then installed immediately in the E-PROM if a Memory

TM 990 LISTING: UNIVERSAL PROGRAMMER FOR ERASABLE PROGRAMMABLE READ-ONLY MEMORIES

```
0010  0000          *This program makes the TM 990/189 University Board a smart terminal which
0020  0000          *receives an incoming RS-232 data stream, converts each character into an
0030  0000          *equivalent hexadecimal digit, then combines each successive group of four
0040  0000          *characters into 16-bit words stored in memory. Data is stored in a four-word
0050  0000          *buffer; each word contains one of the original nibbles sent. Each nibble has
0060  0000          *been shifted to a position in the buffer which corresponds to that nibble's
0070  0000          *position in the original word. The word is reassembled by OR'ing together these
0080  0000          *four words; the assembled word is then stored.
0090  0000
0100  0000
0110  0000
0120                    IDT   'DATAIN'
0130  0890              AORG  >890              Program location
0140  0890              DREG                    Assembler directive
0150        01C0  BUFR  EQU   >1C0              Define work buffer
0160  0890  02E0  STRT  LWPI  >1A0              Initialize work space
      0892  01A0
0170  0894  04E0        CLR   @ >36             Switch to external terminal
      0896  0036
0180  0898  0206        LI    R6, >900          Load alpha adjustment factor
      089A  0900
0190  089C  0203        LI    R3, >200          Define data buffer start
      089E  0200
0200  08A0  0201  ST2   LI    R1, 2             Load work buffer
      08A2  0002
0210  08A4  C081        MOV   R1, R2            Index values
0220  08A6  06A0        BL    @ CLR             Clear work buffers
      08A8  08F2
0230  08AA  C202  GET   MOV   R2, R8            Save multiplicand
0240  08AC  3A01        MPY   R1, R8            Complete shift count
0250  08AE  C009        MOV   R9, R0              and load it into register R0
0260  08B0  2EC4        XOP   R4, 11            Enter ASCII character
0270  08B2  0284        CI    R4, >3900         Test to see if character is numeric
      08B4  3900
0280  08B6  151B        JGT   LPHA              Make adjustment if necessary
0290  08B8  0244  MASK  ANDI  R4, >0F00         Strip character to 4 bits
      08BA  0F00
0300  08BC  06C4        SWPB  R4                Move data to least significant nibble
0310  08BE  0B04        SRC   R4, 0             Shift by count in R0
0320  08C0  C844        MOV   R4, @ BUFR (R1)     and store in buffer
      08C2  01C0
0330  08C4  05C1        INCT  R1                Bump the pointer
0340  08C6  0281        CI    R1, 8             Test for last character
      08C8  0008
0350  08CA  12EF        JLE   GET               Not done. Continue
0360  08CC  0641        DECT  R1                Adjust index pointer
0370  08CE  04C5        CLR   R5                Done, combine hexadecimal characters
0380  08D0  E161  ORIN  SOC   @ BUFR (R1), R5     using successive OR operations
      08D2  01C0
0390  08D4  0641        DECT  R1                Count down index
0400  08D6  0281        CI    R1, 2             Test for end of loop
      08D8  0002
0410  08DA  14FA        JHE   ORIN                and repeat as needed
0420  08DC  CCC5        MOV   R5, *R3+            and then store data words
0430  08DE  0283        CI    R3, >7F0          Check for end of buffer
      08E0  07F0
0440  08E2  1301        JEQ   EXIT                and quit if out of memory
0450  08E4  10DD  NEXT  JMP   ST2               Continue in an endless loop
0460  08E6  0560  EXIT  INV   @ >36             Go back to on-board terminal
      08E8  0036
0470  08EA  06A0        BL    @ >3000           Then go home
      08EC  3000
0480  08EE  A106  LPHA  A     R6, R4            Adjust for alpha
0490  08F0  10E3        JMP   MASK              Return to data stream
0500  08F2  0207  CLR   LI    R7, 8             Set up counter and index
      08F4  0008
0510  08F6  04E7  CL2   CLR   @ BUFR (R7)       Clear a work buffer
      08F8  01C0
0520  08FA  0647        DECT  R7                Count down
0530  08FC  1BFC        JH    CL2                 and repeat as needed
0540  08FE  045B        RT                      Then return
0550        0890        END   STRT
```

335

Expansion Module[3] is used. This module allows nearly automatic operation, with only 14 keystrokes needed to initiate the 90-second (1 kilobyte) programming cycle. As for control tasks, this scheme has worked well for students attending courses in 9900 assembly language, where the word processor of TI's 990/302 software development system is the system log that generates assembly listings. In transfer tasks, in-house development systems serving as extra text-entry stations are both flexible and inexpensive. This scheme will allow the addition of at least five such systems for the price of the original two included with the word processor. □

References
1. James Nicholson, "Upgrade your EPROM programmer," EDN, Nov. 5, 1980, p. 90.
2. H. S. Mazumdar, "Build a simple PROM programmer," EDN, June 20, 1980, p. 100.
3. Memory and I/O Expansion Module—User's Guide, George Goode & Associates Inc., 1st edition, Jan. 1980.

Home computer displays inverse Laplace transforms

by Michael A. Wyatt
Honeywell Inc., Avionics Division, St. Petersburg, Fla.

Expanding upon the Gaver algorithm[1] for estimating the time value of a Laplace transform as described by Kinchu Woo,[2] this program provides the time-domain display of such functions on the popular Apple II personal computer. Utilizing the graphics capabilities allows a user to view a time-domain response almost as it would be seen on an oscilloscope. Also included is a feature not found on most scopes—the ability to scan the display with a cursor in order to extract measurements at any given point on the plot.

Taking the given Laplace transform, which is entered between lines 250 and 1000 of the floating-point Basic routine, the program finds the function's corresponding time value from:

$$f(t) = (\ln 2/t)\sum_{i=1}^{10} V_i F(i \ln 2/t)$$

where the set of constant V_i coefficients is entered at lines 160–170 and where it is assumed that $f(t)$ is suitably bounded and well defined. Once the time increment, t_i, is specified at line 145, the program performs the inversion (lines 200–1170) until $t_f = 279 t_i$, where the number 279 is the number of points that may be resolved on the Apple's screen. Lines 2000–3030 then automatically scale and plot the previously calculated data points using the Apple's high-resolution graphics.

The cursor-scanning subroutine is implemented between lines 4000 and 7020. The user need only position the cursor on the curve by means of the left (L) and right (R) keys on the operator console. The value of the function will be displayed as a function of time at the lower f left portion of the screen as $f(t) = A$, where t will be expressed in seconds and A is a scalar quantity (a number).

Consider the example of the classic second-order system that is excited by a step function:

$$F(s) = \omega_n^2/s(s^2 + 2\zeta\omega_n s + \omega_n^2)$$

with $\zeta = 0.4$ and $\omega_n = 1$. Following initialization of the program, the routine will flash a message on screen instructing the user to enter the desired transform, to which the user responds: $F(n) = 1/s(s^2 + 0.8s + 1)$. Alternatively, he may specify the value of ζ and ω_n first and then enter the general (symbolic) equation.

TIME	CALCULATED	PLOTTED
0.1	0.0048653	0.0048661
0.2	0.0189122	0.0189238
0.5	0.1076671	0.1077150
1.0	0.3599150	0.3580033
2.0	0.9270844	0.9404786
5.0	1.0760872	1.0773478
10.0	1.01564972	0.9849304

PROGRAM FOR DISPLAY OF INVERSE LAPLACE TRANSFORMS ON APPLE II

```
10   TEXT : HOME : VTAB (8)
20   DIM K(11),V(281),F(11),Y(281)
30   PRINT "          LAPLACE INVERS
     ION FOR
40   PRINT "         TIME-DOMAIN D
     ISPLAY"
50   VTAB (20)
60   GET U$
70   HOME : VTAB (8)
80   PRINT " ENTER EQUATION IN S B
     ETWEEN 250 & 1000"
90   VTAB (10)
100  PRINT "         USE F(N) AS VA
     LUE OF EQUATION"
110  VTAB (16)
120  INPUT "      HAS EQUATION BE
     EN ENTERED? Y/N";A$
130  HOME : VTAB (8)
140  IF A$ = "N" THEN 50000
145  INPUT "          INPUT TIME I
     NCREMENT";T: HOME
150  PRINT "           INVERSION COM
     PUTING"
160  K(1) = 1 / 12:K(2) =  - 32.08
     333333:K(3) = 1279.000076:K(
     4) =  - 15623.66689:K(5) = 8
     4244.16946:K(6) =  - 236957.
     5129
170  K(7) = 375911.6923:K(8) =  -
     340071.6923:K(9) = 164062.51
     28:K(10) =  - 32812.50256:K(
     11) =  LOG (2)
200  FOR I = 1 TO 279
210  F(0) = 0:S = 0
220  FOR N = 1 TO 10
230  S = N * K(11) / (T * I)
250  F(N) = 1 / (S * (S * S + .8 *
     S + 1))
1100 F(N) = F(N) * K(N) + F(N - 1
     )
1110 NEXT N
1120 V(I) = F(10) * K(11) / (T *
     I)
1130 IF V(I) < PHIGH THEN 1150
1140 PHIGH = V(I)
1150 IF V(I) > PLOW THEN 1170
1160 PLOW = V(I)
1170 NEXT I
2000 REM ----PLOT ROUTINE-----
2010 HGR
2020 HCOLOR= 7
2030 IF PLOW < 0 THEN 2200
2050 Y(1) = INT (156 - 155 * ((V
     (1) - ABS (PLOW)) / (PHIGH -
     PLOW)))
2100 FOR I = 2 TO 279
2110 Y(I) = INT (156 - 155 * ((V
     (I) - ABS (PLOW)) / (PHIGH -
     PLOW)))
2120 NEXT I
2130 GOTO 3000
2200 Y(1) = INT (156 * ABS ((PL
     OW - V(1)) / (PHIGH - PLOW))
     ) + 1
2210 FOR I = 2 TO 279
2220 Y(I) = INT (156 * ABS ((PL
     OW - V(I)) / (PHIGH - PLOW))
     ) + 1
2230 NEXT I
3000 HPLOT 1,Y(1)
3010 FOR I = 2 TO 279
3020 HPLOT  TO I,Y(I)
3030 NEXT I
4000 REM -----CURSOR ROUTINE-----
4010 I = 125
4020 GET C$
4030 IF C$ = "L" THEN 5000
4040 IF C$ = "R" THEN 6000
4050 IF C$ = "S" THEN 50010
4060 GOTO 4020
5000 REM ---DECREMENT CURSOR---
5010 I = I - 1
5020 IF I >  = 1 THEN 5040
5030 I = 1
5040 HCOLOR= 4
5050 HPLOT I + 1,Y(I + 1) - 1 TO
     I + 1,Y(I + 1) + 1
5060 HCOLOR= 7
5070 HPLOT I + 1,Y(I + 1)
5080 HPLOT I,Y(I) - 1 TO I,Y(I) +
     1
5090 GOTO 7000
6000 REM ---INCREMENT CURSOR---
6010 I = I + 1
6020 IF I <  = 279 THEN 6040
6030 I = 279
6040 HCOLOR= 4
6050 HPLOT I - 1,Y(I - 1) - 1 TO
     I - 1,Y(I - 1) + 1
6060 HCOLOR= 7
6070 HPLOT I - 1,Y(I - 1)
6080 HPLOT I,Y(I) - 1 TO I,Y(I) +
     1
7000 HOME : VTAB 22
7010 PRINT "          F(";I * T;")=
     ";V(I)
7020 GOTO 4020
50000 PRINT " ENTER EQUATION IN
     S BETWEEN 250 & 1000
50010 END
```

The return key is then depressed and the program will query if the equation has been entered. Depressing the Y key to indicate yes (or N to indicate no) triggers analysis of the equation.

As seen, the results of the tabulated analysis compare favorably with the analytical (mathematical) result for the time-domain equivalent, which is:

$$f(t) = 1 - [1/(1-\zeta^2)^{0.5}]e^{-\zeta\omega_n t}\sin[\omega_n(1-\zeta^2)^{0.5}t + \theta]$$

where $\theta = \arctan[(1-\zeta^2)^{0.5}/\zeta]$. To exit the program, the user simply depresses the S key on the console.

Modification of the basic program for other machines (TRS80, Pet, and so on) should pose no serious problem. The plot and cursor function will not be so easily implemented, however, and so this section of the program must be rewritten accordingly. □

References
1. D. P. Gaver, "Observing Stochastic Processes and Approximate Transform Inversion," *Operational Research*, Vol. 14, No. 3, 1966, p. 444–459.
2. Kin-chu Woo, "TI-59 inverts Laplace transforms for time-domain analysis," *Electronics*, Oct. 9, 1980, p. 178–79.

'Surgical' program speeds 6909 debugging process

by Ralph Tenny
George Goode & Associates Inc., Dallas, Texas

Without a precise move program, the traditional fix for many problems that arise during debugging is to jump to

6809 DEBUG PATCH PROGRAM

```
                   0D30    TEMP1    EQU      $D30       REGISTER Y STORAGE
                   0D32    TEMP2    EQU      $D32       REGISTER U STORAGE
                   0D34    TEMP3    EQU      $D34       REGISTER D STORAGE

                           *THIS ROUTINE WILL MOVE A BLOCK OF DATA TO ANOTHER
                           *LOCATION.  PASS THE BLOCK LENGTH (BYTES) IN TEMP3,
                           *THE SOURCE ADDRESS IN TEMP1, AND THE DESTINATION
                           *ADDRESS IN TEMP2.  ALLOWANCE MADE FOR BUFFER OVERLAP.

0000 34    76              MVBLK    PSHS     D,X,Y,U    SAVE REGISTERS

                           *ALLOW FOR POSSIBLE BUFFER OVERLAP

0002 FC    0D30                     LDD      TEMP1      GET START ADDRESS
0005 B3    0D32                     SUBD     TEMP2      GET DISTANCE BETWEEN BLOCKS
0008 27    15                       BEQ      EXIT2      SAME ADDRESS, WHY BOTHER?
000A 2D    16                       BLT      REV        MOVE CODE FROM BOTTOM FIRST

                           *NOTE: THIS MOVE ALLOWS UNWANTED CODE TO BE
                           *OVERWRITTEN.  USE WITH CARE!

000C FC    0D34                     LDD      TEMP3      GET NUMBER OF BYTES TO MOVE
000F 10BE  0D30                     LDY      TEMP1      ALSO START ADDRESS
0013 FE    0D32                     LDU      TEMP2      AND DESTINATION START
0016 AE    A1              B1       LDX      ,Y++       LOAD TWO BYTES
0018 AF    C1                       STX      ,U++       AND PUT THEM DOWN
001A 83    0002                     SUBD     #2         COUNT THE OPERATIONS
001D 24    F7                       BHS      B1         LOOP UNTIL DONE
001F 35    76              EXIT2    PULS     D,X,Y,U    RESTORE REGISTERS
0021 39                             RTS                 AND GO HOME

                           *THIS MOVE ALLOWS CODE TO BE OPENED UP TO INSERT
                           *ONE OR MORE OP CODES FOR A PATCH.

0022 FC    0D34            REV      LDD      TEMP3      GET NUMBER OF BYTES TO MOVE
0025 F3    0D30                     ADDD     TEMP1      AND POINT TO BOTTOM OF BUFFER
0028 1F    02                       TFR      D,Y        LOAD SOURCE POINTER
002A FC    0D34                     LDD      TEMP3      GET BYTE COUNT AGAIN, THEN
002D F3    0D32                     ADDD     TEMP2      FIND END OF TARGET BUFFER
0030 1F    03                       TFR      D,U        LOAD DESTINATION POINTER
0032 FC    0D34                     LDD      TEMP3      ONE MORE TIME!
0035 AE    A4              B2       LDX      0,Y        LOAD TWO BYTES AND
0037 31    3E                       LEAY     -2,Y       POINT TO NEXT LOAD
0039 AF    C4                       STX      0,U        STUFF THE DATA AND
003B 33    5E                       LEAU     -2,U       POINT TO NEXT TARGET
003D 83    0002                     SUBD     #2         COUNT THE PASSES
0040 24    F3                       BHS      B2         LOOP UNTIL DONE
0042 20    DB                       BRA      EXIT2      THEN BLOW THE JOINT

0 ERROR(S) DETECTED
```

a patch area elsewhere in memory, do the missing operation, and jump back to the instruction that follows the jump's takeoff point. But the ideal solution is to insert one or two instructions inline in the program. With this routine, a gap can be opened in the existing program by means of a simple keyboard entry to three locations and a simple jump statement elsewhere in memory. The missing instructions can then be keyed in and tested. The new code still has to be recompiled, but the recompilation is one that matches code that has been verified.

A normal block-move program would scramble the last part of the program, since these routines almost always move the top byte first. Thus, the first byte to be moved would land on the fourth byte downstream of the patch, thereby obliterating part of the program. To overcome this problem, this routine tests the addresses in buffers TEMP1 and TEMP2. If the code is to be moved down, the program jumps to its last section. This section peels off code from the bottom of the section being moved, so that there is just enough space to open the required gap. Thus the last byte to be moved will be the byte at the start of the patch.

The program is used by entering the address of the first byte to be moved in TEMP1, the address of the last byte to be moved in TEMP2, and the number of bytes from the address in TEMP2 to the end of the program in TEMP3. Then the program is run. Inspection of memory will show that the code that resided between the specified addresses has been duplicated and that the rest of the program has been moved down in memory accordingly.

For the reverse problem, unwanted code that must be deleted, the usual debugging method is to replace an offending instruction with a no-op code. Instead, this program takes the first byte after the unwanted code and replaces the first byte to be deleted with it.

Note that if either operation happens to change the distance between the start and end of a relative branch, the branch displacement must be recomputed.   □

Very efficient 8080 program multiplies and divides

by Jerry L. Goodrich
Pennsylvania State University, University Park, Pa.

Making an appearance for the third time in this section of *Electronics*, an 8080 program is being presented that can compute 32-by-16-bit division and 16-by-16-bit multiplication. However, this subroutine betters its immediate predecessor's divide and multiply execution times [*Electronics*, March 27, 1980, p. 156] by 75% and over 60%, respectively. Also, this program's 8-by-16-bit multiplication subroutine surpasses all others with an execu-

8080 PROGRAM FOR 32-BY-16-BIT DIVISION

LABEL	SOURCE	CODE	COMMENT
DIV:	MOV	A, L	; CHECK FOR OVERFLOW
	SUB	C	
	MOV	A, H	
	SBB	B	
	RNC		; RETURN ON OVERFLOW
	MOV	A, B	; 2'S COMPLEMENT BC
	CMA		
	MOV	B, A	
	MOV	A, C	
	CMA		
	MOV	C, A	
	INX	B	
	CALL	LOOP	; DIVIDE INTO HIGHEST-ORDER 3 BYTES OF DIVIDEND
; LOOP DIVIDES 3-BYTE DIVIDEND BY 2-BYTE DIVISOR			
LOOP:	MOV	A, D	; MOVE THIRD BYTE TO BE DIVIDED INTO A
	MOV	D, E	; SAVE LOWEST-ORDER BYTE DIVIDEND OR HIGHEST-ORDER BYTE QUOTIENT
	MVI	E, 8	; LOAD LOOP1 COUNTER
LOOP1:	DAD	H	; SHIFT DIVIDEND LEFT
	JC	OVER	; JUMP IF DIVIDEND OVERFLOWED HL
	ADD	A	
	JNC	SUB	
	INX	H	; CONVEY CARRY IF THERE
SUB:	PUSH	H	; SAVE HIGHEST-ORDER 2 BYTES OF DIVIDEND
	DAD	B	; SUBTRACT DIVISOR
	JC	OK	; JUMP IF NO BORROW
	POP	H	; UNSUBTRACT IF BORROW
	DCR	E	; UPDATE LOOP1 COUNTER
	JNZ	LOOP1	; LOOP UNTIL DONE
	MOV	E, A	; PUT BYTE OF QUOTIENT IN E
	STC		
	RET		
OK:	INX	SP	; CLEAN UP STACK
	INX	SP	
	INR	A	; PUT A 1 IN QUOTIENT
	DCR	E	; UPDATE LOOP1 COUNTER
	JNZ	LOOP1	; LOOP UNTIL DONE
	MOV	E, A	; PUT BYTE OF QUOTIENT IN E
	STC		
	RET		
OVER:	ADC	A	; FINISH DIVIDEND SHIFT, PUT 1 IN QUOTIENT
	JNC	OVERSUB	
	INX	H	; CONVEY CARRY IF THERE
OVERSUB	DAD	B	; SUBTRACT DIVISOR
	DCR	E	; UPDATE LOOP1 COUNTER
	JNZ	LOOP1	; LOOP UNTIL DONE
	MOV	E, A	; PUT BYTE OF QUOTIENT IN E
	STC		
	RET		

8080 PROGRAM FOR 16-BY-16-BIT MULTIPLICATION

LABEL	SOURCE	CODE	COMMENT
MULT:	MOV	A, E	; LOAD LOWEST-ORDER BYTE OF MULTIPLIER
	PUSH	D	; SAVE HIGHEST-ORDER BYTE MULTIPLIER
	CALL	BMULT	; DO 1-BYTE MULTIPLY
	XTHL		; SAVE LOWEST-ORDER BYTES PRODUCT, GET MULTIPLIER
	PUSH	PSW	; STORE HIGHEST-ORDER BYTE OF FIRST PRODUCT
	MOV	A, H	; LOAD HIGHEST-ORDER BYTE OF MULTIPLIER
	CALL	BMULT	; DO SECOND 1-BYTE MULTIPLY
	MOV	D, A	; POSITION HIGHEST-ORDER BYTE OF PRODUCT
	POP	PSW	; GET HIGHEST-ORDER BYTE OF FIRST PRODUCT
	ADD	H	; UPDATE THIRD BYTE OF PRODUCT
	MOV	E, A	; AND PUT IT IN E
	JNC	NC1	; DON'T INCREMENT D IF NO CARRY
	INR	D	; INCREMENT D IF CARRY
NC1:	MOV	H, L	; RELOCATE LOWEST-ORDER BYTES OF SECOND PRODUCT
	MVI	L, 0	
	POP	B	; GET LOWEST-ORDER 2 BYTES OF FIRST PRODUCT
	DAD	B	; GET FINAL PRODUCT LOWEST-ORDER 2 BYTES
	RNC		; DONE IF NO CARRY
	INX	D	; OTHERWISE UPDATE HIGHEST-ORDER 2 BYTES
	RET		
; BMULT PERFORMS A 1-BYTE BY 2-BYTE MULTIPLY			
BMULT:	LXI	H, 0	; ZERO PARTIAL PRODUCT
	LXI	D, 7	; D = 0, E = BIT COUNTER
	ADD	A	; GET FIRST MULTIPLIER BIT
LOOP1:	JNC	ZERO	; ZERO-SKIP
	DAD	B	; ONE-ADD MULTIPLICAND
	ADC	D	; ADD CARRY TO THIRD BYTE OF PRODUCT
ZERO:	DAD	H	; SHIFT PRODUCT LEFT
	ADC	A	
	DCR	E	; DECREMENT BIT COUNTER
	JNZ	LOOP1	; LOOP UNTIL DONE
	RNC		; DONE IF NO CARRY
	DAD	B	; OTHERWISE DO LAST ADD
	ADC	D	
	RET		; AND RETURN

tion time less than half that of its larger counterpart. The program works with 3 bytes of the dividend or product at a time to improve the execution time, which is the most appropriate measure of program efficiency.

The divider program works much like ordinary long division of a four-digit decimal number by a two-digit number. The divide routine of the program stores the dividend in registers HL-DE and keeps the most significant digits in HL. The divisor is for the operation placed in register pair BC. The quotient from the division builds up in DE with the remainder placed in HL. For a quotient that is longer than 16 bits, the carry flag is cleared in order to indicate an overflow. The worst-case execution time for the divide routine is 1,745 clock cycles.

In the multiplication routine, the multiplicand is stored in BC and the multiplier in DE. The result appears in registers DE-HL with the most significant digits being placed in DE. The execution time is 1,023 clock cycles.

For the 8-by-16-bit multiply routine, the 8-bit number is placed in register A and the 16-bit number in BC. The result then appears in A-HL with the high-order byte being placed in A. The worst-case execution time for this subroutine is 424 clock cycles. □

Hardware-software integration eases E-PROM programming

by P. R. Ramraj
ISRO Satellite Centre, Bangalore, India

This hardware and software marriage offers the RCA Cosmac development system a built-in programmer for an 8-K-byte electrically programmable read-only memory. Object files can now be easily transferred to E-PROM with this swapping program. The hardware interface has read-and-program modes of operation that may be selected by flipping a switch. In addition, address information and data are entered by a single command while

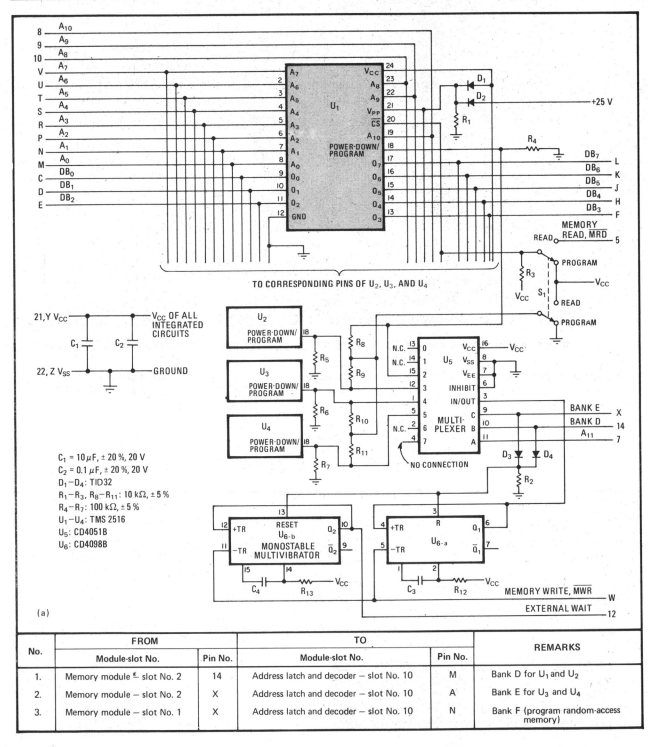

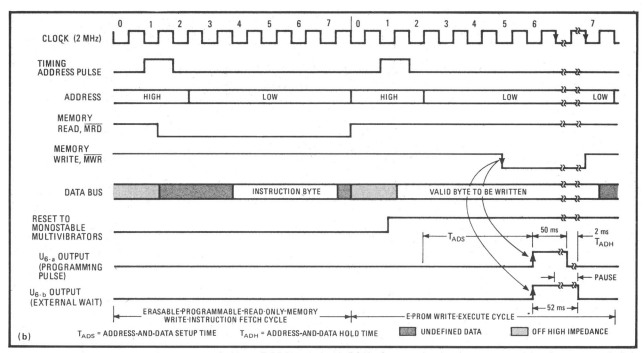

Integration. This hardware interface links an 8-K-byte E-PROM card with RCA's Cosmac development system, giving the system a built-in programming facility. The circuit uses four 2-K-byte E-PROMs U_1 through U_4 to form memory banks D and E, situated in slot 2. The timing diagram (b) depicts the E-PROM write cycle. Monostables U_{6-a} and U_{6-b} generate 50- and 52-ms pulses, respectively.

SWAPPING PROGRAM TO TRANSFER DATA FROM RANDOM-ACCESS MEMORY TO ERASABLE PROGRAMMABLE READ-ONLY MEMORY		
0000	;	0001 ..SWAPPING PROGRAM TO TRANSFER DATA.
0000	;	0002 ..SOURCE ADDRESS (START & END) & DESTINATION
0000	;	0003 ..ADDRESSES (START) ARE AVAILABLE IN THE LIST.
0000	;	0004 LIST = #FF30 ..LIST ADDRESS.
0000	;	0005 ORG #FF00 ..PROGRAM STARTS @ FF00.
FF00	7100;	0006 DIS, #00 ..DISABLE INTERRUPT ; P = X = 0.
FF02	F8FFB0F830A0;	0007 LDI A. 1 (LIST) ; PHI RD;LDI A. 0(LIST) ; PLO RD
FF08	;	0008 ..LIST ADDRESS LOADED IN REGISTER RD.
FF08	0DBA1D0DAA;	0009 LDN RD;PHI RA;INCREMENT RD;LDN RD;PLO RA
FF0D	;	0010 ..SOURCE START ADDRESS LOADED IN REGISTER RA.
FF0D	1D1D1D;	0011 INCREMENT RD;INCREMENT RD;INCREMENT RD
FF10	;	0012 ..REGISTER RD TO POINT DESTINATION START ADDRESS.
FF10	0DBB1D0DAB;	0013 LDN RD;PHI RB;INCREMENT RD;LDN RD;PLO RB
FF15	;	0014 ..DESTINATION START ADDRESS LOADED IN REGISTER RB
FF15	2D2D;	0015 DECREMENT RD;DECREMENT RD ..REGISTER RD TO POINT SOURCE END ADDRESS.
FF17	0A;	0016 RDSORC:LDN RA ..READ FROM SOURCE (RAM).
FF18	5B;	0017 STR RB ..WRITE IN DESTINATION (EPROM).
FF19	;	0018 .."PAUSE" INITIATION BY HARDWARE INTERFACE.
FF19	ED8AF33A24;	0019 SEX RD;GLO RA;XOR;BNZ UPDATE
FF1E	;	0020 ..END "LO" SOURCE ADDRESS?
FF1E	2D9AF332211D;	0021 DECREMENT RD;GHI RA;XOR;BZ*;INCREMENT RD
FF24	;	0022 ..END "HI" SOURCE ADDRESS?
FF24	1A1B3017;	0023 UPDATE: INCREMENT RA;INCREMENT RB;BR RDSORC
FF28	;	0024 ..UPDATE SOURCE & DESTINATION ADDRESS.
FF28	;	0025 ..LIST AVAILABLE AT FF30.
FF28	;	0026 ORG#FF30;,#0000, #1FFF ..SOURCE START & END ADDRESS.
FF30	00001FFF;	0026
FF34	D000;	0027 ,#D000 ..DESTINATION START ADDRESS.
FF36	;	0028
0000		
;;		

the E-PROM is being programmed.

Slot 2 on the hardware interface for the 8-K-byte E-PROM card accommodates the E-PROM by using E-PROMs U_1 and U_2 to form memory bank D and E-PROMs U_3 and U_4 to form bank E. Slot 1 holds the swapping program (see table). Once the program's source and destination are specified within the program, an object file is developed and transferred to the E-PROM.

The hardware interface provides a read-and-program mode of operation with the help of switch S_1. Multiplexer U_5 selects the E-PROM during the read mode and routes a pulse with a 50-millisecond duration (b), which is generated by monostable U_{6-a} for the addressed E-PROM during the program mode. A dc voltage of 25 volts is applied only in this mode of operation.

During the program mode, the monostable reset pulse is created, and the memory write pulse ($\overline{MWR}$) initiates a pause cycle through the monostable U_{6-b} for 52 ms. An extra 2 ms is tacked on to the cycle to ensure a sufficient hold time for address and data bits. The maximum clock frequency is 2 megahertz.

Programming is accomplished by first loading the swapping program into the random-access memory in locations $FF00_{16}$ through $FF27_{16}$ and using the UT20 command. The program immediately collects the source and destination address and executes the transfer. Whenever bank D or E is addressed and a byte transfer has been made, a pause occurs. After an object file is edited, the contents of the E-PROM are transferred to the RAM for modifications, and the edited file is then transferred back to the E-PROM at bank D or E. □